INDUSTRIAL BIOCHEMISTRY

By

Dr. G.P. Garg

DISCOVERY PUBLISHING HOUSE PVT. LTD.

NEW DELHI-110 002

First Published-2010
Reprint : 2012
ISBN: 978-81-8356-645-2

Published by:
DISCOVERY PUBLISHING HOUSE PVT. LTD.
4831/24, Ansari Road, Prahlad Street
Darya Ganj, New Delhi-110002 (India)
Phone: 23279245, 43764432 • Fax: 91-11-23253475
E-mail: parul.wasan@gmail.com • dphbooks@rediffmail.com
web: www.discoverypublishinggroup.com

Printed at:
Dynamic Printers,Delhi

Preface

Biochemistry today has made spectacular progress in unraveling the mysteries of animate nature. This progress has allowed us to gain deeper insight into the principles of vital activity and has to a very significant extent stimulated the development of applied disciplines, especially medicine. The present title *"Industrial Biochemistry"* is intended for those who wish to understand living organisms, especially man. Biochemistry is essential for this purpose, but it would be almost impossible for a student to survey on his own the massive body of existing knowledge, constantly augmented by a remarkable torrent of brilliat discoveries. The purpose of the book, then is to organize our knowledge into something that can be comprehended in a relatively short time and still convey a reasonable complete picture of the chemical structure and function of man. Readability without sacrifice of coverage has been a prime goal. An important device in gaining that goal is to keep attention constantly focused on function, with repeated use of rationalization to show that the chemical facts are not isolated, but part of a whole.

Biochemistry has two major goals as a fundamental science, it treats the vital functions from the standpoint of physical chemistry, and as an applied discipline, it points out practical applications for the wealth of scientific knowledge it has acquired. The dual purpose has been, as far as possible, take into account in the writing of this book. We have also tried to summarize our pedagogical experience in teaching biochemistry to students specializing in medicine and pharmacy. The material of this book has been organized according to the principle of functionally in order to trace the close relationship between the functions of the living organism and the structures of its constituents molecules as well as the chemical and physico-chemical process in which they are involved.

The aim of this book is to present a core of biochemical knowledge that is desirable for undergraduate and postgraduate students and also those involved in the field of medical, microbiology, biotechnology and pharmaceutical. Every attempt has been made to keep abrest of the advances in the subject and at the same time to include the fundamentals.

To make the work more comprehensive and informative, the author has consulted many authoritative books, research journals, abstracts, monographs etc. He is grateful to all those great scholars whose work are cited or substantially reproduced.

There can be no claim to originality except in the manner of treatment and much of the information has been obtained from the books and scientific journals available in the different libraries.

The author expresses his thanks to his friends and colleagues whose continue inspirations have initiated him to bring out this book.

The author expresses his gratitude to Mr. Wasan and Staff of M/s Discovery Publishing House Pvt. Ltd., for their whole hearted co-operation in the publication of this book.

In the mean time, the author will remain sincerely responsible for any shortcomings of the book and be grateful to the readers for their suggestions and constructive criticism for the continuous betterment of the book. He takes this opportunity to appeal to the readers to send their suggestions straightway to his publisher.

Author

CONTENTS

CHAPTER 1

Introduction

Sterilisation is the action of eliminating microorganisms from a medium. Sterility is the absence of any detectable and viable microbes in a culture medium or in the gas phase. Sterilisation is a process that destroys all living organisms, spores and viruses in a pressurised vessel at high temperature. In the food and dairy industries, sterilisation is commonly used to preserve food products. At the laboratory scale, huge steel vessels with live stream at 105 kPa (15 psig) are commonly used for 20-30 minutes. This is a closed system known as an autoclave; therefore it is batch sterilisation. The temperature is raised to 121 °C when air is initially flashed out and all live streams are replaced. Wet steam is usually used for effective autoclaving. The high temperature and long duration may kill all living microorganisms, spores and viruses.

When media is prepared, to eliminate bacterial and fungal contaminants it must be heat treated at high pressure. Even at high temperatures the fungal spores may survive if only heat is used. Therefore, media is autoclaved at 121 °C and 105 kPa. In fact, an autoclave is used like a pressure cooker, which is convenient and economic autoclaving equipment. Overheatmg the prepared media may have negative impact, causing the media pH to be unstable.

Acid pH is very sensitive to overheating. Overheating in media containing sugars cuses the carbohydrate to be caramelised. This media may have reverse impact, reducing bacteriological performance. Gelatinous media or any nutrient agar at acidic pH is hydrolysed. This is due to the acidic condition of catalytic activities or excess amounts of protons breaking down the solid media and extra sugars forming. This may cause substrate inhibition on growth of organisms on solid slant media.

BATCH STERILISATION

Batch sterilisation uses steam to eliminate living organisms. Heat losses, heating and cooling are major steps, and it is a time-consuming process. It requires air to be evacuated replaced with steam. At first the chamber is flashed with pressurised steam. With this technique we should get rid of the existing air. Steam sterilisation is performed in a jacketed vessel by supplying steam and maintaining the set pressure and temperature at a constant level for a fixed duration. Batch sterilisation wastes energy and overcooks the medium. There is no conservation of energy; therefore the process may not be economical for implementation at a large scale. Batch sterilisation is common at bench and laboratory scales. Batch media sterilisation is done in an autoclave.

Basically, it is a huge steam cooker. Steam enters the jacket of the surrounding chamber. When the pressure from the steam has been built up in the jacket, a venting valve for the outlet of chamber air is closed and the inlet valve allows the steam to enter the chamber. The pressure of the chamber is increased to 105 kPa. (15 P_{sig}) At this point the sterilisation time begins to count down. Usually 15-30 minutes are used, depending on the type and volume of media. For instance, sterilisation of one litre of media at 121 °C requires 20 minutes. Larger volumes may require longer retention times for sterilisation. There are recommended holding times of 20, 10 and 3 minutes for 121, 126 and 134 °C, respectively. The high pressure in a closed container allows the temperature to exceed 121 °C.

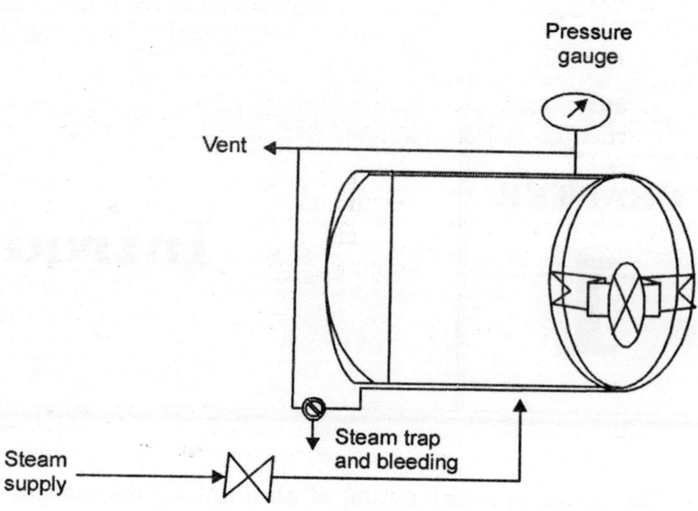

Fig. 1.1. Pressure steam sterilizer (autoclave).

CONTINUOUS STERILISATION

One of the methods for continuous sterilisation of medium for fermentation is the direct use of live steam by injection of steam into the medium. The heat exchanger is eliminated. The medium stays in a loop for a predetermined holding time until the entire medium is sterile. The problem with directly injecting steam is dilution of media since it is initially cold. However, it has better heat economy, since it comes from substituting heat exchangers for direct steam injection. To utilise all the supplied heat, preheaters or heat economisers are used. Instead of having a cold water stream to cool the sterile media, the lower temperature, unsterile media stream takes the primary heat from the warm stream, cooling the sterile media.

Continuous sterilisation has a holding coil for detention long enough to kill all the microorganisms. The medium from a make-up vessel flowing through the exchanger is held in the coil, and then goes through the heat exchanger, heating more unsterile medium while become cool itself, as it is collected in a sterile fermenter. Presents the sequence of piping media goes through,

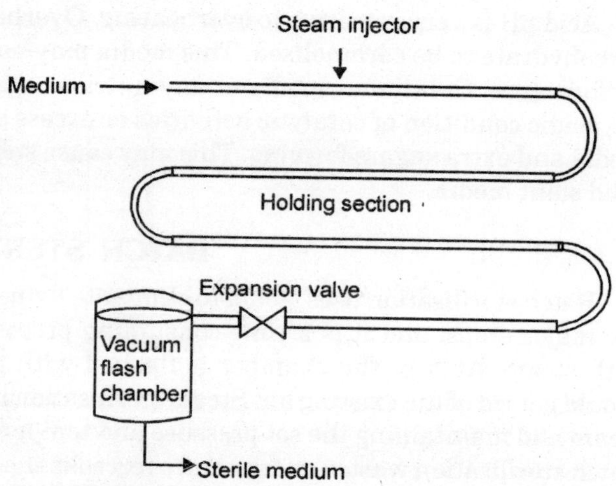

Fig. 1.2. Continuous sterilization.

along with steam. Finally, the media is collected in a vacuum flash drum. Heat economisers are important for large-scale plant unit in continuous sterilisation. For a small-scale bioreactor use of direct steam injection is much simpler to operate.

A heat exchanger is then needed with cooling water to bring the medium back to ambient temperature. Therefore, jacketed vessels are commonly used with cooling water. There are advantages and disadvantages to batch and continuous sterilisation. The energy savings are related to the use of an economiser and direct stream or indirect sterilisation. In a largescale system economy and role decide which one is the most suitable process to be implemented. For each case the specific design needs to be evaluated in terms of the thermal efficiency and fixed costs involved.

HOT PLATES

In directing steam heating, continuous sterilisation is preferable. The medium passed through La preheater may reach about 90 °C; then quite fast sterilisation may take at 140 °C. Fig. 1.3 shows a counter-current hot-plate heat exchanger is normally used for continuous sterilisation. Hot-plate heat exchangers are extremely efficient and easily maintained if any fouling or scale deposition takes place. Normally fouling is serious problem, occuring by deposition of proteins on the hot surface of the exchanger, which causes reduction in the overall heat transfer coefficient. Plate exchangers are easily separated and cleaned. Special care must be taken when the media contains any agglomerated particles. Correlation is required to calculate the suitable residence time for sterilisation.

The choice of a suitable sterilisation process is based on economics and cost of the heat exchangers to reduce energy consumption. In a suitable exchanger, the larger the heat transfer coefficient, the greater the energy recovery.

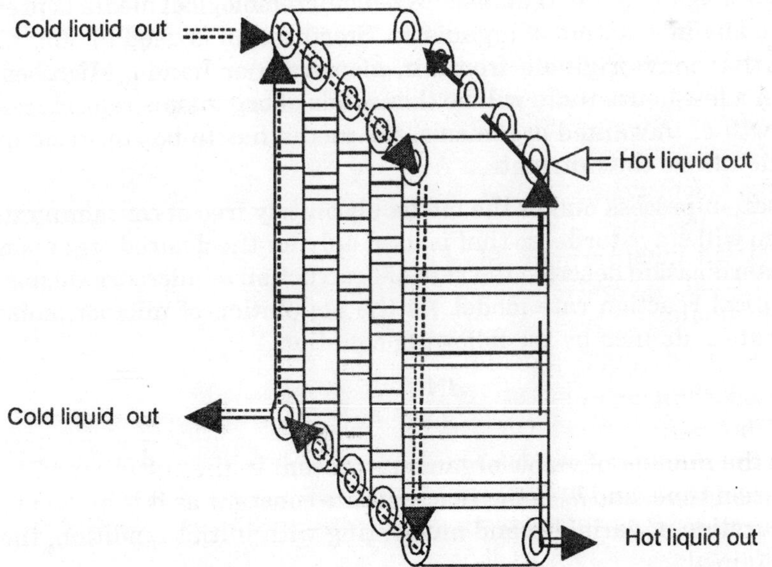

Fig. 1.3. Counter current hot plate heat exchanger.

HIGH TEMPERATURE STERILISATION

Fast sterilisation is performed at high temperature. High-temperature sterilisation requires short holding times. This technique is used in the fast preparation of nutrient media for industrial bioprocesses and in pasteurising milk. A short retention time may favour media with heat-sensitive proteins and cause less damage to the biochemical composition of the media than more prolonged times at lower temperatures. The ability of high temperatures to perform rapid sterilisation is related to activation energies. That is affected by how fast bacteria are killed in an elevated temperature.

On the other hand, long sterilisation at high temperatures may destroy the protein and biochemical composition of the media. Short-duration sterilisation at high-temperatures are more lethal to organisms and less chemically damaging than are longer sterilisation processes at lower temperatures. Sterilisation at high temperature is recommended for 3 minutes at 134 °C. It is preferable to 20 minutes at 115 °C in conventional operation.

STERILISED MEDIA FOR MICROBIOLOGY

Sterile media are used for pure culture. The media used to culture microorganisms depend on the living conditions of the microorganisms. Media compositions must be identified based on the needs of the organisms for carbon, nitrogen and phosphorus. Nutrients such as protein and sugars are added at wide ranges of pH. The acidic, neutral and alkaline conditions are defined based on differentiation of microbes in their biochemical reactions. Colour indicators are used to observe any transition of pH. Buffer is used in the media to maintain constant pH. Growth factors and growth stimulants are also used to accelerate microbial growth and to maintain exponential growth.

The most common media in the liquid phase is nutrient broth; in the solid phase, nutrient agar is used to propagate microorganisms. When microbiological media is prepared, it has to be sterilised before any inoculation of organisms. Sterilisation is used to eliminate any microbial contamination that may originate from air, glassware or hands. Microbes propagate quite quickely: within a few hours there will be thousands of organisms reproduced in the media. To control the growth of unwanted organisms, the media has to be sterilised quickly before any microbes start to utilise the nutrients.

The sterilisation process makes the media absolutely free of contaminants and guarantees that the medium will stay sterile, so that is used only for the desired organisms. The kinetics of culture media sterilisation describe the rate of destruction of microorganisms by steam using a first-order chemical reaction rate model. As the population of microorganisms (*N*) decreases with time, the rate is defined by the following equation:

$$-\frac{dN}{dt} = \boldsymbol{k_d N} \qquad \text{...(1.1)}$$

where N_0 is the number of viable organisms present in the culture media, t is the retention time or sterilisation time, and k_d is the reaction rate constant as it is known for a specific death rate. Using separation of variables and integrating with initial condition, the following useful expression is obtained:

$$N(t) = N_0 e^{-k_d t} \qquad \text{...(1.2)}$$

where N_0 is the number of viable organisms present in the media at the starting point before sterilisation. Now take the natural log; it is reduced to a linear model:

$$\ln \frac{N(t)}{N_0} = -k_d t \qquad \text{...(1.3)}$$

The graphical presentation of the equation shows a straight line with a negative slope for k_d. As the death rate constant follows Arrhenius' law, the death rate constant is temperature dependent. The value of k_d is about 0.02 min^{-1} at 100 °C, the death rate constant increases by 10-fold at 110 °C and 100-fold at 120 °C.[2]

$$k_d = k_0 e^{-E/RT}$$

where k_0 is the death rate constant at a reference temperature also known as Arrhenius' constant, R is the gas constant, T is the absolute temperature, E is the activation energy, which is 60-70 kcal per mole for microorganisms, 100-150 kcal per mole for spores, and 30-40 kcal per mole for media with vitamins and protein solution.[2]

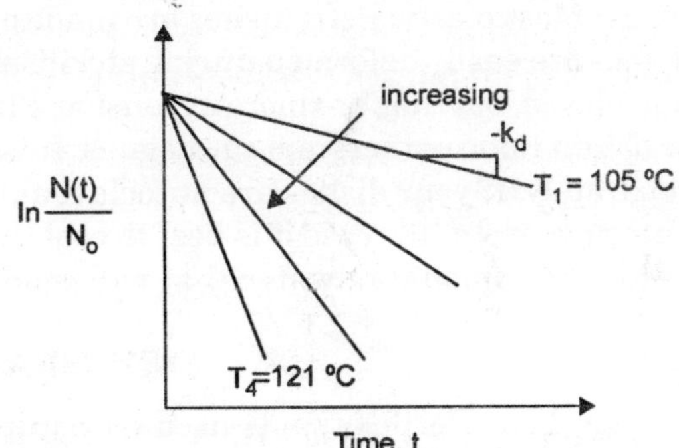

Fig. 1.4. The loss of cell viability at various temperatures.

The linear model for the loss of cell viability at various temperatures. As the temperature increases from 105 to 121 °C, the value for the slope of the line increases. This means that the number of viable cells at a fixed time of sterilisation will drastically decrease as the temperature increases by 16 °C.

Sterilisation of Media for Stoke Cultures

To clear slant tubes, culture agar solution is prepared. To have clear agar solution, it has to boil. Agar is a polysaccharide. It is not soluble in cold water, so heating helps to dissolve it in water. Once the solution is dispensed in the tube, it goes for sterilisation. Agar is in the solid state at room temperature, about 35 °C.

Sterilisation of Bacterial Media

Often, sterilisation of culture media is done in an autoclave at temperatures between 121 and 134 °C. It is good to know the damage caused to the medium by heating it. Heating the culture media, which contains peptides, sugars, vitamins, minerals and metals, results in nutrient destruction, either thermal degradation or reaction between the components of the media. Protein in the media is denatured by high temperature, or sugars are easily caramelised. During the heat treatment, toxic compounds are formed, which may retard or inhibit the microbial growth. For minimal damage to the ingredients of the media, it is important to optimise and minimise the heating and holding time of the sterilisation process, respectively.

Sterilise Petri Dishes

There are several ways to handle Petri dishes. The standard dishes are 15-20 mm deep and 12-15 cm in diameter. They are normally available in glass and also transparent plastic. Petri dishes are individually wrapped. The media is separately sterilised while we add the sterilised media to the Petri dishes in front of a flame. It is recommended to use 20-25 ml of medium in each Petri dish. These are all called culture dishes. The media contains about 1.5-2% agar. After the media cools off it solidifies. The Petri dishes with the nutrient agar should be sterilized and stored in a refrigerator. It is common to sterilise the media and Petri dishes separately. When the medium is cooled to about 55 °C, in front of a flame or in a laminar flow chamber, lift the lid of the dish enough to pour about 25 ml of the medium to the desired depth and lower the lid in place.

It is best to gently move the Petri dish in way that spreads a thin layer of agar uniformly without any air bubbles. Distribution of media in the Petri dishes should be done in front of a flame. Most plastic Petri dishes are made of polystyrene and are not autoclaveable. Plastic Petri dishes are easily deformed during sterilisation at high temperature. Some plastic dishes can be autoclaved, but they are more expensive. Please follow the instructions given by the manufacturer or obtain information from catalogues. It is recommended to autoclave glass dishes and medium separately. If your dishes are autoclaveable, you may dispense the agar into the dish and then autoclave it. In this case it is best to cool the dishes in the autoclave or in the pressure cooker to reduce the amount of water that will condense on the cover of the dish.

DRY HEAT STERILISATION

Dry heat sterilisation is used for equipment that can withstand high temperature and dry heat but cannot withstand wet or steam autoclave. This method is often used for glassware as it dries and sterilises in one operation. The pipets must be wrapped in dustproof aluminum foil or placed in metal pipette cans. The can lids are removed during heating and replaced after sterilisation, that is before any dust can get in the can. Disposable items are not recommended for dry heat sterilisation. This method may only be good for permanent reusable glass pipettes. Normal laboratory glassware must first be washed and cleaned. It has to be rinsed with deionised water.

The clean glassware is sterilised in an oven set at 200 °C for 1–4 hours. It is suitable to cover glassware with aluminum foil to maintain aseptic conditions after removing the glassware from the oven. If aluminum foil is not available, special heat-resistant wrap paper can be used. The sterile glassware must be protected from the air, which has micro-flora, or any contaminants. Avoid the use of any plastic caps and papers. Detach any labelling tape or other flammable materials, as they are fire hazards.

STERILISATION WITH FILTRATION

Certain media components are susceptible to heat; it is going to be denatured if it is heated. Therefore they must be added to the media after autoclaving. To do so, it is necessary to carry out filtration, the components using a 0.22 μm pore size filter that is appropriate to the solvent used. Filters are normally available from Whatman, or Fisher Scientific. It is recommended you consult filter experts or suppliers about the solvent and the special application you are intending to implement for the desired filters for the process. Normally filtration of media is used instead of sterilised media with an autoclave. To filter sterilised media, filter the water using a 0.45 μm pore size filter, before using a 0.22 μm pore size filter for final sterilisation.

MICROWAVE STERILISATION

Rapid sterilisation of media is achieved by using microwave ovens. Most plant tissue culture media can be sterilised using a microwave, although it may not be suitable with a medium containing complex additives like oatmeal.

ELECTRON BEAM STERILISATION

Electron beam sterilisation is a high-voltage potential established between a cathode and an anode in an evacuated tube. The cathode emits electrons, as a cathodic ray or electron beam. A high intensity of electrons is produced. These electrons are accelerated to extremely high velocities. These accelerated electron intensities have great potential as a bacteriocide. Most electron beams operate in a vacuum. As a result the unwanted organisms in the media vanish and the media is sterilized. The electron accelerator equipment producing the high-voltage beam also has various applications in the medical field and surgical supplies. Electron beam sterilisation is a successful technology used for sterilisation of disposable medical appliances and devices with a wide range of densities. The electron beam inactivates microorganisms that cause destruction of biomolecules, and results in death of microbes by an indirect chemical reaction. Irradiation by gamma rays is a similar mechanism. Advanced electronics precisely control the use of electron beams in the sterilisation of medical devices. Electron beam sterilisation results in less material degradation than with gamma irradiation. Medical products are sterilised in the original shipping containers, saving lots of time and maintaining integrity of the original package.

CHEMICAL STERILISATION

Chemical agents are used to sterilise heat-sensitive equipment. Chemical solutions are used as a suitable method for sterilising long pipettes and glassware. Normally at pilot scale in the absence of life, steam and chemical agents are often used. Application of an oxidising agent such as 10% chlorox for 20 minutes or longer proves the system operates without any contamination. Excess amounts of chemical agent have to be removed; otherwise organisms are able to grow in a toxic environment. The bleach (sodium hypochlorite) then needs to be removed by rinsing with sterile water or rubbing with ethyl alcohol, 70% ethanol-water (pH = 2) or an alcohol solution of 70% *iso*-propanol without recontaminating the glassware or tools. Ethanol is commonly used for cleaning surfaces. Use of bleach on metal devices is not suitable as it corrodes metals rapidly.

Ethylene oxide in the gas phase is commonly used. This is a special chemical effectively used for column bioreactors. It is a volatile compound and strong oxidizing agent. It boils at ambient temperature, therefore the solution of ethylene oxide (liquid phase) must be stored in a refrigerator (4°C). An excellent oxidizing agent such as a 3% sodium hypochlorite is used for chemical sterilization of equipment.

Tyndallization

John Tyndall, a famous British scientist, noticed that a period of about 24 hours at ordinary room temperatures is usually sufficient to enable dormant and heat-resistant spores to germinate and grow into the vegetative state in which they are sensitive to heat. He devised a process of sterilization based on these observations. It consists in steaming for a few minutes at 100° C, on three or four successive occasions, separated by 24-hour intervals at a temperature favorable to spore germination. The intervals permit the dormant, resistant spores to become active, vulnerable vegetative cells, readily killed by 100° C. This process renders an infusion sterile, whereas one single continuous boiling for one hour may not, since many spores remain in their dormant and resistant state during this time. This intermittent process is sometimes put to practical use in the home canning of foods. A disadvantage of the method is the time required to achieve complete sterilization; an advantage is that it requires no special apparatus. In some fluids, such as water, spores may not grow out promptly. Also, if the material is freely exposed to air, anaerobic spores may not germinate and may survive the process. If not freely exposed to air, aerobic spores will not grow out freely.

Table 1.1. Boiling point at various altitiudes

Location	*Altitude**	*Boiling Point***
New York City	0	100
Chicago, Ill	589	98.9
Denver, Col	5,280	94.3
Fort Laramie, Wyo	7,390	92.2
Tahoe, Nev	10,000	89.1

*Feet above sea level.

Degrees C.

* Feet above sea level.

** Degrees C.

Compressed Steam: Autoclaving

Anyone familiar with the operation of a home pressure cooker is familiar with an *autoclave,* because the cooker is a simple form of autoclave. In the autoclave, be it a small and simple home pressure cooker or an apparatus large enough to fill a room and fitted with various gauges, pipes, valves, clocks and wheels, the object of both is alike: to heat the articles to be sterilized by means of steam under considerable pressure.

Steam under pressure is hotter than boiling water or free-flowing steam such as is used in tyndallization. The higher the steam pressure, the higher the temperature will be.

It must be remembered that it is the compressed steam (moisture, hydration) that sterilizes and not compressed air (dry and usually not as hot as steam).

For example, pure steam at 15 pounds pressure has a temperature of 121° C. If the steam is mixed with an equal amount of air, at the same pressure the temperature is only 110° C, while if the mixture is two thirds air the temperature is only 109° C.

Steam *hydrates* and thus promotes coagulation; air does neither. Steam, being water vapor, also produces *hydrolysis* at autoclave temperatures. Dry air cannot do this at any pressure or temperature. In autoclaving, therefore, as in using a pressure cooker, a valve is left open for the escape of all air before the steam pressure is allowed to rise.

The actual amount of water present as steam in an autoclave is small, and articles soon dry off after removal, especially if removed from the autoclave while hot.

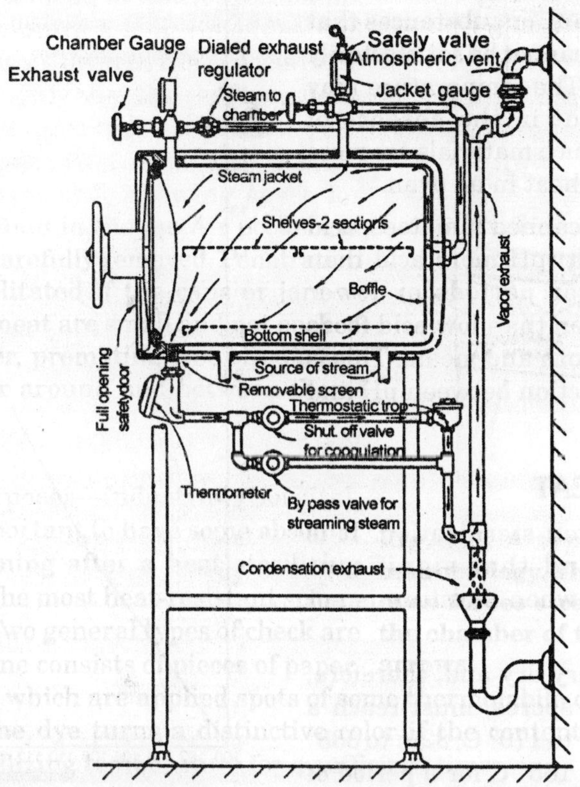

Fig. 1.5. Diagrammatic illustration of steam jacketed autoclave. Steam enters the *jacket,* a double-walled shell, at two places just beneath the cylinder. It passes out the top through a pipe to which are attached: a wheel valve admitting the steam to the inner *chamber,* a *safety valve,* a gauge showing *pressure in the jacket.* The steam enters the inner chamber at the right of the diagram, filling the upper portion. Its pressure registers on the *chamber gauge.* It may be allowed to escape rapidly by the *exhaust valve.* If this is closed, steam pushes the cooler air in the lower portion out at the bottom (left), where the *thermometer* registers proper temperature only when the air is gone and is followed by the hot steam. The escaping steam may be allowed to flow out without building up any pressure if the *by-pass valve* is fully opened. If the bypass valve is closed and the *shut-off valve* is opened, steam passes through the *thermostatic trap* where the heat shuts off all but a pinhole opening. This causes pressure to build up in the chamber, yet prevents stagnation by permitting a constant minute flow of steam through the apparatus.

The *thermometer* on the autoclave is the important guide to the process of autoclaving, not the pressure gauge. However, the latter, as well as a steam-escape or safety valve, is essential to safety.

The common practice in autoclaving fluids or freely exposed surfaces such as those of dishes and instruments is to apply 115° to 125° C (10 to 20 pounds' pressure) for 20 minutes. The pressure must be allowed to subside *slowly* after the heating is over or *superheated* fluids in open (cotton-plugged) vessels will boil over. Any large, solid masses must be heated a longer time to allow for heat penetration. Packages must be spaced so as to allow free circulation of steam. Substances such as mineral oil or petrolatum, sand, or any dry objects in tight jars, or substances that are impervious to moisture, cannot be satisfactorily sterilized in the autoclave. The temperature may rise as high as 125° C, but in the absence of moisture this is ineffective. Such materials are more effectively sterilized by dry heat in an oven.

Acid materials, such as canned tomatoes, acid fruits, pickles or sauerkraut (pH more acid than 4.5), require much shorter periods or lower temperatures for sterilization than low acid foods (pH 6.5 to 7) such as milk, corn and meats. This is because of the synergistic action between pH and temperature.

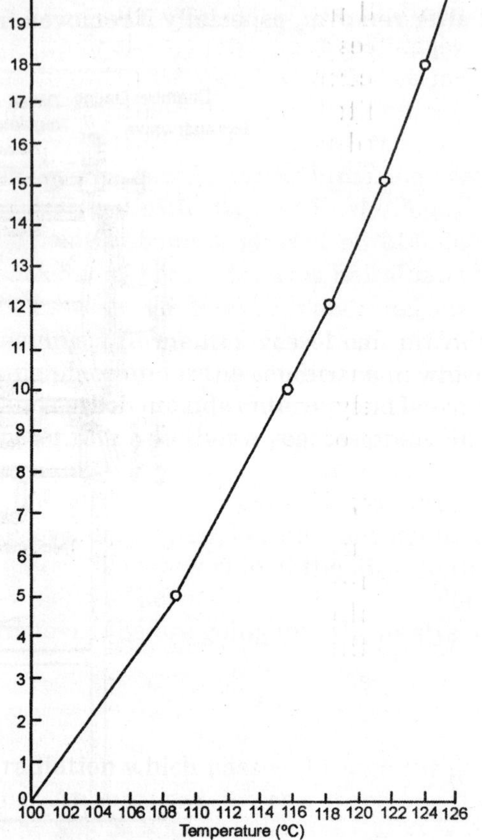

Fig. 1.6. Relation between temperature and pressure of pure steam.

DRY HEAT

Dry heat is used in oven sterilization. It is necessary to bear in mind that significant coagulation does not occur when moisture is not present.

Articles in ovens are very dry and, therefore, in order to be freed of live spores, must reach a very high temperature (165° to 170° C; 329° to 338° F). It is customary to apply 165° C for a period of two hours. This accomplishes no coagulation but, what is more effective, slight charring.

A home oven can easily be used for sterilization. A "moderate" temperature (330° F) is satisfactory, and the heating should be allowed to proceed for two hours *after* reaching that temperature. Paper wrappings should be slightly browned but not brittle; muslin or string should be faintly yellow due to the heat.

Only dry articles not injured by baking (glassware, bandages, instruments, mineral oils, petrolatum, talcum powder and the like) may be thus sterilized. Solutions containing water, alcohol or other volatile substances will, of course, boil away and be ruined.

TIME AND CONTROLS

Thorough Heating Necessary

In any process of disinfection or sterilization by heat it is absolutely essential that the object be heated through and that the center of the object be held at a killing temperature long enough to destroy the bacteria.

Thus, in home canning, a quart jar of spinach may be held in free steam (100° C) for five to ten minutes and when grasped with the hand will feel very hot, yet the center may be well below a microbicidal temperature. Large masses of nonfluid materials like quart jars of canned vegetables and roast meat, in which the contents cannot circulate, require a long time (one and one-half to two hours), even in the autoclave, to be heated so thoroughly that the center reaches a sporicidal temperature. Pieces of meat or vegetables to be sterilized in jars by heating should be loosely packed, allowing space for circulation of the fluid in the jar. Air pockets in the jars should be carefully removed. Penetration of the heat is facilitated if the cans or jars or pieces of roasting meat are small and not packed too closely together, promoting free circulation of steam or hot air around and between them.

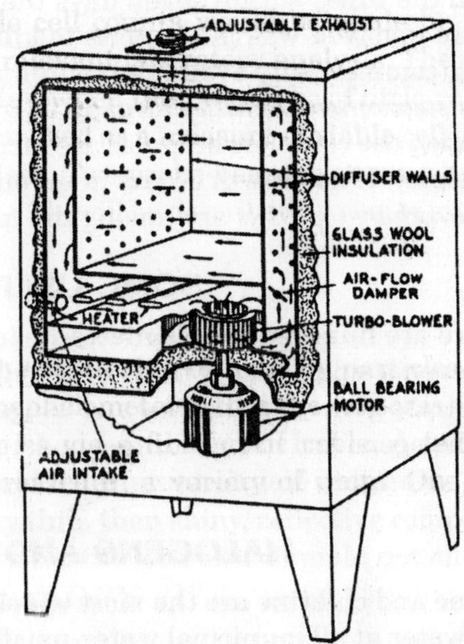

Fig. 1.7. A type of sterilizing oven for microbiology. The motor at the bottom of the oven forces streams of hot dry air through the chamber of the oven, as indicated by the arrows.

Sterility Check

For many purposes—industrial, hospital, research—it is important to have some absolute means of determining after a heat sterilizing process that even the most heat-resistant spores have been killed. Two general types of check are commonly used. One consists of pieces of paper, gauze or thread to which are applied spots of some thermolabile dye. Inserted in the materials being sterilized, the dye turns a distinctive color if the contents of the sterilizer have been subjected to a sterilizing temperature for a sufficient time.

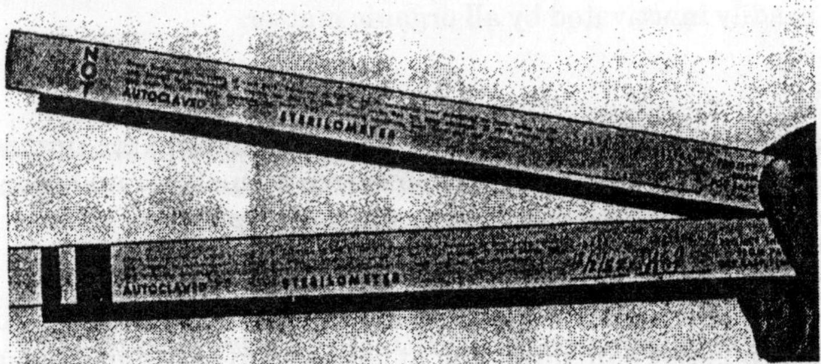

Fig. 1.8. One type of device for indicating proper autoclaving. Before heating the word "NOT" appears clearly as on the upper indicator. After partial heating the leftmost black bar appears as seen on the lower indicator. Not until autoclaving is complete both as to time and temperature is the word "NOT" obliterated by a second dark bar on the lower indicator.

A more direct check (though more time-consuming) consists of strips of filter paper having on them the very heat-resistant spores of *Bacillus subtilis*. This is a harmless aerobic bacterium. The spore-impregnated strips are inserted (enclosed in sterile envelopes) inside materials being sterilized. Afterward the strips are removed to tubes of culture medium and incubated. If growth occurs in the tubes within seven days then it is evident that the sterilization process in which they were involved was inadequate to kill the spores. In a modification of this process, sealed ampules (commercially available) containing culture medium with dormant spores of harmless *Bacillus stearothermophilus* already in it are inserted in the materials being sterilized. After sterilization the complete ampules are incubated at 55° C (since *B. stearothermophilus* is a strict thermophile). If no growth occurs in seven days it is proof of effective sterilization, since spores of *B. stearothermophilus* are among the most heat-resistant of known living things.

SOME USEFUL DISINFECTANTS

There are hundreds of disinfectants in the market and it would obviously be impossible to discuss even a small part of them here. A few commonly used representatives of several classes will illustrate general principles and usages. For convenience we may group the disinfectants as follows: halogens and halogen compounds, compounds of heavy metals, phenol and its derivatives, alcohols, detergent disinfectants, microbicidal gases and radiations.

HALOGENS AND HALOGEN COMPOUNDS

Iodine and chlorine are the most widely used of this group. Chlorine gas is used to disinfect filtered water at all municipal water-purification plants and many sewage-disposal plants. It is usually handled in tanks or tubes like oxygen. It is applied to drinking water in a final concentration of about 1.0 part per million.

A more convenient form of chlorine for the individual and household user is calcium or sodium hypochlorite. Solutions (5 per cent) of sodium hypochlorite are purchasable in ali grocery stores under various names (*e.g.,* Clorox). They have a multitude of household and sanitary uses. They depend for their effectiveness probably on their liberation of free chlorine. Directions for use of these solutions are on each bottle. Four points need to be remembered in their use: the chlorine tends to evaporate from the solution; they give a bad odor to the hands; chlorine is poisonous; and it is readily inactivated by all organic matter.

Chloride of Lime

Chloride of lime (0.5 to 5 per cent aqueous solution) is excellent for similar purposes and is inexpensive. The ordinary chloride of lime of commerce is unstable and soon loses most of its free chlorine. Many organic chlorine compounds which liberate their chlorine more slowly are very effective. *Azochloramid* is one of these; *dichloramine toluol* is another. The odor of chlorine may be objectionable.

In addition to the above substances, there are many new proprietary chlorine compounds on the market based on principles given above.

Lodine

The most actively antimicrobial of the halogens is iodine. However, it is not commonly used for the same large-scale purposes as chlorine because of its physical properties and its cost. Formerly used widely as a local antiseptic in a 7 per cent tincture (alcoholic solution), this strength is far too caustic and poisonous for local application and does more harm than good. The 2 per cent tincture is much better and is the most generally useful for small cuts and abrasions. If the alcohol causes too much pain, the tincture may be diluted with two to four volumes of water.

An aqueous solution containing approximately 1.85 per cent of iodine and 2.2 per cent of potassium iodide is excellent but has a high surface tension.

Lodophors

Nonirritating, nonstaining, virtually odorless and very effective solutions of organic compounds of iodine in which the iodine is loosely combined with a surfactant organic *wdophor* (Gr. *phor* = bearer or carrier) are available commercially. The iodophor releases the iodine slowly and also lowers the surface tension of the solution. Some idea of the amount of iodine remaining in such solutions is readily gained by comparing the diminishing brown color with that of the original full-strength solution. This is not possible with colorless chlorine solutions. Examples of such iodine products are Wescodyne, Iodide and Betadine. In such forms iodine may be used for a variety of purposes, from clinical uses to sanitization of dishes, with aqueous solutions containing 200 parts per million. Under certain conditions of concentration and temperature such solutions may be sporicidal. They appear to kill bacteria, molds and some viruses quickly. Iodine is a much more effective germicide than chlorine.

Iodine, like chlorine, combines indiscriminately with organic matter; its solutions are unstable and, like chlorine, it is poisonous.

COMPOUNDS OF HEAVY METALS

The first scientifically designed, specifically acting drug was based on the heavy metal arsenic. Arsenic was compounded with an organic group to diminish its poisonous properties. Prior to 1907, Paul Ehrlich and his coworkers had synthesized and tried many such compounds in the treatment of spirochetal diseases, notably syphilis, without success. The 606th compound ("SOS" for "606" or Salvarsan or arsphenamine), discovered in 1907, was very effective and, with various modifications such as neoarsphenamine, was used for years in the treatment of syphilis and other spirochetoses. It fell into disuse after 1943 when it was found by Mahoney et al., in the U. S. Public Health Service, that penicillin is much more active against the same organisms and almost nontoxic.

Other metals in common use as microbicides today are iron and zinc, previously discussed, and mercury, silver and copper. All act mainly by coagulative action except in very low concentrations (*e.g.,* 1: 10,000 and lower) in which they appear to have specific molecular toxic effects (see *oligodynamic action* and *ionic effects*).

Mercury compounds were first used in the fifteenth century specifically to treat syphilis. More recently zinc in a chelated form called *Zinc Omadine* has been shown to have potent microbicidal properties, as well as the sodium salt of *Omadine.*

Bichloride of Mercury

Long used as a general disinfectant and as an antiseptic, bichloride of mercury (mercuric chloride) is now much less used than formerly. For general disinfection it may be prepared in dilutions of 1: 1000 to 1:5000. However, it is very toxic to tissues, very poisonous internally and corrosive to materials, especially metals (hence its older name of *corrosive sublimate).* In many situations it fails because it forms a thick coating of coagulum on the outer surface particles of mucus, tissues and feces, and thus protects the microorganisms within. As described previously, in dilutions of 1:10,000 or more its action is more bacteriostatic than coagulative. This action may be reversed by sulfur compounds unless coagulation has occurred.

Organic Molecules

Attempts have been made, following Ehrlich's tead with arsenic, to decrease the toxic, corrosive and irritating qualities of mercuric disinfectants by incorporating mercury in complex organic molecules. This has yielded a number of products that are less toxic and irritating than bichloride of mercury, among them *Mercumchrome, Merthiolate* and *Metaphen*. In the presence of organic matter they are more effective than bichloride of mercury. Mercurochrome is often used for local, superficial application for disinfecting cuts, wounds and skin. Some of these organic mercury compounds are used as preservatives in biological materials. Other useful organic mercurials are *phenylmercuric nitrate,* very effective and of sufficiently low toxicity for external use, and *ammoniated mercury,* long used as a 10 per cent ointment for external wounds and fungal infections of the skin. These mercurials all mainly act microbistatically, and their action, like that of bichloride of mercury, is reversed by thioglycollate and hydrosulfuric acid. All mercury compounds are poisonous internally to some degree, and mercury is absorbed through the skin.

Silver Nitrate

Silver nitrate (1 per cent aqueous solution) is used principally for application to infants' eyes at birth to prevent infection. Silver is also used as a local disinfectant in the form of organic colloidal preparations such as Argyrol or Protargol, in which form it is nonirritating and may be used in adult eyes at 20 per cent concentration. It is not for use in infants' eyes in place of silver nitrate. The use of silver nitrate (1 per cent aqueous) or an equivalent, approved antibacterial in the eyes of infants at birth is required by law in most states as a safeguard against blindness of the newborn *(ophthalmia neonatorum)*. However, antibiotic treatment may be satisfactory in some situations.

Copper Sulfate

Copper sulfate is used chiefly to control the growth of algae in open water reservoirs and as a fungicide in garden sprays *(Bordeaux mixture)*. Organic compounds of iron are also used as fungicides in garden sprays.

PHENOL AND ITS DERIVATIVES

Phenol (C_6H_5OH) (carbolic acid) in the pure state is not commonly used as a disinfectant because it is expensive and because there are more than a dozen derivatives that are not only less costly but more effective. All act similarly to phenol (*i.e.,* mainly by coagulation, partly by little understood lytic and toxic effects). Unlike the halogens, which are indiscriminately effective against a wide range of microorganisms, susceptibility to phenolic disinfectants varies greatly. Some organisms use phenol as a food.

Surface (Residual) Disinfection

Phenol and, more especially, many of its derivatives are surface tension reducers and are commonly used in solutions containing additional surfactants to improve wetting properties.

Since they are surfactants themselves, the phenolic disinfectants tend to be adsorbed in thin, and more or less durable, films on inert surfaces to which they are applied, resulting in what is termed *surface,* or *residual,* disinfection. As compared with the volatile halogens or alcohols their action is thus prolonged, often for many hours. This is an especial value of certain preparations of phenol derivatives for surgical and hospital hand washes.

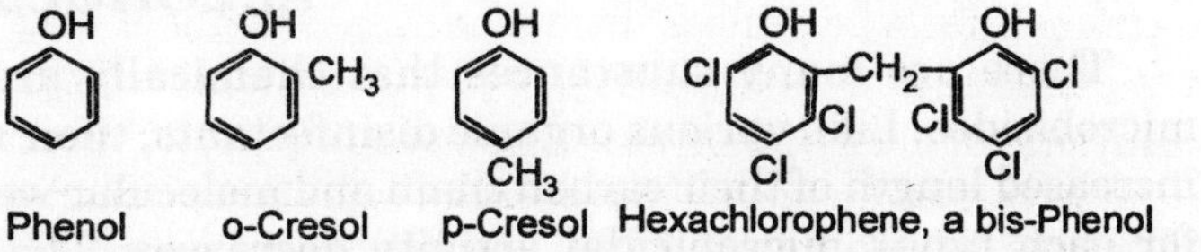

Phenol o-Cresol p-Cresol Hexachlorophene, a bis-Phenol

Fig. 1.9. Structures of phenol, representative cresols and one commonly used bisphenol (hexa-chlorophene).

Types of Phenolic Disinfectants

Phenolic disinfectants are of many types, common among which are: the *cresols,* the molecules of which are like phenol but have methyl groups attached; and the *bis-phenols,* the molecules of which consist of two phenolic groups joined directly or through some other radicle.

Crude Cresols

Crude cresols form colloidal (milky) suspensions in water and are therefore especially effective as disinfectants, because each colloidal droplet consists of *concentrated* cresol. This is true of many colloidal disinfectants. There are many excellent cresol preparations on the market. Some have a clinging odor. Most of them contain a surface tension reducer in addition to the disinfectant. They are commonly used for disinfecting floors, furniture, barns and stables in from 1 to 5 per cent concentrations. Some may also be used in temporary contact with human tissues in 0.5 to 1.0 per cent strengths. Good examples are Lysol and saponated cresol solution, *U.S.P.* The preparation of such solutions is an important industry.

Bis-Phenols

Like the cresols, bis-phenols have surfactant properties. In order to increase their solubility in water they are commonly used in solutions with other surfactants or detergents. In general they are less toxic than the cresols. Among well-known bis-phenols is one combined with chlorine (one of the class of *halogenated bis-phenols),* called Hexachlorophene (G 11). Because it is a surfactant it provides good residual surface disinfection of the skin, and it is not irritating if used properly. It is incorporated especially in surgical soaps and hand washes (Gamophen, pHisohex and Hex-O-San).

Orthophenylphenol

A similar compound orthophenylphenol, is used in mixtures such as *O-syl. Chlorothymol* and *Chlorohexidine* are other halogenated bis-phenols which are similar in general properties. All are broadly germicidal, and each has distinctive properties that adapt it to particular situations. Some can be used effectively as irrigations in surgical wounds in 1 :1000 to 1:10,000 concentrations.

The activity of all phenolics is markedly depressed in the presence of extraneous organic matter, and they are highly microbistatic.

ALCOHOLS

There are many substances that chemically are alcohols: many of these are effective microbicides. Like various organic disinfectants, their microbicidal activity is directly related to increased length of their carbon chain and molecular weight; beyond their critical point (different for each type): microbicidal activity decreases. The two alcohols most commonly used for disinfection or sanitization are ethyl ("grain") alcohol and isopropyl ("rubbing") alcohol. They are mild disinfectants and are nontoxic on external application. Methyl or "wood" alcohol and its vapor are extremely poisonous and should never be used in any bodily contact whatever.

Ethyl alcohol is a surface tension reducer, a lipid solvent and a coagulating agent, properties that should contribute to microbicidal potency. However, it is also a potent dehydrating agent, a property that interferes greatly with its coagulating power and probably also with its other antimicrobial properties. Because of its dehydrating power it is best used as a 70 per cent aqueous solution. Stronger concentrations (95 or 100 per cent "absolute") are much less effective. As a coagulating agent it tends to combine nonspecifically with extraneous organic matter and can produce a thick, protective coating of coagulum around microorganisms. As a lipid solvent it is useful as a *cleansing* agent ("disinfectant"?) for skin prior to hypodermic injections. It adversely affects viruses and cells with envelope lipids. As a surface tension reducer it is a potent adjuvant of other disinfectants such as phenolics and quaternaries. Similar considerations apply to isopropyl alcohol.

DETERGENT DISINFECTANTS

As just mentioned, not all detergents are disinfectants. However, several compounds combine the properties of detergent, surface tension reducer and disinfectant.

Soaps

Among these versatile compounds are the common soaps; potassium or sodium esters of higher fatty acids (stearic or glyceric). They are actively lethal for certain species of microorganisms: *Treponema pallidum,* the syphilis spirochete; *Diplococcus pneumoniae,* a cause of pneumonia; some streptococci; and a few other organisms. Unfortunately, with these few exceptions, soaps are relatively ineffective as general disinfectants. However, soaps are potent surface tension reducers and consequently are good emulsifying and foaming or "sudsing" agents. Foaming results from the fact that, since the surface tension is greatly weakened, the surface can be extended almost indefinitely as foam or as droplets in an emulsion; compare with mercury. When soapy water is applied to skin or instruments, microorganisms are carried into the emulsified grease droplets and removed by rinsing. Detergency is the chief virtue of neutral soaps. Soluble soaps are precipitated by acids, ions of heavy metals, and by Ca^{++}. Hand soap is often used as a carrier for phenolic disinfectants, as previously discussed.

Synthetic Detergents

As generally understood, this term comprises a group of potent surface tension reducers that are not typical soaps. They are synthesized commercially and much used for household and commercial cleansing and laundry purposes. They are available in "large, economy" packages in all grocery stores. Most of them are long-chain, alkyl-benzyl-sulfonates. Few are markedly microbicidal. They are *anionic* compounds (see following paragraphs).

Another group of synthetic compounds, classed as detergents because they are strong surface tension reducers, are much more important as disinfectants and sanitizing agents. The first was discovered in 1936 by Domagk (Nobel Prize winner), who described the properties of dodecyl-dimethyl-benzyl-ammonium chloride, now familiar as *Roccal* or *Zephiran*. This is a quaternary alkyl ammonium salt, and is representative of a number of similar compounds that have since been developed commercially as disinfectants or detergents. As a group they are commonly called *quaternaries* or "quats." Basically, they are ammonium halides (for example, ammonium chloride) in which the hydrogen atoms have been replaced by various alkyl groups.

$$\left[R-\overset{R}{\underset{R}{N}}-R \right]^{-}Cl\ (^{-}Br,\ etc.)$$

A. General formula of alkyl ammonium halides

$$\left[CH_3-(CH_2)_{11}-\overset{H_2}{C}-\overset{CH_3}{\underset{CH_3}{N^+}}-CH_2-C_6H_5 \right]^{-}Cl$$

B. Alkyl-dimethyl-benzyl-ammonium chloride; a cationic, microhicidal detergent

$$\left[CH_3-(CH_2)_{11}-O-\overset{O}{\underset{O}{S}}-O^- \right]^{+}Na$$

C. Sodium lauryl sulfate; an anionic mirobicidal detergent.

Fig. 1.10. Forms of microbicidal detergents.

Chemical Structure and Activity

In general, the more effective of these detergent-disinfectant compounds are those in which at least one of the alkyl groups contains a chain of 12 to 16 carbon atoms, whereas the less effective are those of lower or higher molecular weight. This represents a common observation previously mentioned; *i.e.*, the germicidal efficacy of many organic compounds is greatly affected by chemical structure, especially length of carbon chains.

Cationics, Anionics and Nonionics

There are three types of microbicidal detergents: (1) those in which the surfactant portion is the *cation* (*e.g.*, quaternary alkyl ammonium halides those in which the surfactant portion is the *anion* (*e.g.*, complex soaps like sodium lauryl sulfate; and (3) those that do not ionize. These compounds are classified as *cationic, anionic* and *nonionic* detergents, respectively. The cationic compounds appear to be the most generally effective as disinfectants and are what is generally meant by the term "quaternary." Soaps are anionic detergents. The nonionic compounds are relatively weak disinfectants. They are sometimes used with anionic compounds in household detergents as adjuvants.

Mechanism of Action

Since detergents like the quaternaries and sodium lauryl sulfate are generally alkyl compounds of chlorine or sodium, they are bipolar, *i.e.*, they exhibit both fat-soluble (lipophilic) and water-soluble (hydrophilic) sides. When dissolved in aqueous culture media, the longer alkyl chains tend to be oriented toward, and to dissolve in, the cell lipids. The negatively charged residue (chlorine or sodium ion) is oriented away from the cell and is drawn toward the surrounding aqueous fluid. The surface tension reducer molecules are systematically arranged side by side at the cell-aqueous medium interface, as closely placed as possible.

Since the reducer molecules at any surface have a lower mutual attraction than the water molecules in which they are dissolved, the surface tension of their solution is lowered. Lipids upon which detergents in aqueous solution become oriented tend to become divided into smaller and smaller particles, each coated with the surface tension reducer (*i.e.,* to be *emulsified);* detergents thus characteristically "cut" (remove) grease. The lipophilic property of quaternaries and alkyl sulfates is thought to disrupt the lipids, and therefore the whole structures, of bacterial cell walls and membranes, especially of those rich in lipids like many gram-negative bacteria. Disruption of the cell walls and membranes would permit outward passage of vital intracellular components and passage inward of the surface tension reducer. Electron micrographs show the destruction of bacterial cell walls and inner structures by antimicrobial substances which reduce surface tension. Probably these detergents also act in part by occluding the bacterial surfaces by being adsorbed there. In use, soaps and other anionic compounds should not be mixed with cationic compounds, because the oppositely charged molecular complexes neutralize each other and the agents are precipitated or inactivated. For the same reasons hard, acid and iron-rich waters tend to inactivate them.

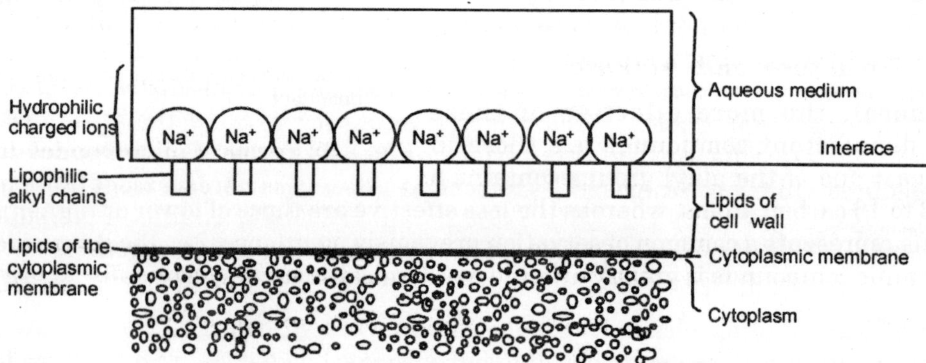

Fig. 1.11. Orientation of polar groups of an *anionic,* surface-active detergent, in this case sodium lauryl sulfate [$Na^+(C_{12}H_{25}OSO_{,1})^-$], in relation to the surrounding aqueous medium and a cell having lipids in its cell wall or cytoplasmic membrane. The hydrophilic group (here represented by Na^+) orients at the fluid-cell interface; the hydrophobic group (here represented by the lauryl sulfate group$^-$) is oriented in the lipids of the cell wall or membrane.

At the air-water interface the hydrophilic groups orient themselves within the surface water while the hydrophobic alkyl groups are at the air-water interface, weakening the intermolecular attraction between water molecules at the surface; *i.e., lowering the surface tension* of the water. Accumulation of molecules of a surface-active, polarized agent at any interface between solids and fluids or between immiscible fluids is called *adsorption.*

A very similar relationship would be that between a cell and a *cationic* detergent of the class of quaternary ammonium compounds; *e.g.,* cetyl pyridinium chloride (Ceepryn):

$$\left[\bigcirc N—C_{16}H_{33} \right]^+ Cl^-.$$

Unlike many other disinfectants, quaternaries retain much of their activity in the presence of considerable organic matter. The quaternaries tend to be microbistatic or slowly lethal rather than to kill instantly. When the effectiveness of quaternaries as disinfectants is being measured, their marked microbistatic action must be offset by the use of special inactivators to allow the microorganisms to grow in the subculture. *Tinctures* of quaternaries appear to be very effective indeed against tubercle bacilli, and some are effective against molds, but they are of limited value against certain viruses.

Since they are emulsifying agents for lipids quaternaries should be effective against viruses that have lipid-rich components, *e.g.*, envelopes, and there are data in support of this view. Opinions differ concerning other viruses. Pi-cornaviruses are resistant. Quaternaries are not sporicidal. They are much used for disinfection of skin and for purposes of sanitization in restaurants and hospitals.

MICROBICIDAL GASES

Attempts to cure and prevent disease with vapors and smokes date from the earliest human records, when "aromatics," smudges, perfumes and the like were endowed by popular supposition with almost miraculous, and generally quite factitious, powers to drive away evil spirits and "noxious humors."

More objective studies in the last few decades have yielded several exceedingly valuable results in the form of definitely sporicidal, and therefore *sterilizing*, gases: formaldehyde, ethylene oxide and beta-propiolactone. Chlorine and sulfur dioxide, formerly used, are too destructive and poisonous for general purposes. As sterilizing agents, gases have the advantages that destructive heat and aqueous solutions are not needed and that they can be applied in large volumes in buildings. On the other hand they affect only exposed surfaces, except those made of porous or permeable materials, and they must be applied under conditions of controlled temperature and humidity. Such sterilizing agents are often referred to as *sterilants*.

Formaldehyde

This very irritating gas (formula: HCHO) is generated by heating para-formaldehyde or concentrated solutions of formaldehyde. Unless very carefully applied at raised temperatures, formaldehyde tends to polymerize as paraformaldehyde or *paraldehyde* in a thin white film over all surfaces, and hence has definite limitations to places not containing valuable furniture or equipment. Its action is to some extent microbistatic and reversible with sulfite ions. In some solutions it is definitely sporicidal and therefore sterilizing but, like bichloride of mercury, it tends to form protective coatings of coagulum around microorganisms if extraneous organic matter is present. For best results a relative humidity of about 70 per cent and a temperature of about 22° C, are requisite. Formaldehyde vapor has low penetrability into such materials as blankets. It is much less used now than formerly.

Ethylene Oxide

This substance (formula: $CH_2 \cdot O \cdot CH_2$) is liquid at temperatures below 10.8° C (51.4° F), its boiling point. It is relatively inexpensive and is handled in metal tubes or bottles like other compressed gases. In the pure state it is very toxic, irritating and explosive. For use it is mixed with other substances: ethylene oxide, 10 per cent and carbon dioxide, 90 per cent (sold as Carboxide); ethylene oxide, 20 per cent and carbon dioxide, 80 per cent (sold as Oxyfume); ethylene oxide, 11 per cent and halogenated petroleum, 89 per cent (sold as Cryoxcide). Each preparation has special advantages in certain situations.

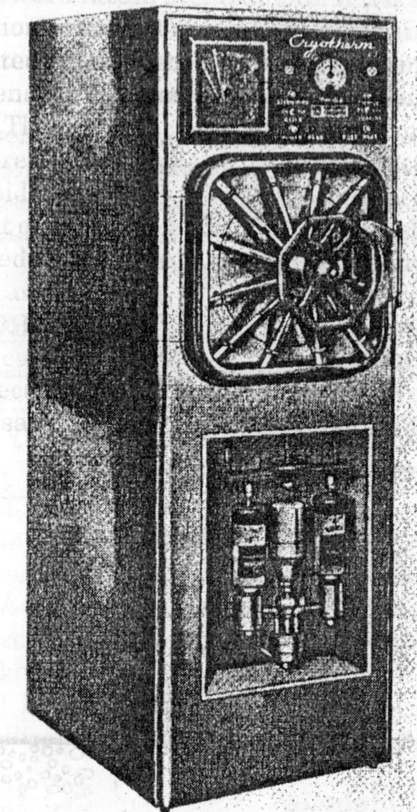

Fig. 1.12. Autoclave arranged for sterilization by sporicidal ethylene oxide gas (Cryoxcide). Below the steel door are seen the containers of gas and a water inlet so arranged that the vapors can be mixed and humidified (much as gasoline, air and water vapor are mixed in a modern automobile carburetor) before admission of the sterilizing vapor mixture to the chamber of the autoclave. Above the door are the dials controlling sterilizing temperature and time, vacuum pump, gas and humidity inlet and a maximum and minimum recording thermometer. Operation is almost completely automatic.

The desired mixture is generally applied in autoclaves or other closed containers where humidity, temperature and pressure may be controlled. The concentration of the gas is generally expressed in milligrams of pure ethylene oxide per liter of space. A typical use situation is 500 mg. of gas per liter at about 58° C (130° F) and a relative humidity of 40 per cent for four hours. If one factor is varied the others must be varied accordingly. For example, if the concentration is doubled, the time may be halved. At a relative humidity of 30 per cent the action of the gas is about 10 times as fast as at 95 per cent relative humidity.

In using ethylene oxide in an autoclave, a typical procedure involves seven steps, all of which can be carried out by fully automatic equipment now available for the purpose:

(1) Withdraw almost all air with a vacuum pump.

(2) Admit a controlled volume of steam or water vapor.

(3) Admit ethylene oxide in required concentration.

(4) Adjust the temperature as indicated and hold constant.

(5) After the required time turn off heat.

(6) Take out all gas with vacuum pump.

(7) Admit sterilized air to normal pressure.

Ethylene oxide is a very effective general microbicide and kills bacterial endospores. It is therefore a sterilizing agent. It acts by alkylation or introduction of an organic group in place of a hydrogen atom, in various compounds in protoplasm. It does not injure even delicate organic materials that are injured by heat and aqueous solutions. It penetrates well, does not condense on objects and soon dissipates from all materials.

It can also be used in the liquid state in 1 per cent concentration to sterilize unstable organic fluids, such as bacteriological culture media, by mixing fluid ethylene oxide with them at 10° C. After a few hours the ethylene oxide is driven off by a short incubation at 37° C.

Beta-Propiolactone (BPL)

In pure form at 20° C this is a liquid with a sweetish but very irritating odour [formula: $(CH_2)_2 \cdot CO$]. Unlike ethylene oxide, it is not explosive. It may be stored at 4° C, but is more stable in sealed glass containers at –25° C. Its aqueous solutions are very unstable. It kills most microorganisms, including bacterial spores, and is therefore a sterilizing agent. It probably acts by alkylation, as does ethylene oxide. In fluid form it has been used to sterilize vaccines, tissues used for grafting and other delicate biological materials that could not be sterilized in any other way. Like ethylene oxide, it must be used in closed chambers where temperature and humidity can be accurately controlled. A relative humidity of 70 to 80 per cent and a temperature of about 25° C are optimal, with a final concentration of beta-propiolactone of about 2 to 4 mg. per liter of air and a sterilizing time of 2 to 3 hours.

When properly applied, beta-propiolactone is much more active than either formaldehyde (HCHO) or ethylene oxide. It is noncorrosive, and does not condense on surfaces as does formaldehyde. However, beta-propiolactone lacks the penetrating power of ethylene oxide and is more like formaldehyde in this respect. It also requires higher relative humidities. It is generally recommended to replace formaldehyde for use in buildings.

Since they are so effective, gaseous disinfectants such as ethylene oxide, formaldehyde, and beta-propiolactone have tremendous potential value in combating massive and widespread dissemination of infectious agents such as might be employed in biological warfare.

Toxic and Destructive Properties

Experience shows that prolonged and repeated exposure to any chemically active disinfectant, or any sterilant, may cause tissue irritation or destruction, and may damage articles containing certain unstable organic and inorganic substances such as magnesium (instrument housings) and some plastics. Users must always be alert to this possibility.

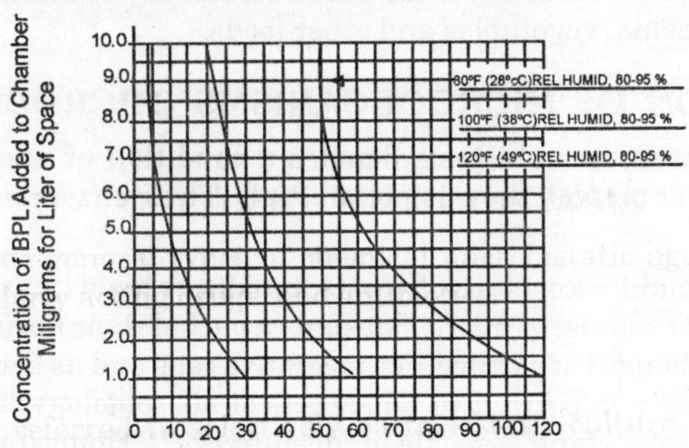

Fig. 1.13. Relation between concentration of beta-propiolactone and time required to kill spores of *Bacillus globigii* on filter paper strips. Curves are shown for three different temperatures at a relative humidity of 80 to 95 per cent.

They must also be alert to toxic properties, especially dangerous potentialities of newly developed disinfectants. For example, with continued study of beta-propiolactone, data have appeared suggesting that it may be carcinogenic as well as irritating. The toxic or carcinogenic properties of a disinfectant do not necessarily militate against its use; otherwise, such substances as chlorine, formaldehyde, bichloride of mercury and phenol, and radiations such as ultraviolet, x-rays and gamma rays would be banned. However, proper precautions must be taken in handling such agents to prevent ingestion or undue contact.

RADIATIONS

Ultraviolet Light

As described, ultraviolet light is actively microbicidal when applied properly. It has very slight powers of penetration. Exposure must be direct, intense and sufficiently prolonged. Eyes must be protected, as they can be severely damaged. It is sometimes used in hospitals, especially in operating rooms, reportedly with good results, to kill microorganisms on instrument tables, in the air and on floors and walls. Ultraviolet light is used on a large scale to prevent growth of molds in meatpacking houses, bakeries and other commercial and industrial establishments.

Another application is in the treatment of biological fluids, such as blood plasma and vaccines, to kill contaminating viruses. The rays are directed on fluid that is made to flow in a micro-thin film that ultraviolet rays can penetrate. The wave length commonly used in disinfectant applications of ultraviolet light is 2537 A, because it is most practical. Shorter waves, such as 1849 A, are much more lethal but are difficult to handle and apply. Ultraviolet kills spores as readily as vegetative cells.

Cathode Rays

A recent development in radiation sterilization, cathode rays (electron beams) are expensive to apply, but are being developed on a commercial scale. They can be made to penetrate thin metal, paper or plastic sheets and have been used successfully to sterilize sealed packages for surgical use, packaged meats, vegetables and other foods.

INHIBITION OF MICROORGANISMS (MICROBISTASIS)

As previously indicated, microbistasis denotes a condition of microorganisms in which, although they are alive (or *viable),* they do not multiply. Their enzymic or other vital functions are in a static condition.

In one sense the dormant ascospores of yeasts, conidia or molds, cysts of some protozoa and the endospores of bacteria represent microbistasis as a part of their normal life cycles. However, microbistasis is usually thought of as affecting vegetative cells and as being produced artificially. Microbistasis is exceedingly important in all aspects of microbiology. There is even published discussion of preserving important(?) people in "deep-freeze" ("Homostasis"?).

Physical Methods

As mentioned elsewhere, there are several physical methods of inducing microbistasis. Most of them are fundamentally methods of depriving the cell of liquid water: desiccation; immobilizing intracellular water by changing it into ice; drawing most of the water out of the cell by immersion in a fluid of high osmotic pressure; and combinations of these.

In addition, mild irradiations with ultraviolet light and x-rays are often microbistatic instead of lethal. Their microbistatic action is not clearly understood but it is not related to dehydration. Apparently, oxidations are among the most important reactions that occur.

Reversibility

The action of microbistatic agents is *characteristically reversible;* that is, the organisms can be reactivated if the inhibiting action has not been so prolonged that the cells have gradually died off. We have seen in foregoing chapters how the microbistatic effect of irradiations may be reversed by visible light and certain chemicals, and how bichloride of mercury poisoning may be "cured" by glutathione. Tubercle bacilli dried in a vacuum were found fully viable and virulent after 17 years; diphtheria bacilli and scarlet fever streptococci, after 25 years. Microbial spores (natural microbistasis) will survive for decades. Viruses are commonly held unaltered for months or years frozen in carbon dioxide ice at –76° C. Cattle breeders and others preserve sperm by freezing. In all cases the totally inactive cells can be reactivated.

MICROBISTASIS AND DISINFECTION

The distinction between chemical microbistasis and disinfection is interesting but entirely arbitrary; a matter of semantics. The differentiation depends on what one means by killing "quickly" (*i.e.,* disinfection) and what one may mean by killing "slowly" (*i.e.,* microbistasis). Even under the most effective conditions of microbistasis, some of the weaker cells die fairly soon; presumably the others are not actually immortal and die eventually, even though only after decades or centuries. As applied to bacteria, one could say that death within a time less than the maximal for one generation (*i.e.,* one cell-fission) under optimal growth conditions might be regarded as more than mere prevention of multiplication and, therefore, true killing or disinfectant action. If death were delayed beyond one generation-time, one might say that multiplication had been delayed or that the organisms had been in a static condition for at least a short time. This last situation could represent true bacteriostasis. There are many obvious difficulties in making such a distinction. The term bacteriostasis (or microbistasis) is, however, widely used and usually means "long" delayed death (*i.e.,* the cells die only after some hours, days or decades), rather than in, say, 10 to 30 minutes.

CHEMICAL MICROBISTASIS

We may note three general types of chemical microbistatic agents: certain aniline dyes; sulfonamide and similarly acting drugs; and antibiotics.

Aniline Dyes

We have already referred to selectively bacteriostatic culture media containing dyes. A representative dye used in such media is crystal violet, commonly used in the Gram stain. Since it is a basic dye, it has a marked affinity for acidic constituents of the bacterial cell, notably nucleic acids. With these the dye forms unstable salts that (in the absence of iodine) soon dissociate, and are either disposed of by the cell or can be removed by washing the cell with a harmless solvent such as water. Cells stained in this way with crystal violet, especially gram-positive cells, are in a static condition. They will not grow in media containing such dyes. They can be quickly reactivated by washing.

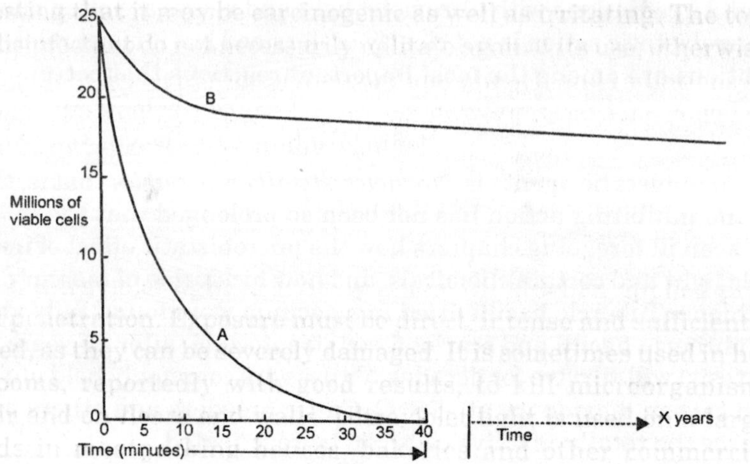

Fig. 1.14. Comparison of disinfectant action with microbistasis. The microorganism concerned has a generation time of 40 minutes at 37° C in broth at pH 7.2. Curve *A* shows complete sterilization within one generation time resulting from action of an applied microbicidal agent such as heat or disinfectant. Curve *B* shows initial rapid diminution of numbers caused merely by the action of the microbistatic agent such as freeze-drying. Bacteriostatic action is characterized by very slow decline in numbers.

On the other hand, gram-negative cells in contact with crystal violet are not stained and generally grow well on media containing such dyes as crystal violet, basic fuchsin or methylene blue-eosin. Such dye-containing media are therefore commonly used for *selective* bacteriostasis; especially to isolate gram-negative bacteria from materials contaminated with gram-positive bacteria.

Sulfonamide Drugs

The sulfonamides are excellent examples of bacteriostatic agents. They were discovered by Domagk (Nobel Prize winner) about 1936. Sulfonamide drugs are derived from the same basic material (coal tar) as aniline dyes. All are derivatives of sulfanilamide like basic dyes they are effective mainly (with a few notable exceptions: *Nesseria, Shigella, Proteus, Klebsiella* and *Escherichia)* against gram-positive bacteria.

Sulfonamide Drugs and Metabolite Antagonism

The true sulfonamide drugs are 'those that readily release the sulfanilamide residue in the living cell. Sulfanilamide so closely resembles paraaminobenzoic acid (PABA) that it can displace PABA in the absolutely essential vitamin, folic acid. This important vitamin is necessary to a co-enzyme (CoA) involved in synthesis of amino acids and hence of proteins. Such absolutely essential specific substances, necessary to the completion of a particular functional part of the cell, but not used as sources of energy or bulk, are sometimes called *essential metabolites* or *micronutrients.* In the present discussion PABA is considered to be an essential metabolite. When we compare the diagrams of the molecules of sulfanilamide and PABA, the resemblance between the major parts of the two is made clear.

As shown in the diagram, the folic acid molecule consists of three functional groups: pterin, PABA and glutamic acid. If any form of a sulfonamide drug is present in a sufficient amount, part of it goes into the folic acid complex in place of PABA. It is said to be a *metabolite antagonist.* In this case it *antagonizes,* or is *competitive* with, the metabolite PABA. The sulfonamide is therefore often called a *competitive inhibitor.* Because the structure of the sulfonamide is analogous to that of PABA, the drug is also called a *metabolite analogue.* But no form of sulfonamide is capable of carrying on the functions of PABA. The sulfonamide is, in effect, a molecular imposter.

The enzymic functions of the microbial cell depending on folic acid are therefore promptly stopped by sulfonamides. Since all of the functions of the cell are closely interdependent, the whole cell mechanism stops also. The cell is then in a state of microbistasis. The production of microbistasis in this way by chemicals or drugs is the basis of modern *chemotherapy*. There are many substances besides sulfonamides (*e.g.,* amino acids, certain ions [see ion antagonism], organic acids and esters, purines, pyrimidines and sugar derivatives) that, because of similarity of structure, can act as metabolite analogues. They antagonize or displace ions or molecules of vital cell structures such as proteins, enzymes and coenzymes or nucleic acids. In the nucleic acids nucleotide analogues can produce mutations.

Reversal of Metabolite Antagonism

It is important to note that, like most other microbistatic reactions, the enzyme blockage produced by sulfonamide is not fatal and may be reversed in several ways—for example, by adding enough PABA to displace and exclude the sulfonamide; by supplying new folic acid ready-made in the medium; by supplying, in the culture medium, the product that would have been formed by the blocked enzyme (or any subsequent enzyme in the chain); or by supplying the finished product of the entire enzymic series, in this case certain amino acids.

DRUG RESISTANCE

The cell itself may be capable of bypassing or overcoming the effects of metabolite antagonists. For example, let us suppose that in a parallel series of enzyme reactions enzyme D is blocked by some microbistatic drug. However, let us assume that some of the substance usually made by the blocked enzyme D is produced as a side reaction or intermediate stage in the function of another enzyme, H. This product becomes available to E and thus the blockage at D is bypassed. Such a cell is more or less *drug-resistant,* depending on how much of the product needed by E is available from H.

Such enzymic rearrangements and bypassings *(alternate metabolic pathways)* are quite common. From the clinical standpoint, sulfonamide resistance of an infecting organism means that the patient must receive some other drug.

Resistant Mutants

It has already been shown that among any large number of microbial cells in a given pure culture, mutant forms may be found. Among the large numbers of bacterial cells of a given species infecting a patient there are, therefore, likely to be mutants with enzymic arrangements that will be unaffected by a microbistatic drug such as sulfonamide. In the presence of the drug these mutants will thrive, while all the other microbial cells are killed off. We thus have bred up a sulfonamide-resistant strain of the infecting organism which is dangerous to the patient and to those who contract the resistant organism from him.

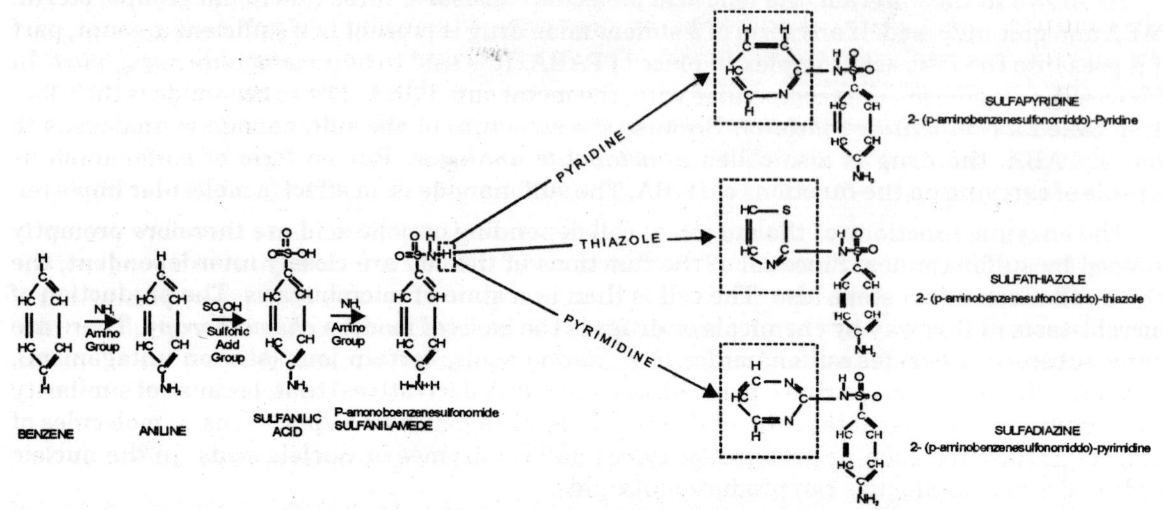

BENZENE → Amino Group → ANILINE → Sulfonic Acid Group → SULFANILIC ACID → Amino Group → P-amonoboenzenesulfonomide SULFANILAMEDE

PYRIDINE → SULFAPYRIDINE 2- (p-aminobenzenesulfonomiddo)-Pyridine

THIAZOLE → SULFATHIAZOLE 2- (p-aminobenzenesulfonomiddo)-thiazole

PYRIMIDINE → SULFADIAZINE 2- (p-aminobenzenesulfonomiddo)-pyrimidine

Fig. 1.15. How the sulfonamide drugs are derived from benzene. First, an amino group is added, yielding aniline (a base from which dyes are synthesized). The addition of a sulfonic acid group to aniline yields sulfanilic acid. The addition of another amino group yields sulfanilamide. The addition to this, in place of one hydrogen atom, of various other chemical groups yields other sulfanilamide drugs, three of which are shown in the diagram.

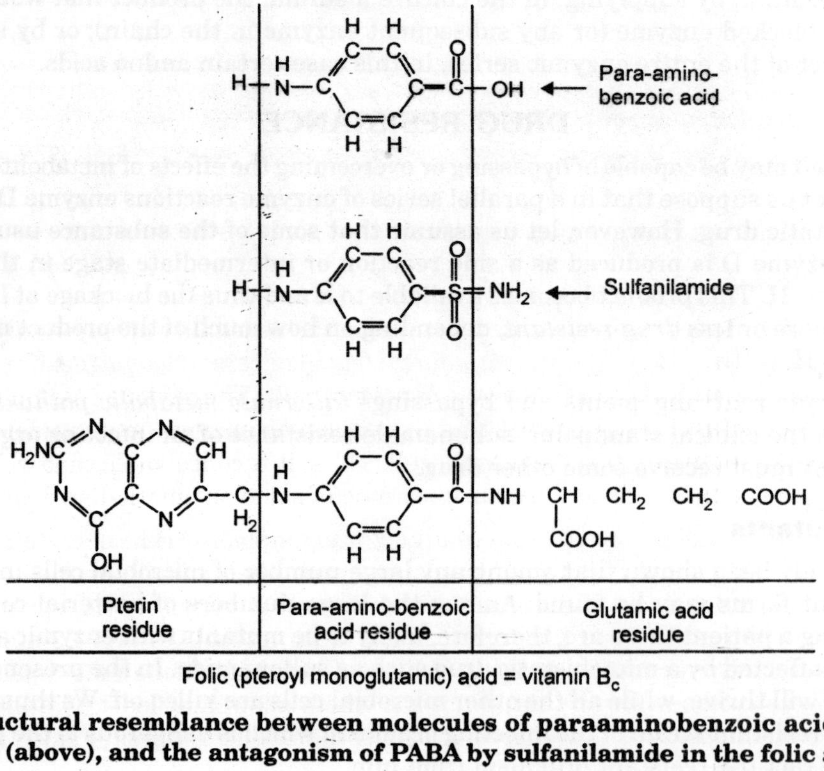

Para-amino-benzoic acid

Sulfanilamide

Pterin residue | Para-amino-benzoic acid residue | Glutamic acid residue

Folic (pteroyl monoglutamic) acid = vitamin B_c

Fig. 1.16. Structural resemblance between molecules of paraaminobenzoic acid (PABA) and sulfanilamide (above), and the antagonism of PABA by sulfanilamide in the folic acid molecule (below).

In some microorganisms that are ordinarily susceptible to penicillin, notably *Staphylococcus,* penicillin-resistance depends on production of an extracellular enzyme *(ptenicillinase)* which destroys the drug. Mutants that produce such enzymes are able to grow unhampered in the presence of the drug, to the eventual exclusion of the nonenzyme-producing sensitive mutants.

Induced Drug Resistance

In addition to the occurrence of drug-destroying enzymes that appear as a result of mutation, such enzymes appear also as *induced* or *adaptive enzymes* resulting from contact of the cell with the drug. The whole problem of mechanisms of drug resistance, not only in microorganisms but in insect pests (*e.g.,* resistance of mosquitoes and flies to insecticides), is a major problem in chemotherapy and pest control, and an excellent field for profitable research.

Drug Dependence

The appearance of strains that are not merely resistant to drugs but wholly *dependent* on them seems as strange as the emergence of resistant strains. The two are probably not related except that they are two manifestations of the phenomenon of mutation. Drug-dependent mutants obviously could not survive and develop significantly except in an environment containing the drug on which they are dependent. It is only because we have tried to cultivate microorganisms in contact with antimicrobial drugs that these curious mutant forms have been revealed.

CHEMOTHERAPY

Chemotherapy, broadly speaking, is therapy by means of chemicals. The term is generally restricted, however, to the use of drugs that act by metabolite antagonism or enzyme blockage. In order to be truly chemotherapeutic such drugs must act on the parasite in concentrations that have no effect (or at least no serious "side" effects) on the host. Some valuable chemotherapeutic drugs are toxic to the patient as well as to the parasite and so must be used with great care and under medical supervision.

Chemotherapy and Cancer

A very interesting aspect of chemotherapy concerns the possibility of selectively poisoning the cells of *neoplasms* (Gr. *neo* = new; *plasm* = form) (*e.g.,* cancers). Neoplasms are thought to have metabolic mechanisms different from those of normal cells. The obvious possibility presents itself of finding some metabolite antagonist that will poison an enzymic mechanism peculiar to the neoplasm cell and not present in the normal cell. Thus, it is conceivable that we could cure cancer overnight by means of a few tablets or injections of some antibiotic-like drug.

This rosy dream may not be too far from reality. For example, the cells of certain cancers of mice differ from normal mouse cells by synthesizing cyanocobalamin (vitamin B_{12}), a cobalt-containing coenzyme. At least two types of antimetabolites that interfere with synthesis of B_{12} produced cures of the cancer when fed to mice. The antimetabolites are 1,2-dimethyl-4(P-carboxyphenylazo)5-hydroxybenzene and 1,2-dichloro-4-benzenesulfonamide-5-nitrobenzene. Several other drugs have been found that, if not the whole answer, at least point definitely to the possibilities for chemotherapy in research on cancer metabolism. Honor, Fame and Fortune are to be won in this field!

CHAPTER

2 Anaerobic Process

BATCH CULTURE FERMENTATION

The purpose of this chapter is to give an introduction to the methodology of batch fermentation using a small, autoclavable, benchtop fermentor system. Specifically, this chapter should provide an understanding of the following:

(i) the engineering and biological concepts of fermentors;

(ii) the principles behind the instrumentation used with a modern fermentor;

(iii) the assembly and preparation of a fermentor vessel and anciliary equipment for autoclaving;

(iv) how samples are taken from the fermentor vessel during operation;

(v) how simple continuous culture can be accomplished.

THE MAIN COMPONENTS OF A FERMENTOR

The main subdivisions of a bench-scale fermentor are as follows:

(i) base components including drive motor, heaters, pumps, gas control, etc.;

(ii) vessel and accessories;

(iii) peripheral equipment such as reagent bottles;

(iv) instrumentation and sensors.

These components combine to perform the following functions:

(i) provide operation free from contamination;

(ii) maintain a specific-temperature;

(iii) provide adequate mixing and aeration;

(iv) control the pH of the culture;

(v) allow monitoring and/or control of dissolved oxygen;

(vi) allow feeding of nutrient solutions and reagents;

(*vii*) provide access points for inoculation and sampling;

(*viii*) use fittings and geometry relevant to scale-up;

(*ix*) minimize liquid loss from the vessel;

(*x*) facilitate the growth of a wide range of organisms.

The example used to illustrate these points is the smallest, simplest system which displays all these components and attributes, *i.e.* a small, autoclavable, bench-top fermentor. Mention is made of larger, *in situ* sterilizable units and special features which can be added to achieve certain objectives such as containment work. Also, a range of basic applications are illustrated, e.g. mammalian cell culture as a starting point for anyone contemplating the use of small-scale fermentors for these purposes. The specific information provided should be treated with a great deal of circumspection as the parameter values required for any particular microbes or cell line could be completely different.

COMPONENT PARTS OF A 'TYPICAL' VESSEL

The vessel is constructed either as a single-walled cylinder of borosilicate glass with a flat bottom or as a glass-jacketed system which typically has a round bottom. The top plate is made from '316' stainless steel and is compressed onto the vessel flange by an easily released clamping system. A seal separates the vessel glass from the top plate. Port fittings of various sizes are provided for insertion of probes, inlet pipes, exit gas cooler, cold fingers, sample pipes, etc. These work by compressing the sides of the probe/pipe against an O-ring seal. A special inoculation port will have a membrane seal held in place with a collar. Culture can be withdrawn into a sampling device or a reservoir bottle via a sample pipe situated in the bulk of the fermentor fluid. A gas sparger is also fixed into the top plate and this terminates in a special assembly which ensures that incoming air is dispersed efficiently within the culture by the flat-bladed 'Rushton-type' impellers fixed to the drive shaft.

A drive motor provides stirring power to the drive shaft and is usually fitted directly to the drive hub on the vessel top plate. An exit gas cooler works like a simple Liebig condenser to remove as much moisture as possible from the gas leaving the fermentor to prevent excessive liquid losses during the fermentation and wetting of the exit air filter. A narrow platinum resistance (Pt-100) temperature sensor completes the list of minimum essential fittings. Temperature control is either by direct heating using a heater pad or by circulating warm water around the vessel jacket.

If direct heating is used, a cold finger is used to cool the vessel contents. The sensors which fit into the vessel do so either by direct coupling with a thread on the body of the electrode, as is the case in the gel-filled type of pH electrode, or by a special fitting on the vessel top plate, as in the case of the dissolved oxygen electrode (polarographic type). Another system involves the use of a simple compression fitting which holds the body of the electrode, as with the foam probe. In this case, the height is variable and the tip of the probe is above the culture fluid. Illustrates all the main components of a fermentor.

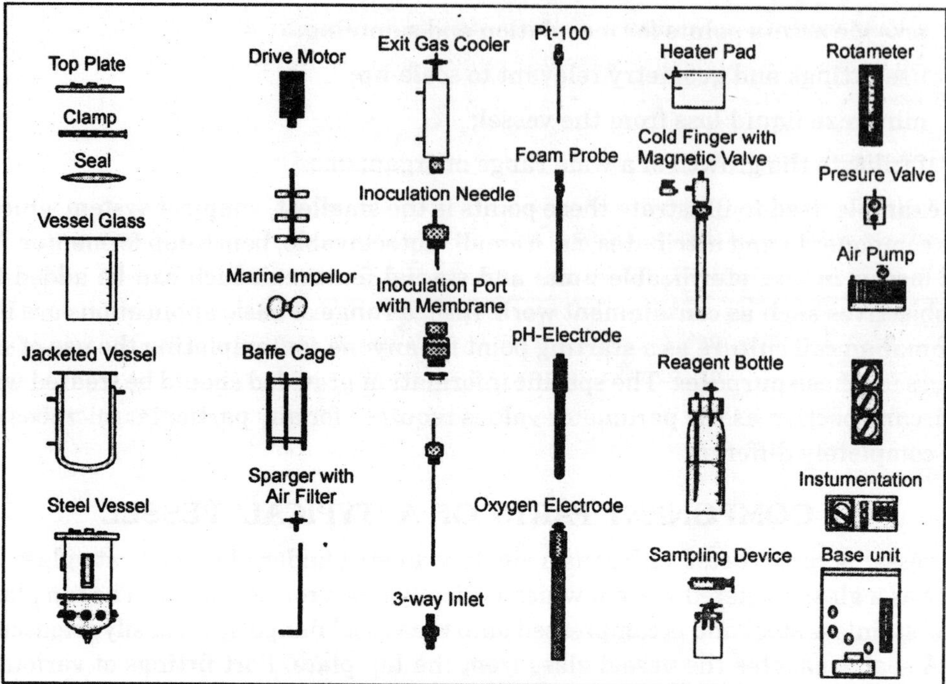

Fig. 2.1. Major Components of a Fermentor.

PERIPHERAL PARTS AND ACCESSORIES

Reagent Pumps

Pumps are normally part of the instrumentation system for pH and antifoam and control. Peristaltic pumps are used and the flow rate is usually fixed with a timed 'shot and delay' feed system of control. Flow rates will depend on the bore of the tubing used and it is a good idea to move the tubing every few days so that one piece does not become worn out by the pump rollers. The peristaltic tubing links the reservoir bottles with the vessel multi-way inlet. The tubing is clamped shut during autoclaving, already connected to the vessel, and opened for active addition of reagent. Alternatively, the bottles can be connected following autoclaving by using an aseptic coupling to join the tubing between the pump and the multi-way inlet. A dip tube goes to the bottom of the bottle. A filter in the top of the reservoir bottle is kept open during autoclaving to prevent a build-up of pressure. For extremely high accuracy of addition, the reservoir bottles are placed on analytical balances which can be used to determine how much reagent has been pumped in a given time.

Medium Feed Pumps and Reservoir Bottles

Medium feed pumps are often variable speed to give the maximum possible range of feed rates. The speed of operation of the pump can be set manually or the whole system put under computer control. The reservoir bottles are usually larger, *e.g.* 5-20 l, but are prepared in the same way as normal reagent bottles. These bottles may have to be changed several times during a long continuous culture experiment, so it is usual to fit an aseptic coupling in the tubing. An effluent pump is often used to remove culture fluid from the fermentor vessel into a storage reservoir bottle.

Rotameter/gas Supply

Gas input, usually air, can be provided by use of a laboratory air supply or from a separate pump (which *must* be oil-free). A variable area flow meter or rotameter is used to control the air flow rate into the fermentor vessel. A pressure regulator valve before the rotameter ensures safer operation. A sterile filter (usually 0.22 μm) is used as a bridge between the tubing from the rotameter and that connected to the air sparger of the fermentor. A second filter on the exit gas cooler stops microbes being released into the laboratory air as the gas leaves the fermentor under a slight positive pressure.

Sampling Device

This allows culture fluid to be removed aseptically during the fermentation at intervals decided by the user. The frequency of sampling and the size of each sample is determined empirically according to the needs of the experiment.

Alternative Vessel Designs

There are alternatives to using a conventional, 'stirred tank' fermentor. Usually, alternative vessel designs are tried when the standard vessel configuration does not allow adequate growth of the organism (*e.g.* animal cells are often disrupted by shear forces in fermentors with turbine impellors) or the scale-up criteria require a different design of bioreactor (*e.g.* production of large quantities of single-cell protein is cheaper on a large scale if air lift fermentors are used to eliminate energy costs associated with a drive system). A number of these special designs are available at the bench/pilot scale of operation to allow small-scale research into the suitability of a particular method.

Air Lift

This vessel design eliminates the need for a stirrer system. A tall, thin vessel is the best shape with an aspect ratio of around 10:1 (height to base diameter). Sometimes a 'conical' section is used in the top part of the vessel to give the widest possible area for gas exchange. Sensors can be mounted in a steel base section, from a collar at the base of the conical section or from the vessel top plate. The culture fluid is both mixed and aerated by a stream of air which enters near the base of the vessel. A hollow pipe or draft tube in the centre of the vessel provides a 'riser' for the air (which is full of bubbles) to move upwards to the top of the vessel. If a very large vessel is used, the hydrostatic head of the fluid provides a pressurizing effect to the lowest region of the culture where the air enters and so increases the dissolved oxygen concentration. The draft tube is usually double-walled to allow heating and cooling using a thermocirculator system. When the aerated culture fluid reaches the top of the draft tube, it 'spills over' and begins to fall towards the bottom of the vessel via the space between the outer wall of the draft tube and the inner wall of the vessel.

A large head space above the top of the draft tube allows for easy gas transfer from the liquid to the gas phase which causes the density/specific gravity of the liquid to increase and so it descends to the bottom of the vessel. The descending liquid returns to the base of the vessel where it is regassed and begins to rise again. A common use of air lift fermentors is the growth of shear-sensitive cells such as plant and animal cultures. Also, the design has been used for the production of large amounts of biomass as single-cell protein.

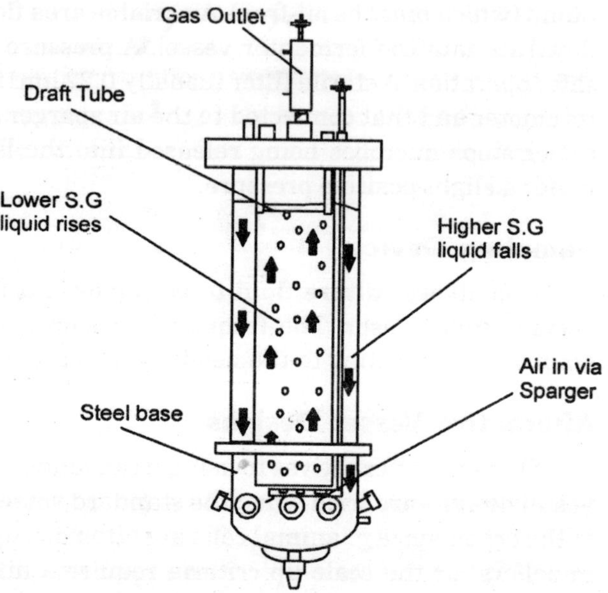

Fig. 2.2. Air lift fermentor.

Fluidized Bed

The microbes/cells are trapped in a physical medium (*e.g.* alginate beads) and held in the vessel by a mesh. Medium is recirculated via a pump and this can be easily adapted to give a continuous/semi-continuous flow to allow the trapped cells to effect chemical changes on constituents of the medium without being washed out along with spent medium. This system is well suited to growth of animal cells on the smaller scale and has large-scale application in effluent/decontamination treatment plants.

Hollow Fibre

This is a similar idea but the cells are now embedded in fibres contained in a cartridge which is bathed in circulating culture medium. This is often used for mammalian cell culture, where anchorage-dependent cell lines can be perfused with oxygenated medium. An extension of this method is to use cartridges containing two different bundles of fibres as a separation 'membrane' between a pair of fermentor vessels and allow dissolved gases and metabolites to be exchanged without cells crossing the barrier to allow, for example, toxicity or interactive studies.

In Situ Sterilizable Fermentors

The use of autoclavable vessels of greater than 5-10 l working volume quickly becomes impractical. For safety and insurance considerations, vessels above this size are usually made of stainless steel (316 L) and are designed to be sterilized *in situ* using house steam or an electrical steam generator (which may be built into the base unit). The heating for the vessel is normally provided via a double jacket which can either be the full length of the vessel or cover just the bottom third. The bottom section contains large (25 mm) port fittings for electrodes and usually has some kind of steam-sterilizable sampling device/harvest valve. The mechanical seal and drive shaft usually enter from the bottom on this size of vessel. As the vessel body is steel, a sight window and a light have to be fitted in order to see the culture. The vessel top plate has port fittings which use a membrane seal and port closure.

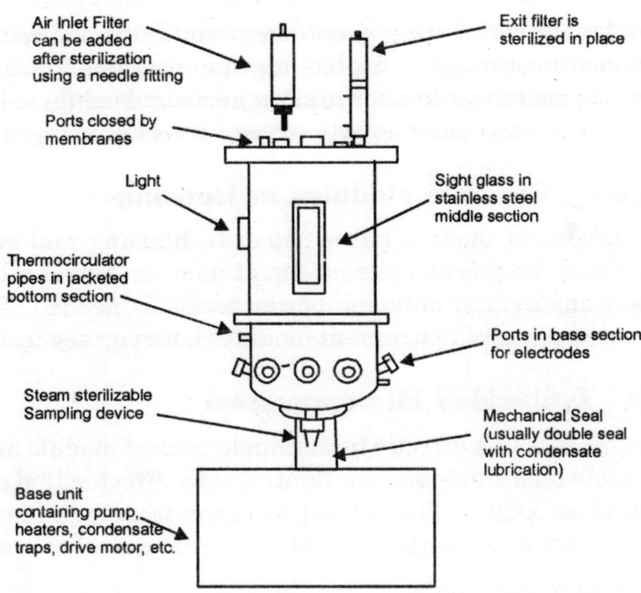

Fig. 2.3. Major components of an *in situ* sterilizable fermentor vessel.

Unlike the bench units, *anything* which dips into the culture liquid or is used as an addition port must be autoclaved separately and pushed through a membrane using aseptic techniques after the vessel has been autoclaved. In some fermentors, this even includes the air inlet filter and connecting pipe to the air sparger. The steam for sterilization is either supplied from an in-house source and used to heat the vessel jacket or is raised electrically from a separate steam generator. The medium in the vessel is heated to 121°C and often supplies the steam for sterilization of the exit gas filter.

Containment

An *in situ* sterilizable fermentor may have to be altered in certain ways if the organism to be cultured is pathogenic or genetically modified. The alterations are designed to contain any release of microbes into the environment by using features such as a double mechanical seal, magnetic coupling, additional air filters, extra foam control systems and special sampling devices. More elaborate precautions include steam traces to all vessel fittings and discharge of any released liquid directly to a tank of disinfectant. There are several different categories of containment (1-3 for pathogens and 1-4 for genetically modified organisms). Requirements differ at each stage and relate to building and working protocols as much as the design of the fermentor. Applications for this sort of technology are medical research and vaccine manufacture.

DIFFERENT TYPES OF INSTRUMENTATION

Fermentor instrumentation can range from simple analogue control modules arranged as a 'stack', to powerful, embedded microprocessors which operate as on-board computers to directly operate the heaters, pumps, valves and other control actuators of a fermentor.

Analogue Controllers - Rack System

Here the individual modules for the measurement and control of each parameter are made to plug into a rack system for compact, flexible instrumentation. A bus system in the racking transmits signals from one module to another to allow measured values to be passed to controllers and for one controller to influence another, *e.g.* oxygen levels altering speed.

Analogue Controllers - Separate Modules in Housings

Each measurement/control module has a separate housing and essentially operates in isolation. The modules may be stacked one on top of another and can be supplied with 'pass through' power connections so that only one power socket is needed. This system allows for gradual building up or swapping of instrument modules between several fermentors.

Digital Controllers - Embedded Microprocessor

The measurement sensors link directly to the single control module and several parameters are displayed immediately on a single screen. Control is by direct action on heaters, valves, etc. (Direct Digital Control or DDC). The microprocessor is permanently embedded in the instrumentation and may even be a single chip. Operation is usually via a simple menu system.

Digital Controllers - Process Controllers

A complete process controller (usually from a production control environment) is added to a housing containing all the signal processing and control actuators for the fermentors. It exerts control in the same way as an embedded controller but is essentially a 'plug-in' component. The controller is usually 'programmed' using simple commands to input set-point values, etc.

Digital Controllers - Direct Computer Control

In this case, there is no external instrumentation or processor between the actuators and the computer. A printed circuit board with operational amplifiers for the probe input signals is the only electronic part which may be present in the fermentor base unit. A special Input/Output (I/O) card is needed for the computer which allows the input values to be accessed by the measurement and control software and sends signals out to operate relays, *e.g.* to turn a heater on or off.

The processing power and speed of modern PCs make them more than capable of replacing separate PLCs or embedded controllers. The advantage with this system is that the computer display and control software is integrated totally with the fermentor. Fewer components means that these systems are usually less expensive. However, the fermentor cannot be used without the computer and there is no backup for the control systems should the computer develop a fault and 'hang'.

COMMON MEASUREMENT AND CONTROL SYSTEMS

Speed Control

Speed control relies on the feedback from a tachometer located within the drive motor determining the power delivered by the speed controller to maintain the speed set-point value set by the user. A digital display shows the actual speed in rpm, as determined by the tachometer signal. A power meter is sometimes included which indicates how hard the motor has to work to maintain the set speed and thereby, indirectly, the viscosity or 'density' of the culture fluid. A DC, low voltage (24-50 V) motor is often used for safety reasons. Speed range is typically from 50 to 1500 rpm (exceptionally up to 2000 rpm).

Where speed is used to control the level of dissolved oxygen, an external signal from the oxygen controller can have an effect on stirrer speed. In this case, an absolute maximum and minimum value for speed can be set on the speed control module to limit the effects of the oxygen controller (which could set either too low a speed and impair mixing or too high a speed and cause excessive foaming)

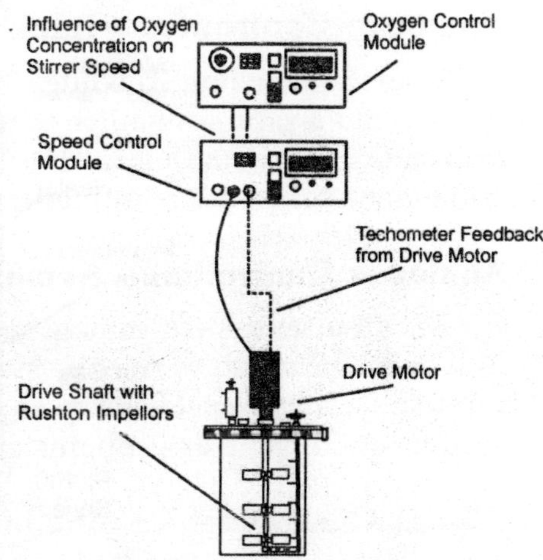

Fig. 2.4. Speed control system.

Temperature Control

A thermocirculation system around a vessel jacket has been chosen as an example here because it is the most complex of all the methods of temperature control. For simple direct systems such as a heater pad, it is simply a matter of fixing the heater, setting the desired temperature and switching on. Cooling is normally via a cold finger and flow of cooling water is controlled via the action of a solenoid valve. The Pt-100 sensor provides the feedback signal which causes the controller to take one of the following actions:

- heat at full power as the actual temperature is some way below set point;
- pulse the heater power as the actual temperature is close to set point;
- turn on the cooling valve as the actual temperature is above set point.

Indicator lights usually show which action the controller is taking at any given moment. A low-voltage DC heater system, *e.g.* 50 W, is often chosen for safety reasons. A circulation pump and pipework are added to the system for water circulation and any heating is indirect, *i.e.* on the water circulating in the vessel jacket. In this case, the connections to cold tap water must be made (securely, using jubilee clips or cable ties) and a drain pipe provided from the overflow point to a sink with a clear fall to the drain, *i.e.* the sink must be the lowest point for the whole length of this pipe. The water should be delivered from the mains at a minimum pressure of 2 bar and a flow rate of greater than 51 min^{-1}. The water hardness should be no more than 50 ppm suspended solids to protect the heating elements from 'furring'.

The vessel must be connected to the circulation loop (normally by rapid coupling connectors and flexible pressure tubing). Water is first supplied by opening a manual valve until the jacket is filled. The manual valve is then used to maintain just a trickle of water going through the jacket to the overflow to compensate for evaporation losses, *e.g.* 30-50 drops per minute. The heating and cooling is controlled in exactly the same way as a directly heated system but only the water in the jacket is affected. The jacket provides a large surface area in contact with the vessel wall for heat exchange.

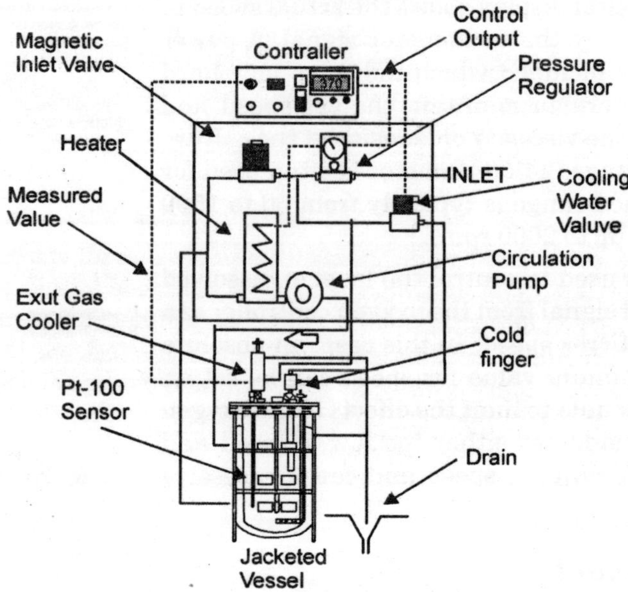

Fig. 2.5. Temperature control system using water circulation.

Good temperature control can be achieved from approximately 5-8°C above the ambient temperature or 5°C above the temperature of the cooling water. Counter-cooling with water ensures stable temperature control when operating near ambient temperatures. Measured range is typically from 0 to 60°C (exceptionally up to 90°C).

Control of Gas Supply

The supply of gas (normally air) to the fermentor vessel is provided from a compressed air supply (oil-free) which may provide air for more than one vessel depending upon its size and output. A pressure regulation valve ensures that air reaches the vessel at a maximum of 1.5 bar. The rotameter controls the actual flow rate of air through the fermentor. This should not exceed 1.5 vessel volumes per minute or droplets of water may become entrained in the stream of gas leaving the fermentor and so wet the exit gas filter, causing it to block.

A valve at the bottom of the rotameter is turned and the indicator 'ball' in the rotameter tube rises or falls in proportion to the valve position. A scale on the tube gives flow rates in ml min^{-1} or h^{-1}. The air passes through the inlet air filter which prevents any microbes from entering the vessel via this path. The end of the sparger is typically a ring with small holes through which the air is forced. The bubbles are immediately broken up and dispersed by the impellors on the drive shaft and the baffles which can be fitted near the wall of the vessel. The use of several impellors ensures all regions of the vessel receive good aeration.

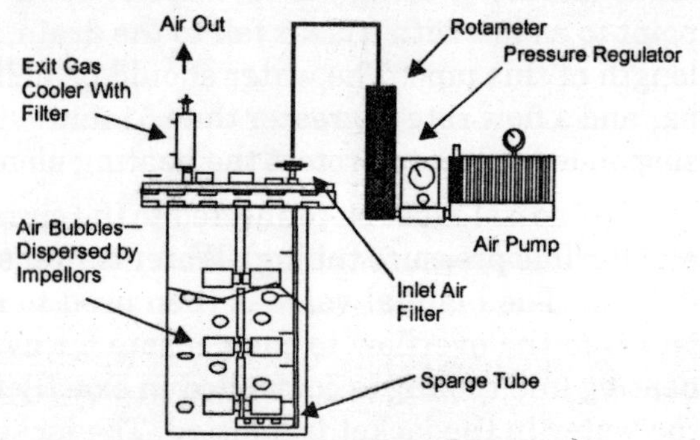

Fig. 2.6. Gas supply system.

A 'head space' of around 20% is normally left between the culture level and the vessel top plate. Sometimes, gas can also be introduced into this region via a short pipe in the fermentor top plate. For certain types of fermentation, *e.g.* mammalian cell culture, a gas mixing station can be used to pre-mix several gases before they are introduced into the fermentor.

Control of pH

pH control is achieved by the addition of either acid or alkali to correct changes in the pH of the culture during growth. The controller uses a pH electrode (either glass or gel type) to sense these pH changes and provide a feedback signal. The pH measurement system is identical to a bench pH meter and is calibrated in the same way *before* the electrode is autoclaved. Temperature compensation during operation may have to be set manually or may be automatic. The calibration normally requires that the temperature is manually set to equal the temperature of the buffer solutions.

Autoclavable electrodes have a limited life (20-50 sterilizations typically). The pumps used to supply the acid and alkali are normally built into the control module or the base part of the fermentor. Measured range is typically pH 2-12. The reagent bottles are connected to the fermentor via transfer lines of silicone tubing. The bore of tubing selected will determine the volume of acid or base added by each activation of the relevant pump by the controller. Selecting the concentration of the acid or alkali will determine how much effect each dose has on the vessel contents.

Normally, solutions of between 0.2 and 0.5 M of acid and base are used as a starting point. Using an ammonium salt for alkali addition can provide an extra nitrogen source for the growing culture. A set-point value is set on the controller and/or an upper and lower limit to provide a 'dead band' range in which the controller is not active. This band should normally be around 0.5 pH units each side of the desired value to prevent overdosing of acid and alkali sequentially. A proportional band adjustment may be present to widen or tighten the range of pH value over which the controller acts and even a switch to give a non-linear control if good control cannot be achieved by any normal adjustments.

Control of Dissolved Oxygen

Dissolved oxygen is one of the most important and yet one of the most difficult parameters to control properly during a fermentation. The electrodes used to measure dissolved oxygen are of two distinct types:

1. Galvanic electrodes, which are generally simple, cheap, slow and have a limited life.
2. Polarographic electrodes, which are generally more complex, accurate, rapid and robust but expensive.

The polarographic electrode is used more commonly and will be the example provided.

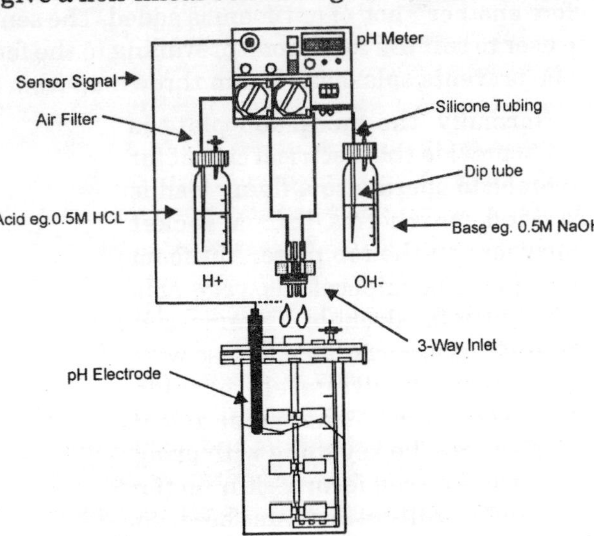

Fig. 2.7. pH measurement and control.

The key point with this type of electrode is that it requires a voltage to polarize the anode and cathode of the detecting cell. This polarization can take between 2 and 6 hours to complete (in an emergency, 40 min is just sufficient for most tasks). During this time, the electrode must be connected to its relevant module which in turn must be switched on. The membrane of the polarographic electrode needs periodic replacement and a special cartridge kit is available from the manufacturer to make this a simple task. The electrode can be 'zeroed' prior to autoclaving using a special zeroing gel provided by the manufacturer.

The electrode is simply inserted into a sachet of gel, left for a short time and then the displayed value adjusted until it reads zero. Alternatively, the electrode can be zeroed after autoclaving by passing nitrogen gas through the vessel for some minutes and then adjusting the zero point. The 100% value is a relative setting made after autoclaving and polarization of the electrode. The air flow and stirrer speed are turned to the maximum needed for the fermentation and the vessel left for some minutes. The display is now adjusted until it reads 100%. Control of dissolved oxygen can be purely by speed control, purely by control of a proportional air valve, or by a combination of both.

Speed adjustment may interfere with good mixing or increase the degree of foaming. Adjusting the air flow rate may also affect levels of foam. The most accurate form of flow control is to use a thermal mass flow control valve which measures and controls air flow based on the cooling effect the gas exerts when passed over a heated element. A simple option is the use of a solenoid valve with a manually set 'bleed' valve to simply vary air supply between a maximum and a minimum flow rate. Range is typically 0-120%.

Antifoam Control

The control of foam is based on its detection by a conductance probe in the vessel head space, which leads to the controller delivering a dose of liquid antifoam reagent via a peristaltic pump. A delay timer ensures the antifoam reagent has adequate time to reduce the foam level before another 'shot' of antifoam is added. The sensitivity of the probe to foam can be adjusted by the user to suit the conditions prevailing in the fermentor. A sheath of inert material around the probe prevents splashes of foam thrown up from the culture from giving 'false positives.

Normally, the metal top plate is used to provide the electrical circuit for the probe to operate so a flying lead is provided which fits into a socket somewhere on the top plate. Antifoam reagents can be mineral oils, vegetable oils or certain alcohols. Commercial preparations are available for use with cultures which ultimately provide materials for pharmaceutical manufacture. The key thing with using oils is that they can form a 'skin' on the surface of the culture and interfere with gas transfer at the liquid/air interface. If foam is allowed to build up unchecked

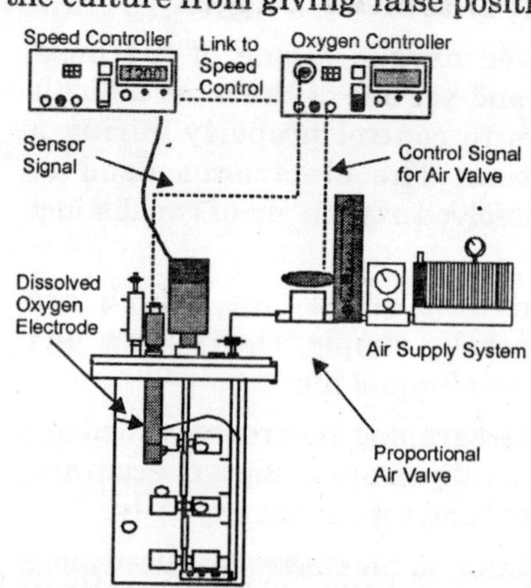

Fig. 2.8. Dissolved oxygen measurement and control.

it can eventually reach the exit gas filter so blocking it and providing a path for contamination.

ADDITIONAL SENSORS

There are several additional process values which can be measured and controlled within a fermentor. The most common are listed below:

Redox

This refers to the reduction/oxidation (redox) potential of a system and can be applied to biological cultures as well as chemical reactions. Redox values are given in millivolts and usually refer to the oxidation or reduction potential of a given chemical relative to a standard hydrogen half-cell (which is defined as having a redox potential of 0.00 V). A half-cell which is more reducing than the hydrogen standard has a negative value and a stronger oxidizing agent will have a positive value. Reducing agents donate an electron to oxidizing agents.

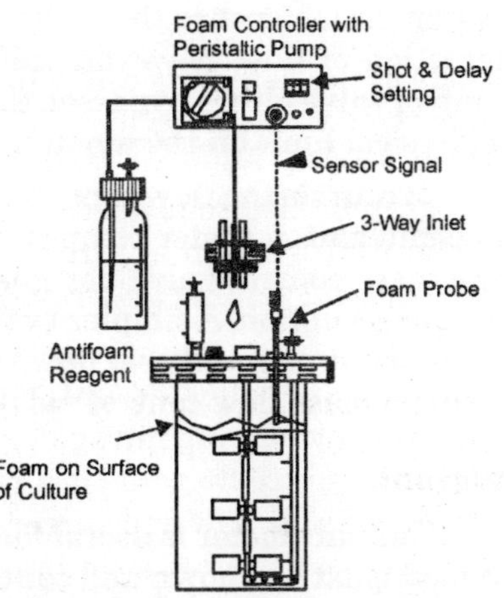

Fig. 2.9. Foam control system.

Depending on conditions (the pH can affect redox potential), this reaction is reversible. The complexity of a typical fermentation broth (in terms of the sheer number of separate reactions going on at any one moment) means that the actual measured value is only of limited importance. However, for anaerobic cultures, the redox electrode provides a very sensitive 'watchdog' for the presence of oxygen. Tiny amounts of oxygen will have a relatively large effect on the redox value. A redox value below -200 mV is a good indicator of anaerobic conditions. The redox electrode is very similar to a pH electrode and the conditions for handling and care are almost identical. An electrical zero can usually be set and shorting out the electrode connections should give a reading close to 0 mV. There is no calibration procedure as such. A check on the electrode can be made using ordinary pH buffers which have been saturated with an excess of solid quinhydrone. The quinhydrone dissociates into equal quantities of benzoquinone and benzoquinol below pH 8.0. The redox potential for this system at pH 7 should be around +285 mV and changes by 59 mV for each pH unit, e.g. pH 9 = 403 mV. Deviations of ±5 mV are acceptable. If an analogue meter is used, a variable offset is usually provided for output to a chart recorder.

Air Flow

This has been mentioned previously in connection with dissolved oxygen control. A simple system would use a magnetic sensor and a variable area flow meter (rotameter) to measure the current air flow and adjust a motorized valve between 0 and 100% open-or closed to control the flow. Alternatively, a solenoid valve could pulse the air flow very rapidly on and off between a maximum and minimum value set manually by the rotameter and a 'bleed valve' respectively. In this case, there is no displayed value for flow rate which could be used for data-logging or control by computer. A more sophisticated system uses a thermal mass flow control valve which combines both the measurement element and the control valve in one compact unit. In this case, measurement of flow is by detecting the change in temperature of a heating element before and after gas passes over it.

The temperature difference is proportional to flow rate and this can be processed to give a measurement signal for logging and/or control. The practical points to consider when using this system are to ensure the valve is the right way round (an arrow on the metal section shows the direction of flow, however, this is sometimes obscured by the valve mounting) and that the control valve either bypasses the rotameter or the rotameter has a maximum value set to act as a physical limit on the amount of gas flow into the fermentor.

Of course, the flow control systems can measure any gas flow, not just air. They can be used in combination under computer control to give a precise gas mix for any total flow rate using one mass control valve per gas and post-blending the mixture. However, this tends to be an expensive option. A simpler system uses solenoid valves to introduce each gas into a pre-mixing chamber at manually set flow rates and then controls the final gas flow into the fermentor with a single mass flow control valve.

Weight

This parameter is useful for continuous culture or fed-batch work where the rate of addition of feed must be known and controlled very accurately. It is possible to mount a whole fermentor on a balance and 'tare' out its weight. However, a more precise approach is to use a system with a small load cell mounted in such a way that only the vessel and its contents are measured. The 'deformation' of the load cell provides an output which can be transduced into an electrical signal. Control is normally by setting a maximum and/or minimum weight such that a feed pump operates whenever the value is less than the maximum.

At the maximum value, a second pump operates to remove culture until the minimum weight is reached. A variant of this system can be used to measure very accurately the amount of reagent, *e.g.* alkali, added to a fermentor. In this case, the reagent bottle is placed on an analytical balance which has a computer output (normally RS232) and the decrease in weight used as a measure of the amount of reagent added.

Pressure

The use of pressure as a control parameter is more common in larger, *in situ* sterilizable fermentors, for the obvious reasons that the vessel is made of steel and is already adapted to working at pressures of several bar. Also, it is typically in the larger vessel that maintaining the dissolved oxygen concentration at a high level presents a problem (an increased pressure increases the amount of oxygen which can be dissolved in the fermentation broth). However, even a small glass vessel can often operate at some overpressure (*e.g.* 1-1.5 bar) but it is vital that this is first confirmed by the manufacturer.

The measurement and control is the same whatever the size and type of vessel, except that the size of the control valve would alter. Measurement is normally provided by a piezo-electric sensor which can be mounted into a port closure and fitted to the vessel top plate. As with a load cell, it is the deformation of the crystal by the internal vessel pressure which gives a proportional electrical signal, typically in the range of 0-2 bar. The control element is a proportional valve which restricts the flow of gas out of the fermentor and thereby creates a 'back-pressure'. A mechanical overpressure valve or burst-disc is an essential safety requirement if overpressure is to be used.

On-line Measurement of Biomass

A direct measurement of the number of organisms in a culture is clearly desirable for the control of feed rate, oxygenation and general process optimization. Normally, measurements of total cell numbers, wet weight, dry weight and viable cell counts would be made by taking samples of the culture at set times and performing the relevant laboratory analysis. The results would be entered some minutes, hours or even days later, when they would be of little value for real-time control. Fluorescence microscopy has also been used as a measure of viable cell density either with a probe or externally through the vessel glass. Two probe-based systems which are reasonably accurate and can be used with bench-scale vessels will be described by way of examples below.

Mettler-Toledo *FCS* Turbidity System

This is actually a nephelometer as it measures the scattering of light by particles in the culture (*i.e.* microbes). The advantage over a standard nephelometer is that the sensor is directly mounted in the vessel and connected to the electronics via a fibre optic cable. A two-point calibration procedure is used and results can be expressed in a variety of units. One of the restrictions of such a system is that if the culture is very thin, then shiny, reflective components close to the sensor could affect the readings. Conversely, too thick a culture would not allow the light scattering to be detected properly. However, a range up to 250 g l^{-1} dry weight is measurable. The probe is resistant to fouling over time.

Capacitance/conductance-based Biomass Monitor

This uses a totally novel measurement system to give viable cell numbers for all types of organisms (bacteria, yeast, filamentous fungi and animal cells). A probe in the vessel uses a radio-frequency electrical field to measure the natural capacitance of living cells with an intact plasma membrane. This build-up of charge is proportional to the cell concentration and specific to different types of cell. The signal is processed and can be ex-pressed as dry weight or concentration in cells ml^{-1}.

An anti-fouling system is used to maintain the ability of the probe to give accurate readings over long time periods. This type of instrument makes possible the control of feed pumps by biomass concentration, and could form the basis for an automated transfer system for delivering a seed culture to a larger vessel after an inoculum of sufficient concentration has been produced in a smaller fermentor.

SIMPLE CONTINUOUS CULTURE

There are a number of different ways of setting up a continuous culture. The method of choice has the virtue of being able to be added to any fermentor vessel without modification or the addition of extra controllers beyond the acquisition of a feed pump and a harvest pump. The feed pump is variable speed but the harvest pump can be fixed speed as long as its speed is greater than that of the feed pump. The harvest pipe, being sited at the surface of the culture, has the disadvantage of taking off only the top of the culture (which may not be representative of the bulk culture) but the advantage of helping to prevent build-up of foam. If foam is a problem then antifoam reagent can be added to the medium feed at a background level, *e.g.* 1 part in 20 000. Samples can be taken by putting a 'Y' piece in the take-off tubing and placing a sampling device in one arm of the 'Y'.

ADDITIONAL ACCESSORIES AND PERIPHERALS

Feed Pumps

A separate feed pump is necessary for fed-batch or continuous applications. The type of pump used will depend on the size of the vessel, the required flow rates, the nature of the additive (*e.g.* viscous, corrosive, etc.) and the speed at which the addition should be made. Often, a feed pump is included as part of the basic fermentor or provision is made for its inclusion as an optional accessory.

Peristaltic Pump

This is the most common type of addition pump used in fermentor applications and is often found in a fixed-speed version for addition of reagents such as acid and alkali. The pump uses rollers to occlude sections of silicone tubing sequentially. This action creates a 'peristaltic' flow of liquid through the tubing (analogous to the movement of food through the body by waves of peristalsis). The more rollers a peristaltic pump has, the smoother the flow. The pumps used for medium feed are typically variable speed units which can accept signals from an external source such as a computer or process controller to allow adjustment of feed rates based on, for example, respiratory quotient (RQ) or optical density (OD). The key advantages of peristaltic pumps are as follows:

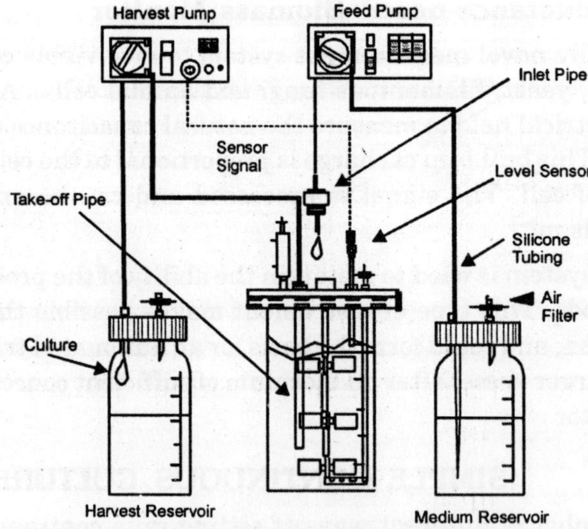

Fig. 2.10. Simple continuous culture.

(*i*) The liquid being pumped is not contaminated by any part of the pump.

(*ii*) The pump is not contaminated by the liquid being pumped.

(*iii*) A wide range of flow rates is possible using tubing of different bores.

(*iv*) The accuracy of delivery is good over a wide range of flow rates.

(*v*) The pump can run safely even when the tubing is empty.

The disadvantages are:

(*i*) The flow is pulsed and cannot 'push' against high backpressure.

(*ii*) The tubing wears over time and flattens, reducing accuracy and increasing the risk of bursting.

(*iii*) Silicone tubing is not suitable for all liquids and alternative tubing material may be too rigid to allow correct peristaltic action.

The problem with tubing wear can be partly overcome by leaving extra length in the tubing when connecting it to the fermentor vessel and moving a new length into the pump head every few days to stop wear in just one small section.

Syringe Pump

This type of pump works on a completely different principle and is useful for fed-batch applications rather than continuous medium feed. In this case, a syringe (of various sizes from µl to 100 ml, typically) is filled with the liquid to be added and secured onto the syringe pump. This is done in such a way that the syringe plunger is linked to a movable piston and the syringe barrel is held firmly to the main body of the pump. A flow rate is set (in mm min^{-1} or directly in ml min^{-1} or ml h^{-1} if the pump can be calibrated).

The piston is then moved at the desired speed by the action of toothed gears linked to a stepper motor engaging with 'teeth' on the underside of the piston. When the syringe is empty, an alarm condition is indicated and the pump automatically stops or, on very sophisticated pumps, a signal can be sent which allows automatic refilling from a reservoir. This type of pump is typically used in medical applications for the precise delivery of drugs. Most now allow for computer control. The advantages of syringe pumps are:

(*i*) Very precise and accurate delivery; non-pulsile flow.

(*ii*) Very low flow rates possible.

(*iii*) Can work against back-pressure.

(*iv*) Can pump very viscous fluids, *e.g.* oils.

The disadvantages are:

(*i*) Must be replaced or refilled regularly.

(*ii*) Range of flow rates not as wide as with peristaltic pumps.

(*iii*) Usually more expensive.

Diaphragm/Positive Displacement Pump

This type of pump is a hybrid of the two previous types in that it uses a flexible membrane and soft tubing to displace the liquid and control its flow but also employs a piston and pump heads of fixed volume. A stainless steel pump head is connected to soft tubing and autoclaved. It is then located onto the pump and held securely. A central piston and two small 'pinch valve' pistons move in and out to push liquid out of the central chamber of the pump head and allow fresh liquid to refill the chamber on each half of the pump stroke. The flow rate is controlled by the frequency of the pump stroke and the size of the pump chamber (which can usually be exchanged). A return spring returns the diaphragm to the correct position after each inward stroke.

A variant of this is to use a 'captive syringe' where the piston moves in and out of the chamber, emptying it fully then drawing back to refill on each stroke.

Advantages of this type of pump include:

(*i*) Accurate delivery.

(*ii*) More choice in type of tubing for corrosive liquids, for example.

Disadvantages are:

(*i*) Less flexibility on flow rate ranges for each pump head.

(*ii*) Stainless steel chamber and diaphragm may not be suitable for some liquids.

(*iii*) Pulsed flow even more pronounced than with a peristaltic pump.

Exit Gas Analysis

Measurement of the amounts of different gases leaving a fermentor vessel can provide valuable information about the metabolic processes taking place under the conditions in which the culture is growing. For example, the ratio of oxygen and carbon dioxide entering and leaving the vessel, or respiratory quotient (RQ), can be used to determine whether a yeast is producing biomass or alcohol. The RQ can be calculated by most fermentation software packages and can then be used as part of a control algorithm so that the flow rate of a feed pump can be altered accordingly. The entry gas does not need to be analysed providing it is air, because the amounts of oxygen and carbon dioxide will be those of the atmosphere. The flow rate of the air into the vessel will need to be accurately measured - normally by using a thermal mass flow controller. The exit gas may need to be conditioned (*e.g.* moisture removed) before going into the analyser, depending on the type of instrument used.

Infrared Carbon Dioxide Analyser

A hot wire is used to generate a source of infra-red radiation which passes through the gas which comes from the fermentor into the sample chamber. An infra-red detector measures the amount of radiation reaching it after some is absorbed by the gas. An optical filter is used to make sure the detector only responds to the gas of interest.

A 'chopper' or rotating shutter is used to allow the detector to see a reference source at regular intervals. This type of detector allows continuous measurement with good sensitivity, accuracy and selectivity. Output signals are normally provided either as analogue signals (*e.g.* 0-10 V) or a string of ASCII characters for printing or transfer to computer software.

Paramagnetic Oxygen Analyser

Oxygen has a particular physical property which this type of analyser can utilize, *i.e.* that it is more susceptible to a magnetic field than other gases. This property is measured using a finely balanced test apparatus suspended in the test chamber of the analyser. A dumb-bell of gas-filled spheres is linked to a support mechanism suspended in a magnetic field created by permanent magnets. If oxygen is present in the test gas, its attraction to the magnetic field will cause the dumb-bell to be displaced. The movement is detected using a mirror in the centre of the balance system which displaces a beam of light shone via the mirror onto a photocell.

The signal from the photocell will be proportional to the amount of oxygen present in the sample. A low flow rate is needed for this system to work properly, so inlet gas is typically pumped into the detector cell with most being discarded through bypass pipework. Once again, suitable analogue and computer outputs are normally provided.

Both infra-red detectors and paramagnetic oxygen analysers need to undergo a calibration procedure before use.

Mass Spectrometer

This represents a step upwards in versatility, capability and also cost! A wide range of gases can be analysed (both in a gas stream and as dissolved gases in the vessel), *e.g.* oxygen, carbon dioxide, argon, nitrogen, ammonia, hydrogen, methanol, ethanol and several other organic volatiles.

The analysis is rapid, so allowing a multi-inlet system to be used which 'shares' the analyser between separate vessels within a bank of fermentors. Mass spectrometers can be magnetic sector, multi-collector or quadropole. Each has its own particular advantages but the magnetic sector can be used to illustrate the principles upon which they all work.

The physical principle involved is the ionization of gas molecules by an electron source, usually a hot filament, followed by their separation in a magnetic field (the quadropole analyser uses a combination of RF and DC electrical fields to do this job) according to their mass/ charge ratio. This separation takes place in a virtual vacuum to minimize collisions prior to sorting. The magnetic field is 'tuned' so that only the ions of interest will be focused onto the detector system - a Faraday Cup. This is a metal plate which generates a tiny electric current whenever a gas ion strikes it. This signal can be amplified and displayed. By re-tuning the magnetic field very quickly (milliseconds) different ions can be detected and measured in a sequential way which is so fast it appears to give an almost simultaneous measurement.

FERMENTOR PREPARATION AND USE

Disassembly of the Vessel

The fermentation is shut down from the control unit and transfer lines plus cable connections removed. Following a fermentation, the vessel should be re-autoclaved, taking care that inlets and outlets are properly prepared. The culture should be disposed of as laid down by departmental safety procedures. The clip/clamps/bolts which retain the vessel top plate are undone until the whole assembly springs free. The top plate can now be lifted upwards away from the glass section of the vessel, taking care that the air sparger, drive shaft/impellors and Pt-100 temperature probe completely clear the vessel. The flanged top section of the glass vessel and seal can now be seen.

Cleaning

The pH and dissolved oxygen electrodes should be removed and stored in suitable reagents according to the manufacturer's instructions. Periodic cleaning and regeneration of the electrodes are also covered by these instructions. Doing this maintenance is very cost-effective. The vessel should be rinsed several times in distilled water to remove any loose culture residues. Cleaning of growths of culture on the vessel walls may require disassembly and light brushing of the glass. At this point, an examination of any chips or cracks in the vessel glass can be carried out and a replacement made if necessary.

Vessels must be stored clean and dry. In use, any spillages of reagents or medium should be wiped up immediately with a damp cloth and not be allowed to dry out. Contact between the top plate and liquids with a high chloride ion content (*e.g.* common salt solutions, hydrochloric acid) should be avoided to prevent corrosion. The pump heads and covers must be thoroughly cleaned if a tube breaks and reagent leaks out.

Preparations for Autoclaving

The vessel seal should be removed and lightly greased with a suitable silicone grease. On replacing the seal, it must be correctly located so that there is no chance of any part lifting or kinking. At this point, the vessel can be filled with medium to a maximum of 80% full (if active aeration is to be used this space is vital for gas exchange). The minimum medium volume is the amount needed to cover the electrodes adequately. The vessel top plate can now be replaced and any clamping ring or bolts tightened firmly. The ports for electrodes have O-ring seals which should be lightly greased with a silicone preparation before autoclaving.

Electrodes normally push directly into the port fitting and the collar is tightened down to compress the O-ring seal. All other fittings such as pipes are fitted in the same way. Ports not in use have 'stoppers' fitted, and their O-ring seals should have a light coating of silicone grease applied. The pH electrode should be calibrated in appropriate buffers *i.e.* pH 7, then either pH 4

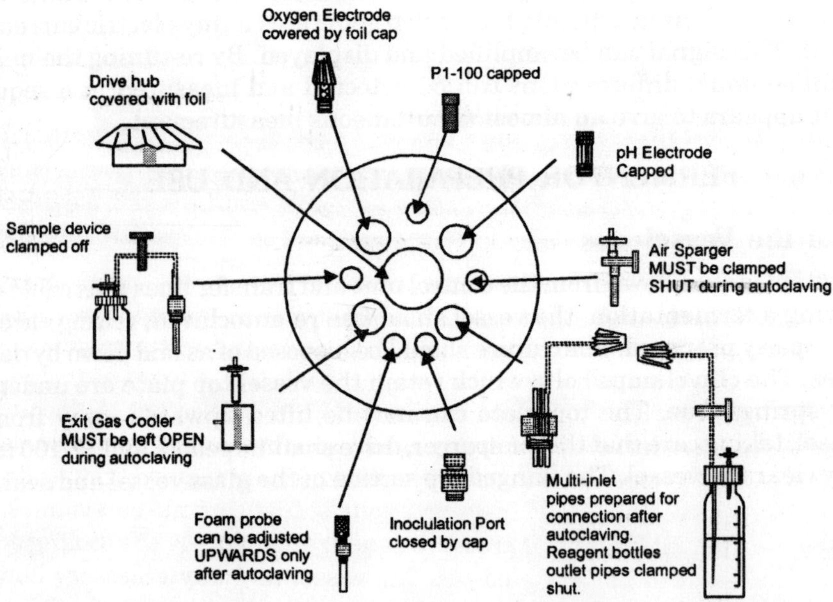

Fig. 2.11. Vessel and accessories prepared for autoclaving.

or pH 9 (or just 4 and 9 for a 'high reference-low reference' calibration of a DDC controller channel). The pH and dissolved oxygen electrodes should be fitted, taking care not to damage them by careless insertion into the port. Both must be tightly capped to prevent moisture getting into the electrical contacts.) For the dissolved oxygen electrode, a cap may have to be improvized from aluminium foil. The Pt-100 temperature sensor must be fitted and capped. If used, the foam probe is fitted so that it is above the liquid level in the head space. Reagent bottles are prepared in a similar way to the fermentor vessel. A membrane is placed over the mouth of the filled bottle and an open-topped cap screwed into place. A short needle is connected to a disposable filter which is then used to pierce the membrane in the bottle. The needle must not dip into the liquid and nothing must block the free passage of air through the filter.

A long piercing needle/pipe is used to puncture the membrane and push it into the liquid as far as possible. This pipe should now have a length of silicone tubing fitted to it which is long enough to reach the peristaltic pump heads of the fermentor instrumentation for which it is intended. The tubing is clamped so that no liquid can escape during autoclaving. A similar procedure is used for sampling and/or harvest bottles except that two short pipes are used so neither dips into the collected culture.

Some preparation is needed at the autoclaving stage for the aseptic coupling of silicone tubing ready for operation. The exit gas cooler should be fitted to one of the wider ports. A short length of silicone tubing should be attached to the top of the air outlet and a small 0.22 mm filter fitted. The air outlet line must be kept open during autoclaving. A short length of silicone tubing must be fitted to the air sparger inlet pipe with a 0.2 μm disposable filter mounted on top. The tubing between the sparger pipe and the filter must be clamped shut during autoclaving. If a port is to be used for inoculation or piercing with a needle, a silicone membrane must be fitted into the empty port and a clamping collar/cap used to hold it in place.

Autoclaving

The vessel and any reagent/sampling bottles already connected by silicone tubing are assembled together on a steel tray or in an autoclave basket. A final check should be made that at least one route is available for air to enter and leave the vessel(s) and that all lines dipping into liquid are clamped closed. If the vessel has top drive and a mechanical seal, the seal must be lubricated (normally with glycerine). It the medium cannot be autoclaved, a suitable volume of distilled water should be used and removed via a sample line before the actual medium and inoculum are aseptically transferred.

A quantity of liquid is certain to be lost during autoclaving (approximately 10%) so either the medium is over-diluted to compensate for this or sterile, distilled water is added afterwards to restore the volume. Some form of indicator such as autoclave tape should be included to provide a warning if the correct sterilization procedure has not been carried out. Autoclaving at 121°C for 30 min is normally considered adequate for vessel sterilization to minimize damage to the constituent chemicals of the medium. However, some work may require temperatures of 134°C for several hours to ensure sterility. If in doubt, your safety committee should be consulted. Also, the autoclave used must have good pressure equalization during the cooling down phase of operation to prevent medium being boiled off. The vessel and any accessory bottles must be allowed to cool completely before handling.

Set-up Following Autoclaving

Firstly, the silicone tubing prepared prior to autoclaving must be connected to the fermentor vessel (if the bottles they are connected to were autoclaved separately). The air sparger is connected to the rotameter by a piece of silicone tubing from the top of the filter to the air outlet of the rotameter. The air sparger line is undipped between the metal pipe and the air filter. The exit gas cooler is connected to the water supply, either directly or via the fermentor base unit.

The tubing for water in, out and drain is connected to the vessel jacket for a water system, and the water turned on so that the vessel jacket is filled. Alternatively, any pads or heater cartridges are connected to the base unit or temperature control module; the cold finger is connected to the water supply. The tubing from the reagent bottles is connected to the multi-way inlet (if necessary), and the silicone tubing from the reagent bottles is located in the relevant peristaltic pump. Any aseptic connections must be made first if the reagent bottles were autoclaved separately from the vessel. The clamps are removed so liquid can flow freely.

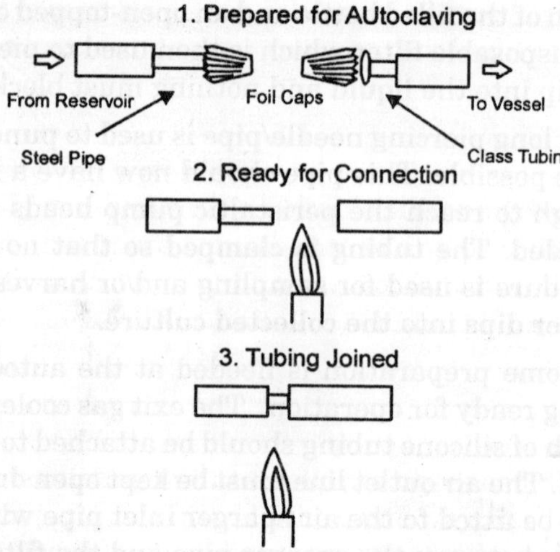

Fig. 2.12. Connecting tubing aseptically.

A manual switch is often fitted which allows the pumps to be primed with liquid prior to use. The drive motor is located onto the top plate (if appropriate), ensuring a good connection is made to the drive shaft. The Pt-100 temperature sensor is connected to the control module and the pH electrode by removing the shorting cap and screwing in the cable. The dissolved oxygen electrode is connected to the appropriate cable (this requires some care, but the connector should lock firmly when it is correctly positioned). Connections to the foam probe are made, usually one wire into the electrode and one on the vessel top plate to make a circuit. The mains electricity to the instrumentation modules is switched on, allowing some hours (minimum of 2, preferably 24) for the dissolved oxygen electrode to polarize properly. This delay is not needed for galvanic electrodes.

Setting the temperature control at this stage will ensure the fermentor is ready to inoculate after calibration of the dissolved oxygen electrode. After polarization the dissolved oxygen electrode is calibrated and the air supply turned on. The maximum stirrer speed to be used and the maximum air flow required on the rotameter are set. After leaving for about 5 min, the 100% level is set. The fermentor is now ready to inoculate.

Inoculation of a Fermentor Vessel

This section assumes that all the set-up procedures listed above have been carried out. The simplest way to inoculate is to have a dedicated port fitted with a membrane which is capped off before autoclaving. The inoculum (which should normally be no more than 5-10% of the total culture volume) is aseptically transferred to a sterile, disposable syringe of a suitable size. The port fitting is removed and held vertically to prevent contamination of the bottom end. The syringe needle is quickly pushed through the membrane and the inoculum transferred into the vessel. The vessel may be actively aerated during this procedure to minimize the risk of a contaminant getting into the vessel (safety considerations permitting).

The syringe needle is quickly withdrawn and the silicone membrane reseals. The port fitting is now replaced. For added security, a couple of drops of 70% ethanol can be placed on the membrane surface before piercing and ignited to provide a thermal barrier. Alternatively, an 'aseptic connection' can be made to an inlet pipe. If the line connected has a 'Y' coupling in it, then the same aseptic connection could be used to introduce medium. This technique is useful if a dedicated inoculation port cannot be provided.

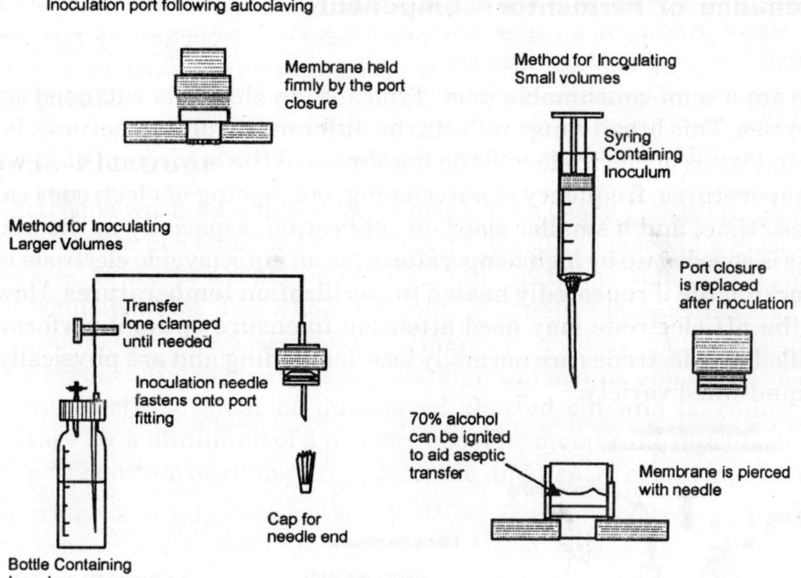

Fig. 2.13. Inoculation of a fermentor vessel.

Sampling from a Fermentor Vessel

All sampling starts with a sample pipe, which should dip into the bulk of the culture liquid. The simplest system is to have a length of silicone tubing connected to the sample pipe (and clamped off until a sample is to be taken) which is plugged and covered at the open end. After autoclaving, the tubing is quickly unwrapped, the plug removed and the open end dipped in a container of 70% ethanol. When a sample is to be taken, a sterile syringe is quickly coupled to the tubing and the sample withdrawn into the syringe. Aeration is stopped during this process to prevent 'surging' of culture into the syringe. Any culture in the transfer line is discarded into a 'kill jar' and the tube replaced in the ethanol. Clearly, this method can be prone to contamination and is not suitable for organisms with any risk to operators. A better approach is to use a sampling device. At its simplest, this is a bottle which has two metal needles/pipes permanently fixed through both the metal and rubber seals of its cap. One pipe has a 0.22 µm air filter connected and the second is linked by silicone tubing (clamped off until a sample is needed) to the sample pipe.

A syringe is fitted to the air filter after autoclaving. The sample device is usually attached to the vessel top plate so that the glass bottle hangs down vertically beneath the cap and a supply of bottles of the same size are autoclaved ready for use. When a sample is to be taken, the clamp on the sample pipe tubing is released, and the syringe pulled back to create a partial vacuum in the sample bottle. Culture will flow into the sample bottle. When the required volume of sample has been taken (never more than 75% of the total bottle volume) the syringe is pushed in to clear the line and the sample tubing is clamped shut again. The sample bottle with the culture is quickly removed and capped with a sterile cap from a new glass bottle which replaces it under the sampling device. It is often preferable simply to discard the first few millilitres of the next samplc by using it as a 'wash' for the transfer line, putting a fresh bottle under the sampling device to actually take the sample.

Routine Maintenance of Fermentor Components

pH Electrodes

pH electrodes are a semi-consumable item. Typically, an electrode will need replacing after 20-50 autoclave cycles. This broad range reflects the different results sometimes found between different brands and types of electrode and the harshness of the environment in which they are used, *e.g.* high temperatures, frequency of autoclaving, etc. Ageing of electrodes can show up as a sluggish response time, and a smaller slope on calibration, especially in the alkaline region. The ageing process is speeded up by high temperatures, so an autoclavable electrode will invariably have a shorter working life if repeatedly heated to sterilization temperatures. However, during its working life, the pH electrode may need attention to ensure optimal performance. In this respect, the gel-filled pH electrodes are normally less demanding and are physically more robust than the glass, liquid-filled variety.

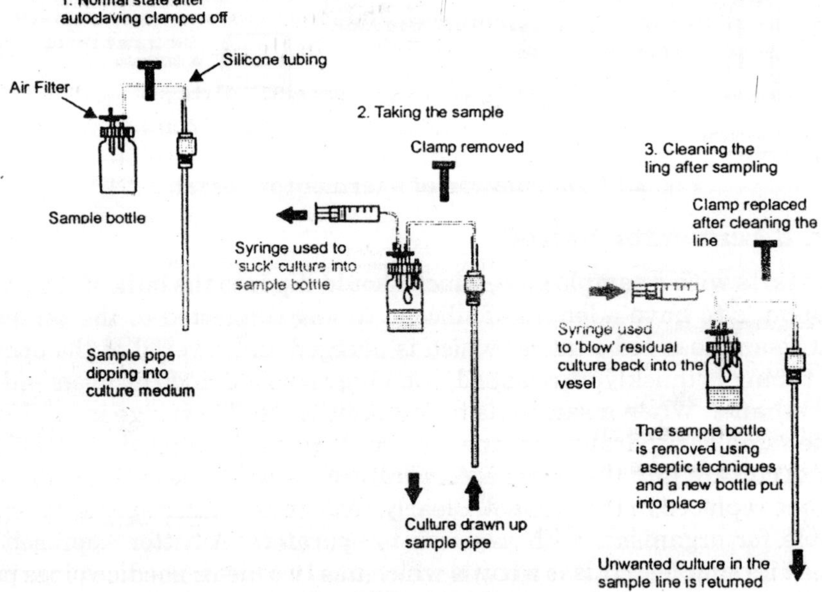

Fig. 2.14. Sampling.

Dissolved Oxygen Electrodes

Galvanic Electrode : This is a simple, self-powered system which relies on a lead anode and a silver cathode which are shorted together. Oxygen is reduced at the silver cathode and an electrical current flows in proportion to its partial pressure. The response time is slower than a polarographic electrode but the initial cost is far lower. The electrode is conditioned for the first time by the autoclaving process. The probe is relatively robust except if the lead anode is exposed to air due to a drop in the level of electrolyte in the probe. Therefore, all air bubbles must be removed when electrolyte is added and probes must always be stored upright with an adequate level of electrolyte and with the electrode shorted out (cap on).

Periodically, the electrode membrane will need replacing. This involves soaking a section of new silicone rubber tubing in chloroform to swell it (safety hazard!). While this is happening, the old tubing and membrane is cut away. A new membrane is stretched tightly across the electrode tip and the silicone tubing placed over it to hold it in place without any wrinkling. The end away from the tip is trimmed off and secured with a little Araldite glue and the electrode tapped gently to remove air bubbles.

Polarographic Electrode

These electrodes work on a different principle (the Clark Cell). The electrode consists of a silver anode/ reference electrode with a platinum cathode encased in a glass rod. The end of the glass rod is shaped to accept a replaceable membrane cartridge. A temperature sensor is included for automatic compensation. The whole assembly is usually encased in a stainless steel housing. A chloride solution is used as an electrolyte. A polarization voltage must be applied to the electrode which can then reduce the oxygen passing through the membrane and generate a current. For this reason, the electrode must be autoclaved shorted out and be connected to a source of polarizing voltage for a minimum of 2 h before use. The membrane must be replaced periodically by unscrewing the bottom section of the electrode and gently removing the old cartridge.

A new cartridge is fitted, having been filled with fresh electrolyte, and pushed onto the inner glass rod. If necessary, the small O-ring in the bottom section of the housing can be renewed before replacing the cartridge. A functional check on the electrode can be performed by setting the reading of a polarized electrode in water saturated with air to 100%. The air is replaced with nitrogen gas and the value should fall to less than 1.5% after about 5 min. If the electrode is now removed into the air, the value should return to 98%+ within 45-90 s.

Changing a Mechanical Seal

Different arrangements of mechanical seal are used for different makes of fermentor. The type described is a good example as it shows many of the features used with the mechanical seals employed on both bench-scale and *in situ* sterilizable fermentors. Seals should be replaced regularly (at least every 6-12 months, depending on use).

MAJOR TYPES OF ORGANISMS USED IN FERMENTATION

A wide range of organisms can be used in fermentation for the production of various chemicals and antibiotics.

Bacteria/yeast/fungi

Either a natural isolate of the culture is made or an uncontaminated 'master' culture is revived from a 'freeze-dried' state or from liquid nitrogen storage. Bacteria, yeasts and filamentous fungi are the usual starting materials. The starting cultures are grown in petri dishes on agar containing the required nutrients and incubated at an optimum temperature, humidity and gaseous atmosphere. If necessary, shake-flask cultures are prepared in an incubator shaker to produce larger volumes of culture for analysis or for use as an inoculum for a fermentor. Flasks are typically filled to 20-25% to provide a large area for gas exchange and are 25 ml to 5 l in total capacity. An inoculum is prepared either from the shake-flask culture or by suspending colonies from growth on solid media in a nutrient/buffer solution. This is aseptically transferred to a fermentor vessel. The inoculum is typically 5-10% of the volume of medium in the fermentor and is often diluted in order to introduce a known number of cells/ml in this volume.

A laboratory scale fermentor can be used to provide larger volumes (typically 1-51 of culture). Growth is normally as a batch fermentation with the following parameters measured and controlled as necessary:

Temperature typically 20—40°C (4-90°C at the extremes) Speed typically 150-1500 rpm (20-2000 rpm at the extremes) pH typically 4-8 (2-12 at the extremes) Dissolved oxygen typically 40-80% (0-100% at the extremes)

In some cases, a continuous culture system may be used to draw off culture of a known cell concentration for long periods. A concentration of a growth-limiting metabolite (chemostat) or an indirect estimate of cell numbers (*e.g.* turbidity, turbidostat) is used to keep the culture density constant for many days/weeks. Samples are taken from the fermentor for counting, culturing on solid medium or biochemical assay for a particular metabolite, sugar utilization, etc. On completion of the culture, the whole culture can be centrifuged, *e.g.* 4000 × g for 15 min to separate. whole cells from the culture supernatant fluid.

Alternatively, a membrane or hollow-fibre system can be used for this initial separation. The separated cells can be washed in buffer, resuspended, and can then be disrupted using a variety of methods, *e.g.* ultrasonic disruption, freeze-thaw, shaking with beads, etc. Supernatant fluids can be further separated and analysed using separation columns, fraction collectors and spectrophotometry. Gel electrophoresis, ELISA assay, etc. are also used. Applications are almost too numerous to mention but obvious categories include brewing (yeasts), antibiotic production (fungi), enzyme production (bacteria) and food manufacture such as yoghurt (bacteria). Environmental uses include removal of toxic wastes, biological control of crop pests and soil regeneration.

Plant Cells

The whole plant is grown by conventional means. A sterile knife is used to cut a small section of root, stern or leaf which is transferred onto solid medium to produce a ball of cells (a thallus). The thallus is separated with a sterile blade and the cells grown in shake-flask culture in an incubator shaker at 20-30°C and 100-200 rpm. The cells are transferred to a fermentor under the following conditions: *e.g.*

- stirrer speed = 100-150 rpm;
- temperature 20-30°C (cooling needed if lights used);
- pH not generally controlled as it tends to cycle naturally;
- dissolved oxygen rarely controlled as it is not generally a limiting factor;
- foaming can be a problem unless controlled.

Cells may be cycled between the shaker and fermentor several times to build up shear resistance. The cell growth is normally too slow for effective continuous culture. For material inside the plant cell a 'Potter' type homogenizer can be used for gentle disruption. Centrifugation at approx. 1000 rpm for whole cells will separate them from the culture supernatant. Higher speeds will be needed to deposit cell fragments. The cellular pellets are discarded or extracted depending on where the required component is located. Plant cell cultures are often used for production of specific biochemicals such as flavinoids and alkaloids used in foods and chemical production. Culture of plant cells for agro-chemical uses is another obvious application.

Mammalian Cell Culture

There are many different cell lines of human, mammalian, avian and even insect origin. For simplicity, a single type, hybridomas, will be used as an example here. Hybridoma cell lines are prepared using cells from mouse spleen which secrete desired antibodies crossed with a continuous cell line grown *in vitro*. The cells are placed in suspension culture in an incubator shaker at 37°C and 60-100 rpm speed. The cells are transferred to a specially adapted fermentor. Large, broad-bladed impellers are used to minimize shear forces and typical stirrer speeds are 20-100 rpm. A gas mixing station can provide an air/carbon dioxide mixture and CO_2 gas can be used for pH control via a solenoid valve as replacement for a peristaltic pump. Oxygen and nitrogen can be introduced for dissolved oxygen control. For many cell lines, the control of foam is vital as its presence can contribute to cell damage.

The cells can sometimes grow freely in suspension but more often are anchored to a mesh, alginate beads or a hollow-fibre system. A special mesh can be connected to the drive shaft to provide a 'spin-filter', or to the sample pipe for taking samples of culture fluid relatively cell-free. Some alternative approaches to growth of mammalian cells are as follows:

(*i*) Air lift. There is no drive motor and it is the pumping of liquid around the draught tube which provides for mixing and gas exchange.

(*ii*) Hollow-fibre cartridges. They remain outside the fermentor vessel but culture fluids can circulate between the fibres and the vessel. All mixing and gas exchange is done in the vessel and the enriched medium passes to the cells anchored to the hollow fibres via a pump.

(*iii*) A gassing basket inside a stirred fermentor. Where cells need good levels of dissolved oxygen but are very sensitive to shear, a large amount of thin-walled silicone tubing supported on a frame can be placed in the vessel to provide a large area for gas exchange. Air is bubbled through the pores in the silicone tubing to give good gas transfer rates without forming large bubbles.

Applications of mammalian cell culture include monoclonal antibody production, culture of viruses for vaccine manufacture and large quantities of specific cells for medical research.

Algae

This category covers a diverse range of microorganisms which differ vastly in their growth conditions and metabolic requirements. Some are halophiles (salt-loving), barophiles (survive under high pressure), thermophiles (grow at high temperatures) and psychrophiles (grow at low temperatures). Most need special culture conditons and so it is best to seek out published literature with specific details. Applications for the use of algae include food additives and chemical production by biotransformation.

MICROBIOLOGY OF INDUSTRIAL FERMENTATION

Microorganisms play a central role in the production of a wide range of industrial chemicals, enzymes and antibiotics. The diversity of microbial processes may be attributed to many factors, including the high surface-to-volume ratio and adaptability to a wide spectrum of carbon and nitrogen sources. The high surface-to-volume ratio supports a very high rate of metabolic turnover, *e.g.* the yeast *Saccharomyces cereviciae* has been reported to be able to synthesize protein several orders of magnitudes higher than plants. On the other hand, the ability of microorganisms to adapt to different metabolic environments makes them capable of utilizing inexpensive renewable resources such as wastes and by-products of the farming and petrochemical industries as the primary carbon source.

Industrially important microorganisms include bacteria, yeasts, moulds and actinomycetes. The rate of product formation in a given industrial process, a significant parameter, is directly related to the rate of biomass formation which, in turn, is influenced directly or indirectly by a whole host of different environmental factors, *e.g.* oxygen supply, pH, temperature, accumulation of inhibitory intermediates, etc. It is, therefore, important that we are able to describe growth and production in quantitative terms. The study of growth kinetics and growth dynamics involves the formulation and use of differential equations and while the mathematical derivation of these equations is beyond the scope of this chapter, it is important that we understand how such equations can be used to further our understanding of microbial growth in general and its impact on product formation (yield) and recovery in particular.

THE GROWTH CYCLE

Now let us consider the basic equation used to describe microbial growth:

$$\frac{dx}{dt} = \frac{ax}{b} \qquad ...(2.1)$$

Equation (2.1) implies that the rate of biomass (x) formation changes as a function of time (t) and that the rate of change is directly proportional to the concentration of a particular factor (a) such as growth substrate or temperature but is inversely proportional to the concentration of another factor (b) such as inhibitors. In the equation above a and b are independent of time t, and the proportionality factor in equation (2.1) can in effect be ignored.

In the early stages of any fermentation process, the increase in biomass is unrestricted and, as such, the pattern of growth follows an autocatalytic first-order reaction (autocatalytic growth) up to a point where either side of equation (2.1) becomes negative, resulting in autocatalytic death. During batch fermentation, a typical pattern of growth curve, otherwise known as the growth cycle, is observed. Clearly a number of different phases of the growth cycle can be differentiated. These are:

I Lag phase
II Acceleration phase
III Exponential (logarithmic) phase
IV Deceleration phase
V Stationary phase
VI Accelerated death phase
VII Exponential death phase
VIII Survival phase

The changes in the specific growth rate (μ) as the organism progresses through the growth cycle.

We shall now describe the metabolic events and their implications in as far as growth, survival and productivity are concerned. Naturally the scenario begins with the first phase of the growth cycle.

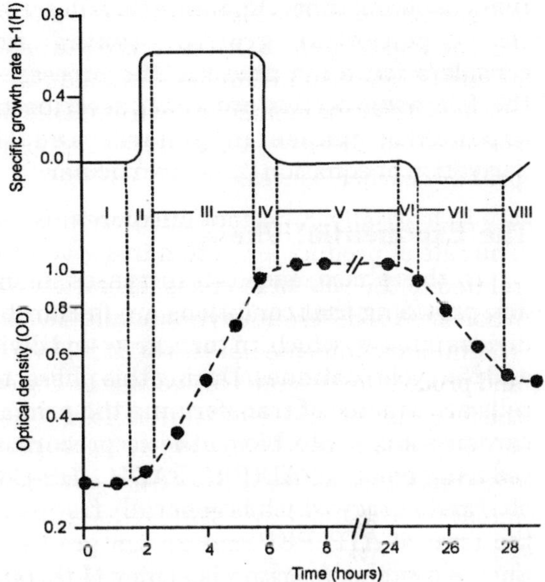

Fig. 2.15. Typical pattern of growth cycle during batch fermentation. The vertical dotted lines and the Roman numerals indicate the changes in specific growth rate (m) throughout the cycle.

The Lag Phase

In this phase the organism is simply faced with the challenge of adapting to the new environment. While adaptation to glucose as a sole source of carbon appears to be relatively simple, competition with other carbon sources, although complex, is resolved in favour of glucose through the operation of two different mechanisms, namely catabolite inhibition and catabolite repression. Adaptation to other carbon sources, however, may require the induction of a particular set of enzymes which are specifically required to catalyse transport and hydrolysis of the substrate, *e.g.* adaptation to lactose, or to fulfil anaplerotic as well as regulatory functions as is the case in adaptation to acetate as the sole source of carbon and energy.

Irrespective of the mechanisms employed for adaptation, the net outcome at the end of the lag phase is a cell that is biochemically vibrant, *i.e.* capable of transforming chemicals to biomass, *i.e.* growth. Entering a lag phase during the course of industrial fermentation is not desirable as it is very costly and as such should be avoided. The question of whether a particular organism has entered a lag phase in a given fermentation process can be determined graphically by simply plotting log n (biomass) as a function of time. Note that the transition from the lag phase to the exponential phase is generally interrupted by an acceleration phase, as described earlier. This difficulty canbe easily overcome by extrapolating the lag phase sideways and the exponential phase downwards as shown, with the point of interception *(L)* taken as the time at which the lag phase ended. What is also interesting about the graph that if one continues to extrapolate the exponential phase downwards, then the point at which the ordinate is intersected gives the number of cells that were viable and metabolically active at the point of inoculation.

The question of whether a lag *(L)* has occurred during the course of the fermentation process and for how long can be easily determined from the following equation:

$$\frac{\log n - \log n_0}{t - L} = \frac{\log 2}{T} \qquad \ldots(2.2)$$

where n is the total number of cells after a given time (t) since the start of fermentation, n_0 is the number of cells at the beginning of fermentation and T is the organism's mean generation time (doubling time). Equation (2.2) describes the exponential growth, taking into consideration a lag phase in the process. In the following section we shall describe the exponential phase in general and the derivation of equation (2.2) inparticular.

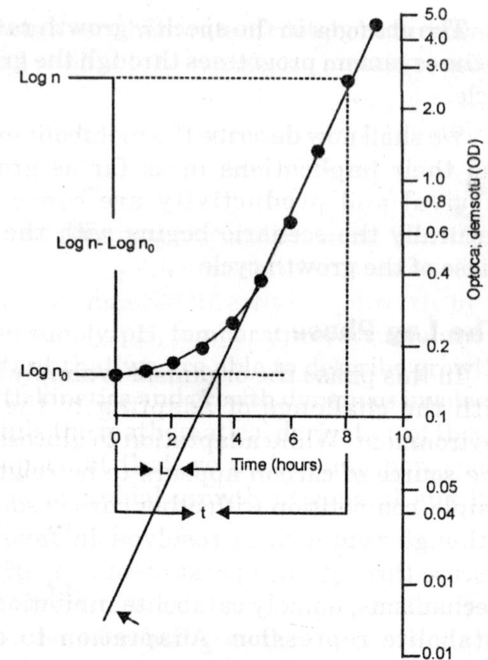

Fig. 2.16. Graphical determination of the lag phase and the number of viable cells at the onset of batch fermentation. As the exponential phase is extrapolated downwards, it intercepts the extrapolated line of the lag phase and the ordinate respectively (see text for details).

The Exponential Phase

In this phase, each cell increases in size, and providing that conditions are favourable, divides into two which, in turn, grow and divide and the cycle continues. During this phase, the cells are capable of transforming the primary carbon source into biosynthetic precursors, reducing power (NAD(P)H, $FADH_2$, HS-CoA, etc.) and energy which is generally trapped in the form of ATP, PEP and proton gradients. Such a complex function is achieved through the activities of various enzymes of central metabolism. The biosynthetic precursors thus generated are then channelled through various biosynthetic pathways for the biosynthesis of various monomers (amino acids, nucleotides, fatty acids, sugars) which, in turn, are polymerized to give the required polymers (proteins, nucleic acids, ribonucleic acids, lipids). Finally, these polymers are assembled in a precise way and the cell divides to give the new biomass characteristic of each organism. The time span of each cycle (cell division) is known as generation time or doubling time but because we generally deal with many millions of cells in bacterial cultures, the term mean generation time is more widely used to reflect the average generation times of all cells in the culture. Such a rate, providing conditions are favourable, is fairly constant. If a given number of cells (n_0) are inoculated into a suitable medium and the organism was allowed to grow exponentially, then the number of cells after one generation is $2n_0$; at the end of two generations, the number of cells becomes $4n_0$ (or 2^2n_0). It follows therefore that at the end of a certain number of generations (Z), the total number of cells equals 2^Zn_0. If the total number of cells, or its log value, at the end of Z generations is known, then

$$n = n_0 2^z \quad ...(2.3)$$

$$\log n = \log n_0 + Z \log 2 \quad ...(2.4)$$

To determine the number of cell divisions, *i.e.* number generations (Z) which have taken place during fermentation, the above equation can be modified to give:

$$Z = \frac{\log n - \log n_0}{\log 2} \quad ...(2.5)$$

If T is the mean generation time required for the cells to double in number and t is the time span over which the population has increased exponentially from n_0 to n, then

$$Z = \frac{t}{T} = \frac{\log n - \log n_0}{\log 2} \qquad ...(2.6)$$

If during the course of a particular fermentation, a lag time has been demonstrated, equation (2.6) can be modified to take account of this observation. The modified equation is as follows:

$$Z = \frac{t-L}{T} = \frac{\log n - \log n_0}{\log 2} \qquad ...(2.7)$$

The above can be rearranged to give the familiar equations governing the determination of the mean generation time (T) as follows:

$$\frac{\log n - \log n_0}{t} = \frac{\log 2}{T} \qquad ...(2.8)$$

$$\frac{\log n - \log n_0}{t-L} = \frac{\log 2}{T} \qquad ...(2.9)$$

While equations (2.8) describes the exponential phase of growth in pure terms, takes into consideration the existence of a lag phase in the process in Fig. 2.2.

The exponential pattern (*i.e.* 2-4-8-16, and so on) demonstrated in Fig. 2.3 obeys equations (2.8) and (2.9) for logarithmic growth. Note the different position of log n_0 to that cited in Fig. 2.2.

Although the use of log to the base 2 has the added advantage of being able to determine the number of generations relatively easily, because an increase of one unit in log *2n* corresponds to one generation, the majority of researchers continue to use log to the base 10. In this case, the slope of the line. The relationship between the mean generation time (*T*) and the specific growth rate (μ) can be described mathematically by the following equation:

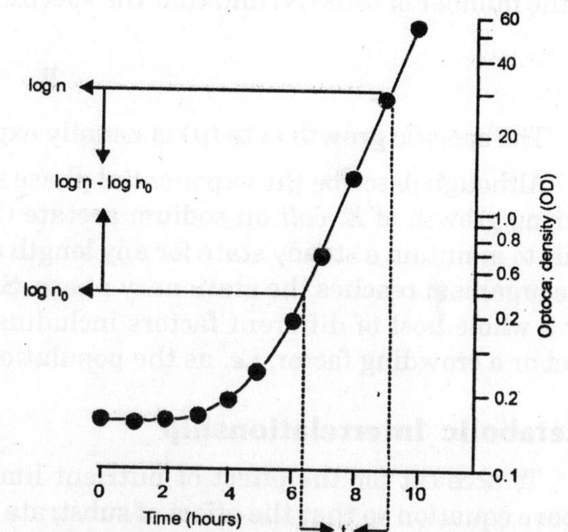

Fig. 2.17. Graphical determination of the mean generation time (*T*) during batch fermentation. Note the different position of log n_0 with respect to that in Fig. 2.2.

$$\ln 2 = \mu T \qquad ...(2.10)$$

or

$$\mu = \ln 2/T \qquad ...(2.11)$$

Because ln 2 = 0.693, either of the above equations can be rearranged to give

$$T = \frac{0.683}{\mu} \qquad ...(2.12)$$

During the course of exponential growth, the culture reaches a steady state and as such the intracellular concentrations of all enzymes, cofactors and substrates are considered to be constant. During this phase, one can therefore safely assume that all bacterial cells are identical and that the doubling time is constant with no loss in cell numbers due to cell death. The rate of growth of a given population (*N*) represents, therefore, the rate of growth of each individual cell in the population multiplied by the total number of cells. Such a rate can be described mathematically by the following differential equation:

$$\frac{dN}{dt} = \mu N \qquad ...(2.13)$$

This equation implies that the rate of new biomass formation is directly proportional to the specific growth rate (μ) of the organism under investigation and the number of cells (*N*). This pattern of growth may be referred to as ***autocatalytic***: a term which has been described earlier. If the above equation describes the exponential phase correctly, then a straight line should be obtained when ln N is plotted as a function of time (*t*). During the exponential phase, it is generally assumed that all cells are identical and as such the specific growth rate (μ) of individual cells equals that of the whole population.

Therefore eqn, 2.13 can be rearranged to take account of the fact that dN/*dt* is proportional to the number of cells (N) and that the specific growth rate (μ) is a proportionality factor to give:

$$\mu = \frac{1}{N}\frac{dN}{dt} \qquad ...(2.14)$$

The specific growth rate (μ) is usually expressed in terms of units per hour (h^{-1}).

Although describe the exponential phase satisfactorily, in some fermentations, as is the case during growth of *E. coli* on sodium acetate (M. El-Mansi, unpublished results), the organism fails to maintain a steady state for any length of time and so μ. falls progressively with time until the organism reaches the stationary phase. Such a drop in growth rate (μ) can be accounted for by a whole host of different factors including nutrient limitations, accumulation of inhibitors and/or a crowding factor, *i.e.* as the population increases in size μ decreases.

Metabolic Interrelationship

To account for the effect of nutrient limitations on growth rate (μ), Monod modified the above equation so that the effect of substrate concentration (limitations) on μ could be assessed quantitatively. Monod's equation is as follows:

$$\mu = \frac{\mu_m S}{K_S + S} \qquad ...(2.15)$$

where S is the concentration of the limiting substrate, μm is the maximum specific growth rate and K_S is the saturation constant, i.e. when S = K_S then μ = u' m/2

During growth in a steady state, the following equations may be used to describe growth using biomass (X) rather than number of cells (N) as a measure of growth

$$\frac{dX}{dt} = \mu X \qquad ...(2.16)$$

$$\frac{dS}{dt} = -\mu\frac{X}{Y} \quad ...(2.17)$$

where X is the biomass concentration, Y is the growth yield (*g* biomass generated per gram substrate utilized). It is noteworthy, however, that during steady state growth, *i.e.* in a chemostat or a turbidostat, the terms used to describe the bacterial numbers (*N*) and biomass (*X*) in equations identical. This is not necessarily the case in batch cultures because the size and shape of bacterial cells vary from one stage of growth to another.

While eqn. (2.15) addresses the relative change in biomass (the first variable) with respect to time, eqn. (2.16) addresses the relative change in substrate concentration (the second variable) as a function of time. Note that following the exhaustion of substrate concentration d*X*/*dt* = dS/*dt* = 0. While either equation (2.16) and (2.17) can surely predict the deceleration and stationary phases of growth, it should be remembered that nutrient limitation is not the only reason for the deceleration of growth and subsequent entry into the stationary phase. In addition to environmental factors, crowding is reported to have an adverse effect on growth rate (μ), *i.e.* growth rate diminishes as the size of the population increases. Although the equation describing the effect of crowding on growth rate (Verhulst-Pearl logistic equation) predicts a sigmoidal pattern of growth and fits rather well with the growth curve for populations of higher organisms, yeasts as well as bacteria.

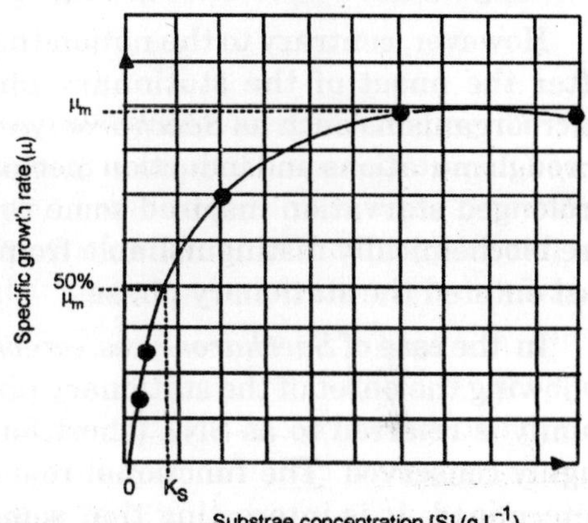

Fig. 2.18. Graphical determination of the maximum specific growth rate (μ) and the saturation constant (K_s). Note that this graph was constructed without taking maintenance energy into consideration. With maintenance in mind the curve should slide sideways to the right in direct proportion to the fraction of carbon diverted towards maintenance.

Stationary Phase and Cell Death

As the exponential phase draws to an end, the organism enters the stationary phase and then the death phase. The equations used earlier to describe growth do not adequately describe cell death. However, if we assume that the kinetics of death are similar to that of growth, then the specific rate of cell death (λ) can be described mathematically by the following equation:

$$\frac{dN}{dt} = (\mu - \lambda)N \quad ...(2.18)$$

Equation (3.18) clearly indicates that if μ is greater than λ, then the organism will grow at a rate equal to μ - λ. If on the other hand μ is less than λ, then the population dies at a rate equal to λ – μ. Under conditions where μ equals λ, neither growth nor death is observed: a situation which is thought to prevail throughout the course of the stationary phase. As the energy supply continues to fall, the equilibrium between λ and μ will finally shift in favour of λ and consequently cell death begins. Microorganisms respond differently to nutrient limitations during the stationary phase.

For example, while *Bacillus subtilis* and other Gram-positive, spore-forming organisms respond by sporulation, *E. coli* and other Gram-negative, non-spore-forming bacteria cannot respond in the same way and as such other mechanisms must have evolved. Although a fraction of the bacterial population die during this phase of growth, a relatively large number remains viable for a long time despite starvation. The ability of cells to remain viable despite a prolonged period of starvation is advantageous, as most microorganisms in Nature are subject to nutrient limitations in one form or another. Our understanding of the molecular mechanisms employed by microorganisms for survival during this phase of growth is rather limited.

However, contrary to the notion that microorganisms enter a logarithmic death phase soon after the onset of the stationary phase, recent investigations have revealed that some microorganisms such as *Saccharomyces cerevisiae* and *E. coli* adapt to starvation, presumably through mutations and induction mechanisms very effectively. The ability of some cells to survive prolonged starvation inspired some researchers to pursue the question of whether such cells are biochemically distinguishable from their predecessors which were actively growing or had just entered the stationary phase.

In the case of *Saccharomyces cerevisiae,* analysis of mRNAs which are specifically expressed following the onset of the stationary phase revealed the presence of a new family of genes. This family is referred to as SNZ (short for snooze) and sequence analysis revealed that they are highly conserved. The functional role of each member of this family, however, remains to be determined. It is interesting that some researchers use the term Viable But Non-Culturable (VBNC) to describe cells in the stationary phase. However, as colony-forming ability is our only means of assessing whether a particular cell is alive or not, other microbiologists argue the case for reversibility, *i.e.* the cell's ability to transform itself from being dormant to being metabolically active.

Recent investigations have revealed that resuscitation of stationary phase cells may be aided by the excretion of pheromones; this has recently been demonstrated in *Micrococcus luteus.* In the case of *E. coli,* an attempt was made to identify the genes which are uniquely turned on in the stationary phase in response to starvation, *i.e.* not expressed during the exponential phase, with the aid of random transposon mutagenesis.

A number of mutants which were unable to survive in the stationary phase were isolated and, subsequently designated as survival negative mutants (Sur^-) and the genes involved were designated as *sur* genes. While some genes, *e.g. sur A,* are required for survival during 'famine' (starvation), others, *e.g. surB,* enable the organism to exit the stationary phase as conditions change from 'famine' to 'feast'. Recent studies have also revealed that starvation induces the production of a stationary phase-specific transcriptional activator (sigma factor) that is essential for the transcription of *sur* genes by RNA polymerase. The physiological function of the *sur* gene products is to downshift the metabolic demands made on central and intermediary metabolism for maintenance energy.

Maintenance and Survival

Maintenance can be defined as the minimal rate of energy supply that is required to maintain the viability of a particular organism without contributing to biosynthesis. The fraction of carbon oxidized in this way is, therefore, expected to end up in the form of carbon dioxide (CO_2). The need for maintenance energy is obvious as the living-state' of any organism, including ourselves, is remote from equilibrium and as such demands energy expenditure. Moreover, in addition to the carbon processed for maintenance, another fraction of the carbon source may be wasted through excretion, *e.g.* acetate, a-ketoglutarate, succinate, lactate, etc. In this context, excretion of metabolites ought to be seen as an accidental consequence of central metabolism rather than by design to fulfil certain metabolic functions. The fraction of carbon required for maintenance differs from one organism to another and from one substrate to another.

For example, maintenance requirements for the lactose phenotype of *E. coli* are greater than those observed for the glucose phenotype as the *lac*-permease is much more difficult to maintain than the uptake system employed for the transport of glucose, *i.e.* the phosphotransferase system (PTS). The metabolic interrelationship between maintenance and growth was first described by Pirt (1965). In his theoretical treatment of this aspect, he formulated a maintenance coefficient (m) based on the earlier work of Monod, and defined the coefficient as the amount of substrate consumed per unit mass of organism per unit time (*e.g.* g substrate per g biomass per h). If s represents the energy source (substrate), then the rate of substrate consumption for maintenance is governed by the following equation:

$$-(ds/dt)_M = mX \qquad ...(2.19)$$

where m is the maintenance coefficient. The yield can therefore be related to maintenance by the following equation:

$$Y = \frac{\Delta x}{(\Delta s)_G + (\Delta s)_M} \qquad ...(2.20)$$

where Y is the growth yield, Δx is the amount of biomass generated, $(\Delta s)_G$ is the amount of substrate consumed in biosynthesis and $(\Delta s)_M$ is the amount of substrate consumed for mainteneace of cell viability.

If maintenance energy $(\Delta s)_M$ was determined and found to be zero, then equation reduces to:

$$Y_G = \frac{\Delta x}{(\Delta s)_G} \qquad ...(2.21)$$

Eqn, (2.21) clearly means that growth yield (Y) is a direct function of the amount of substrate utilized. The growth yield (Y, g dry weight per g substrate) obtained in this case may be referred to as the true growth yield to distinguish it from the Y_G where maintenance $(\Delta s)_M$ is involved. However, if a specific fraction of carbon is diverted towards maintenance energy or survival, then one might conclude that the slower the growth rate (μ), the higher the percentage of carbon that is diverted towards maintenance, the less the growth yield.

On the other hand, however, during unrestricted growth, *i.e.* no substrate limitation, the interrelationship between substrate concentration and maintenance can be assessed from the following:

$$ds/dt = (ds/dt)_M + (ds/dt)_G \qquad ...(2.22)$$

In this case eqn. (2.22) implies that the overall rate of substrate utilization-equals the rates of substrate utilization for both maintenance and growth. With growth rate expressed in the usual way, *i.e.* $dx/dt = \mu x$, and assuming that Y equals Y_G, can be rearranged to give:

$$1/Y = m/\mu + 1/Y_G \qquad ...(2.23)$$

According to eqn. (2.23) if m and $\bar{Y}_G$ are constants, then the plot of $1/Y$ against $1/\mu$ should give a straight line, the slope of which is maintenance (m) and the point at which the ordinate is intercepted is equal to $1/Y_G$.

During the course of fermentation, the minimum rate of energy supply may fall below maintenance requirements due to a shortfall in the supply of phosphorylated intermediates, ATP and/or reducing powers (NAD(P)H, $FADH_2$, etc.). Although such a drop in energy supply may not be fatal, it is likely to have an adverse effect on the fitness of the organism, *i.e.* the cells become less capable of taking advantage of favourable changes in the environment and consequently cells resume growth after a lag period. On the other hand, if the cells were able to grow immediately following inoculation or transfer into another medium, *i.e.* without lag, then the cells must have been left in optimal conditions.

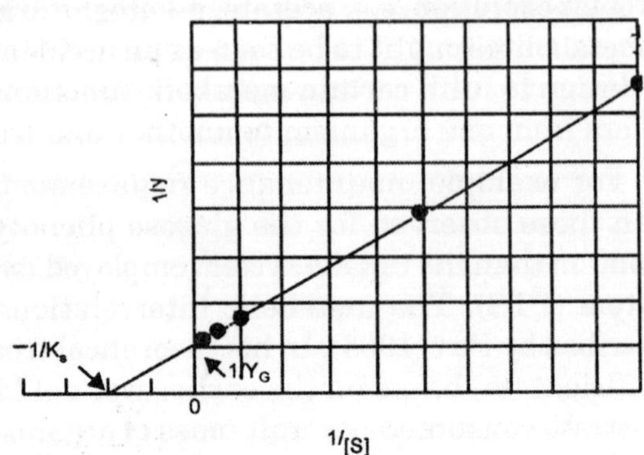

Fig. 2.19. Graphical determination of maintenance (m) and the saturation constant (*Ks*).

However, if the cells were to enter the lag phase prior to growth then it is fair to suggest that the organism was left in a sub-optimal state, *i.e.* the organism must have suffered a drop in energy supply below its maintenance requirements. It follows, therefore, that the physiological state of the cell at the time when energy supply falls below maintenance is very important and that the deeper the shortfall in energy supply, the longer the lag period. A lag period due to the above condition is surely different to that observed when the cells are faced with the challenge of changing phenotype, as is the case when glucose phenotype of *Escherichia coli* is forced to change to lactose phenotype). Individual cells in a given population may respond differently when a given drop of energy supply is exerted. The ability of a given population to respond successfully to 'shift-up' or to survive a 'shift-down' in nutrients appears to be directly related to the intracellular concentration of ribosomes. For example, in comparison with cells grown in rich medium, restricted growth of *Salmonella typhimurium* was accompanied by a 35-fold drop in the concentration of ribosomes. In the yeast *Saccbaromyces cerevisiae*, a drop in growth rate from 0.40 h^{-1} to 0.10 h^{-1} was accompanied by a drop in the cellular concentrations of RNA, DNA and proteins from 12.1%, 0.6% and 60.1% to 6.3%, 0.4% and 45% respectively. Furthermore, it was demonstrated that the cells were capable of 'scaling up' or 'scaling down' the intracellular level of ribosomes in response to the 'shift-up' or the 'shift-down' in nutrients respectively. Recent investigations have revealed that such ability to modulate the cellular content of ribosomes is achieved through a mechanism of autogenous regulation.

To account for such a phenomenon, Nomura *et al.* (1984) proposed a new hypothesis, *translational couplings*, which simply implies that binding to and initiation of translation at the first ribosome binding site of the operon is essential in exposing the ribosome binding sites of all other cistrons and that inhibition or prevention of translation at the first binding site means that all other sites within the polycistronic message will remain unexposed and as such untranslated. The argument over the question of whether smaller cells are less able to cope with environmental changes, *i.e.* nutrient limitations and the drop in energy supply below maintenance, because they contain less ribosomes can now be answered to the satisfaction of everyone as recent research revealed that it is not the number of ribosomes that matter but rather their concentration inside the cell.

Furthermore, the need for the cell size to be relatively large during growth under no limitations is not a reflection of high concentration of ribosomes but rather the need to attain a certain mass before replication of the chromosome can be initiated. The essence of Donachie's discovery is that if initiation of replication is triggered in response to a certain volume of cell mass, then the faster the cell divides, the faster the growth rate, the larger the size of the cell.

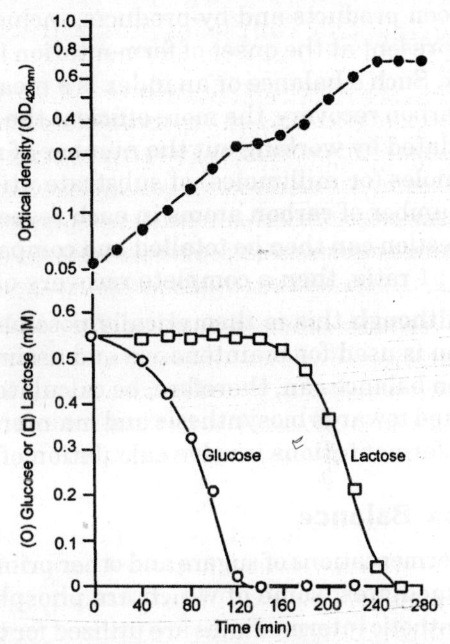

Fig. 2.20. Diauxic pattern of growth and the pattern of sugar utilization during growth of *E. coli* ML30 on equimolar concentrations (0.55 mM) of glucose and lactose.

DIAUXIC GROWTH

The ability of microorganisms to display biphasic (diauxic) growth patterns is well documented. For example, *E. coli* and *Aerobacter aerogenes* display such a pattern when faced with two different substrates, glucose and lactose in the case of the former and glucose and citrate in the case of the latter. This is generally due to the preferential utilization of glucose to either of the competing substrates. This can be explained on the grounds that glucose has been found to exert catabolite inhibition, *i.e.* inhibition of other sugars or organic acids uptake systems, and catabolite repression of the enzymes required for the transport as well as hydrolysis of other sugars, as hydrolysis of other sugars, as is the case in lactose metabolism.

GROWTH YIELD

The relative contribution of each substrate to total biomass formation can be assessed with a good deal of precision providing that growth limitation is only enforced as a result of exhaustion of the substrate in question rather than the accumulation of toxic end-products or adverse changes in pH. The relationship between total biomass formation and the concentration of any given substance can therefore be determined from the plot of log *n* (biomass) as a function of substrate concentration and providing that all other components are in excess, the organism will enter the stationary phase upon depletion of the substrate or the substance in question. This experiment should obviously be done over a wide range of substrate concentrations and from the slope obtained, the yield constant or yield coefficient per unit of the substance in question can be determined. If the unit is, say, one mole of a substrate, then the yield coefficient obtained is referred to as the molar growth yield coefficient. This method is used widely as a biological assay for the determination of vitamins, amino acids, purines and pyridines.

FERMENTATION BALANCES

Carbon Balance

Apart from the fraction of carbon used for biomass formation, the remainder is partitioned between products and by-products, including carbon dioxide. The ratio of recovered carbon to that present at the onset of fermentation is referred to as the carbon balance or carbon recovery index. Such a balance or an index is a measure of efficiency; *i.e.* the higher the index, the higher the carbon recovery, the more efficient the fermentation process. The carbon balance is generally calculated by working out the number of moles (or millimoles) produced of a given product per 100 moles (or millimoles) of substrate utilized. The values obtained can then be multiplied by the number of carbon atoms in each respective molecule. The resulting values for the products in question can then be totalled and compared with that of the substrate. If the values are equal, *i.e.* 1 : 1 ratio, then a complete recovery of carbon into product formation has been achieved.

Although this is theoretically possible, our experience indicates otherwise as part of this carbon is used for maintenance and assimilation to support growth, however slow. A complete carbon balance can, therefore, be calculated for any given fermentation if the fraction of carbon diverted towards biosynthesis and maintenance is determined. In addition to the carbon balance, some fermentations involve calculation of the redox balance.

Redox Balance

Fermentations of sugars and other primary carbon sources give rise to a whole host of different intermediates, some of which are phosphorylated (energy rich) while others are not. While biosynthetic intermediates are utilized for the biosynthesis of monomers, other intermediates are produced in excess and this is balanced by their excretion into the medium. The stoichiometry of product and by-product formations of any given fermentation process can be ascertained by carefully analysing the culture filtrates at different stages. From a physiological standpoint and in order for fermentation to go to completion, the redox balance must be maintained.

EFFICIENCY OF CENTRAL METABOLISM

The efficiency of carbon conversion to biomass and desirable end-products is influenced by many different factors (see Chapters 4, 6 and 7 for more details) of which futile cycling is a major contributor. For example, *Klebsiella aerogenes* has been shown to have an NAD^+-dependent pyruvate reductase, an enzyme which catalyses the formation of D-lactate, and an FAD-dependent D-lactate dehydrogenase which catalyses the conversion of D-lactate to pyruvate. The operation of these two enzymes provides a futile cycle in which NADH,H^+ is oxidized at the expense of reducing FAD, thus bypassing the first phosphorylation site in the electron transport chain. It follows therefore that any changes in the metabolic environment which lead to a transient increase in the intracellular level of pyruvate may lead to the operation of this futile cycle and, in turn, lower the efficiency of central metabolism.

The presence and function of glutamine synthetase and glutammase in ammonia-limited cultures represents yet another possible futile cycle in which the synthetase generates glutamate while the other affects its hydrolysis with a net loss of one molecule of ATP per turn. In addition to futile cycling, flux to metabolite excretion in general and acetate in particular is another significant factor in influencing the efficiency of central metabolism (El-Mansi and Holms, 1989).

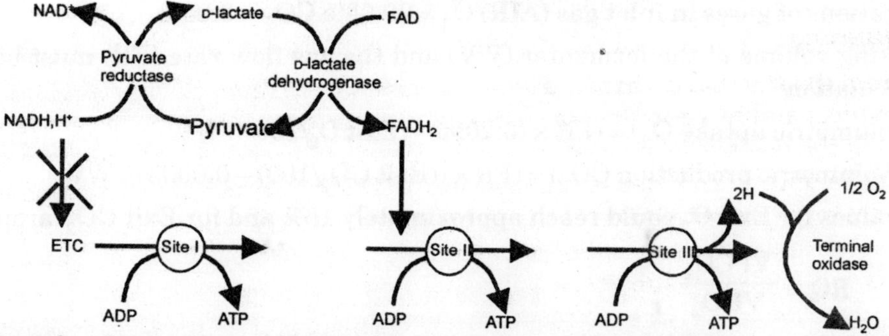

Fig. 2.21. The role of pyruvate reductase and D-lactate dehydrogenase in bypassing the first phosphorylation site of the electron transport chain (ETC) in the oxidation of NADH, H^+.

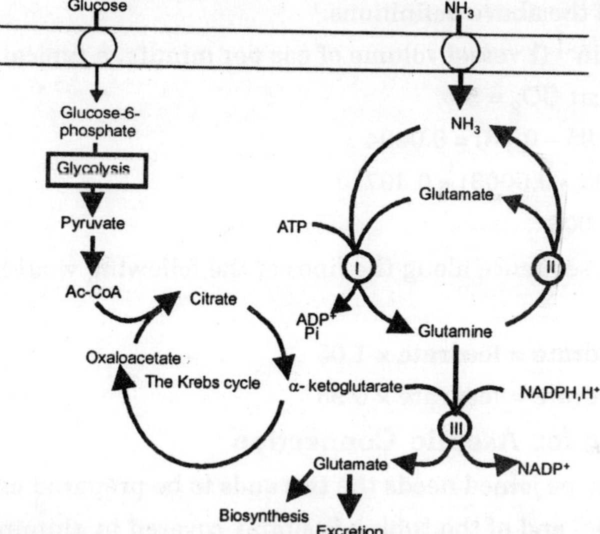

Fig. 2.22. The role of glutamine synthetase and glutaminase in energy dissipation through the formation of a futile cycle.

Furthermore, because acetate, particularly in its undissociated form, has proved to be a potent uncoupler of oxidative phosphorylation, any conditions which promote flux to acetate excretion will, in turn, diminish flux to product formation due to the diversion of a much larger fraction of carbon towards high maintenance energy requirements (El-Mansi, unpublished results).

Calculation of RQ

When data are obtained from an exit gas analysis, it is generally required that the parameter RQ is calculated in order to gain an insight into the metabolism of the organism under investigation. This is normally done by programming a definition into a computer software package for datalogging and control. Here is a generalized outline of the steps needed:

RQ = CPR (Carbon dioxide production rate)/ OUR (Oxygen uptake rate)

CPR = [Flow CO_2 out] × [Conc'n CO_2 out] miny [Flow CO_2 in] × [Conc'n CO_2 in] gmol/O_2/h

OUR = [Flow O_2 in] × [Conc'n O_2 in]-[Flow O_2 out] × [Conc'n O_2 out] gmol/O_2/h

Concentrations of gases in inlet gas (AIR) $O_2 = 20.95\%$ $CO_2 = 0.03\%$

The working volume of the fermentor (WV) and the gas flow rate (FR) must be known to make the calculation:

VUO_2 (Volumetric uptake O_2) = (FR × (0.2095 – (Exit O_2/100))) WV

$VPCO_2$ (Volumetric production CO_2) = (FR × ((Exit CO_2/100) – 0.0003))) WV

Typical values for Exit O_2 could reach approximately 16% and for Exit CO_2 around 5%.

$$RQ = \frac{VPCO_2}{VUO_2}$$

Values of RQ for yeast biomass fermentation, for instance, would need to be kept around 1 for optimal biomass production.

A worked example of the above definitions:

WV = 1 l. FR = 1 l min^{-1} (l vessel volume of gas per minute, a typical flow rate).

Exit O_2 = 16% and Exit CO_2 = 5%.

$VUO_2 = 10/10 \times (0.2095 - 0.16) = 0.0494$

$VPCO_2 = 10/10 \times (0.05 - 0.0003) = 0.497$

$RQ = 0.497/0.494 = 1.006$

To achieve control, a sequence along the lines of the following would have to be set up:

if RQ < 0.95 then feedrate = feedrate × 1.05

if RQ > 1.05 then feedrate = feedrate × 0.95

Preparation of Tubing for Aseptic Connection

Any silicone tubing to be joined needs the two ends to be prepared as follows:

1. The open (female) end of the tubing is simply covered in aluminium foil held in place with autoclave tape. For added security, the open end can be sealed with a short piece of glass rod flattened at one end to make a 'bung'.
2. The other (male) end has a short length of stainless steel pipe pushed into the silicone tubing. The exposed length of pipe is covered with foil which is taped closed.
3. Silicone tubing is connected to any of the inlet pipes intended for use for reagent addition. The tubing must stretch to the peristaltic pump heads when the vessel is in place.
4. The tubing is closed with clamps or any other closure system which will withstand autoclaving (the arterial clamps used in medicine are often employed where a connection must be opened and closed many times, e.g. to a sampling device). The tubing which goes to the sampling device from the top plate must be clamped shut during autoclaving.

Routine Maintenance of pH Electrodes

1. Always store an electrode which is not in use in liquid – preferably 3 M KCl solution, although pH 7 buffer or even fermentation medium will do (in descending order of preference).

2. An electrode which has completely dried out will need regenerating—a procedure which normally involves a brief immersion in a hydrofluoric acid solution (typically supplied by the manufacturer) followed by prolonged soaking in storage solution. This procedure may also regenerate 'aged' electodes for a time. (Note that glass vessels cannot be used for this task due to the effects of the hydrofluoric acid).
3. A build-up of protein material on the electrode membrane can be dealt with by soaking the electrode in a pepsin/hydrochloric acid solution for about an hour.
4. If a discolouration (blackening) of the diaphragm has been caused by sulphides in the medium, soaking for several hours in an HCl/thiourea solution will recover its shiny surface.
5. Never wipe the tip of the electrode with a paper towel as this causes scratching.
6. Glass/liquid electrodes should have the reference electrolyte topped up periodically.
7. To test if the pH meter/cable are functioning correctly, a pH simulator can be used. If such a device is not at hand, a paper clip can be used to short the center pin of the electrode plug and the outside metal ring of the cap. The pH reading should be very close to 7.0.

Replacing a Mechanical Seal

The procedure for replacing a typical single mechanical seal is as follows:

1. Unscrew the top part of the drive shaft and remove. Measure the distance between the bottom of the retaining ring and the inner surface of the top plate. This allows the new seal to be correctly tensioned by setting the same gap at the end of the procedure.
2. Remove any retaining ring by loosening the small screw(s) holding it in place. Disassemble the retaining ring, hard seal, soft seal and bottom O-ring in order. Examine the faces of the static and rotating seals. Scoring of the surface indicates the seal has worn out. Never try to replace the old seal if it looks undamaged - it cannot be reused.
3. Insert a new O-ring into the 'hole' in the vessel plate which leads to the drive shaft bearings. This may be a tight fit but it must seat properly to prevent leaks. Replace the static seal by pushing it into the O-ring to obtain a snug fit. The protective rubber cover over the top face must be removed at this point.
4. Fit the rotating seal onto the drive shaft. Ensure the inner O-ring and metal washer are firmly pushed down into place as the tensioning spring will rest on these. Fit the new tensioning spring into place then push onto this firmly with the retaining ring until the spring is tightly compressed. Measure the gap with a ruler/calipers and only tighten the screw(s) on the retaining ring when the correct distance has been achieved.
5. Replace the drive shaft and gently turn the shaft by hand a small distance to ensure it moves freely. Lubricate with glycerine.

CHAPTER

3

Engineering and Applications of Chimeric Nucleases

Each human cell contains about 3×10^9 base pairs (bp) within its genome. With the first draft sequence of the human genome now available, biologists estimate that there are about 30,000–40,000 different genes within the genome. This is fewer than originally anticipated, but still a huge number. These genes code for all of the human body's proteins. Simple mutations within the coding region of critical genes can lead to the formation of abnormal proteins, resulting in disease phenotypes, premature death, or failure of an embryo to develop. Furthermore, mutations that affect I the regulatory region of genes can result in aberrant gene expression within cells, and give rise to cancer phenotypes. The 'Holy Grail' of the Human Genome Project is 'Gene Therapy', that is, how genes might someday be used, i modified, or even changed to correct human disease. Gene therapy is based on a simple concept : Mutations within the coding or regulatory regions of certain critical genes give rise to disease phenotypes.

Correction of these mutations to normal alleles will result in the reversion of disease phenotypes to normal phenotypes. This is particularly true of monogenie diseases. Thus, gene therapy provides a new paradigm for treating human disease by correcting the causative genetic defect. However, the implementation of this novel concept into clinical practice and therapeutic reality has proven to be extremely difficult; and, the progress has been very slow. This is because targeting a particular defective gene for correction among 30,000 or more human genes within the genome requires exquisite sequence-specificity of the gene therapy vectors, which are the vehicles used to deliver the therapeutic genes into cells. This problem is further compounded by the fact that these genes and their regulatory sequences, which account for about 5% of the human genome, are buried within the context of a 3×10^9-bp DNA sequence.

Current gene therapy vectors lack the required sequence specificity that is necessary for the targeted correction of a defective gene within the genome. As a result, the therapeutic genes are delivered and inserted randomly within the genome that is at many sites other than the location of the genetic defect. Such random insertions of the therapeutic gene at unwanted locations are mutagenic, especially when the insertions occur in critical genes. Over time, these insertions can and do give rise to cancer or other disease phenotypes. Many of the difficulties associated with gene therapy are likely to be overcome if one could insert the corrected version of the mutation at the precise location of the genetic defect within the genome. How then does one achieve targeted correction of a gene defect within cells? Biologists have known for a while that when a defined chromosomal break is introduced at a unique site within a genome, homologous recombination is induced at that site in a large fraction of cells in a population.

The challenge then, so far unfulfilled, is to develop a general means of introducing a double strand break (DSB) uniquely at a given locus in the genome to induce homologous recombination. In this chapter, we discuss the progress towards targeted correction of a genetic defect. This involves two steps : (1) engineering of chimeric nucleases, the molecular tools necessary to make a chromosomal DSB (double-strand break) at a chosen site within the human genome; (2) application of these molecular tools for targeted chromosomal cleavage and correction of a genetic defect by inducing homologous recombination at that locus in cells.

Engineering of Chimeric Nucleases

In order to make a unique chromosomal DSB within a mammalian genome, we need restriction enzymes that recognize DNA sequences of 16 bp or more in length. Such restriction enzyme sites will occur once every 4^{16} (= 4.3×10^9) bp on average, which is about once per human genome. Most bacterial enzymes typically recognize short palindromes that are 4 to 8 bp in length. This means that they will cut DNA on average once every 4^4 to 4^8(=256 to 65,536) bases depending on the restriction enzyme used. For instance, one can expect an 8-bp site to occur about 45,776 times within the human genome. Obviously, bacterial restriction enzymes are not useful for the task at hand, which is to introduce a targeted chromosomal break at a site of our own choosing within the human genome. Over a decade ago, our laboratory set out to engineer chimeric nucleases, the molecular tools necessary for introducing a targeted chromosomal DSB, within a mammalian genome.

The research started with the study of FokI restriction endonuclease, a bacterial Type US restriction enzyme. FokI recognizes the nonpalindromic pentadeoxyribonucleotide, 5′-GGATG-3′:5′-CATCC-3′, in duplex DNA and cleaves 9/13 nucleotides downstream of the recognition site. It does not recognize any specific sequence at the site of cleavage. This property implies the presence of two separate protein domains within FokI: one for sequence-specific recognition of DNA and the other for the endonuclease activity. Once the DNA-binding domain is anchored at the recognition site, a signal is transmitted to the endonuclease domain, probably through allosteric interactions, and the cleavage occurs. We reasoned that one may be able to swap the FokI recognition domain with other naturally occurring DNA-binding proteins that recognize longer DNA sequences or other designed DNA-binding motifs to create chimeric nucleases. In this way, Type US enzymes like FokI are probably ideal candidates for engineering novel sequence specificities.

Functional Domains in FokI Restriction Endonuclease

Several laboratories including ours have independently cloned the FokI restriction-modification system from *Flavobacterium okeanokoites*. These labs have also reported overproduction and purification of FokI endonuclease. As a first step, we probed the domain structure of FokI by limited proteolysis using trypsin. Our studies on proteolytic fragments of FokI endonuclease revealed an N-terminal DNA-binding domain and a C-terminal domain with nonspecific DNA cleavage activity. Waugh and Sauer (1993, 1994) showed that single amino acid substitutions decouple the DNA-binding and strand scission activities of FokI endonuclease.

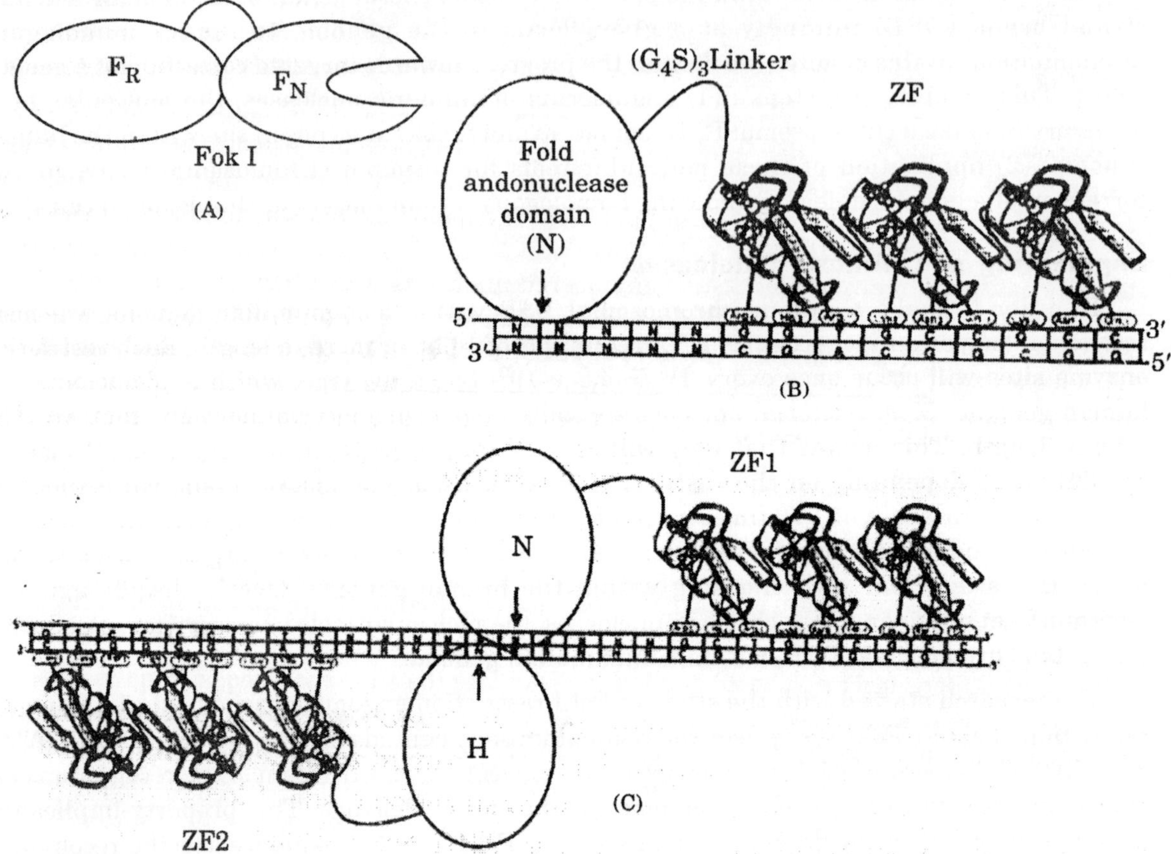

Fig. 3.1. A Schematic representation of FokI endonuclease and ZFN. A Domain structure of FokI; B ZFN bound to its 9-bp cognate site; and C A pair of ZFN with different specificities bound to their target sequence. Since ZFN requires two copies of the 9-bp recognition sites in an inverted orientation in order to dimerize and produce a DSB, it effectively has an 18-bp recognition site. F_R FokI recognition domain; F_N or *N* FokI cleavage domain; *ZF1* and *ZF2* zinc finger proteins.

The mode of DNA-binding by FokI was analyzed by DNA foot-printing methods. These studies showed a lack of protection at the cleavage site of DNA by FokI. Wah et al. (1997, 1998) reported the crystal structures of FokI and FokI-bound to its cognate site, which confirmed the modular nature of FokI endonuclease. The endonuclease domain appears to be sequestered in a "pig-gyback" fashion by the recognition domain. This is consistent with the DNA foot-printing analysis. Thus, the crystal structure of FokI endonuclease is in complete agreement with the model derived from rigorous biochemical studies.

Chimeric Nucleases

The modular nature of FokI endonuclease suggested that it might be feasible to engineer chimeric nucleases by fusing other DNA-binding proteins to the cleavage domain of FokI. The obvious next step was to create novel chimeric nucleases by simply swapping the DNA recognition domain of FokI with other DNA-binding proteins that recognize longer DNA sequences. This indeed proved to be the case. Several novel fusion nucleases were generated in our lab by linking the nonspecific FokI cleavage domain to other DNA-binding proteins.

The latter include the three common eukaryotic DNA-binding motifs namely the helix-turn-helix motif, the zinc finger motif and the basic helix-loop-helix protein containing a leucine zipper motif (bzip). Our lab reported the creation of the first chimeric nuclease by fusing the *Drosophila Ubx* homeodomain to the FokI cleavage domain. This was followed by the fusion of zinc finger motifs and the N-terminal 147 amino acids of the yeast Gal4 protein respectively, to the cleavage domain of FokI. Such engineered nucleases were shown to make specific cuts very close to their expected recognition sequences in vitro. Since then other labs have reported making FokI nuclease domain fusions with various other DNA-binding proteins. FokI fusions have been used to identify high-and low-affinity binding sites of transcription factors in vitro, to study recruitment of various factors to the promoter sites in vivo using the method called protein position identification with a nuclease tail (PIN-POINT) assay and to analyze Z-DNA conformation-specific proteins.

Chimeric nucleases now form a novel class of engineered nucleases in which the non-specific cleavage domain of FokI is fused to other DNA-binding motifs. The most important chimeric nucleases are those based on zinc-finger DNA-binding motifs.

Zinc Finger Binding and Specificity

The modular structure of zinc finger domains (ZF) and modular recognition by zinc finger proteins make them the most versatile of DNA recognition motifs for designing artificial DNA-binding proteins. Each zinc finger consists of about 30 amino acids and folds into a ββα-structure, which is stabilized by the chelation of a zinc ion by the conserved Cys_2His_2 residues. Each finger typically recognizes a 3-bp DNA sequence by inserting the α-helix into the major groove of DNA. Binding of longer DNA sequences is achieved by linking several of these zinc finger motifs in tandem. Each finger, because of variations of certain key amino acids in the α-helix of one-zinc finger to the next, makes its own unique contribution to DNA-binding affinity and specificity.

Because they appear to bind as independent modules, zinc fingers can be linked together in a peptide designed to bind a predetermined DNA site. Although more recent studies suggest that there might be a synergistic interaction between adjacent zinc fingers and that the zinc finger-DNA recognition is more complex than originally perceived, it still appears that zinc finger motifs will provide an excellent framework for designing DNA-binding proteins with a variety of new sequence-specificities. In theory, one can design a zinc finger for each of the 64 possible triplet codons, and using a combination of these fingers, one could design a protein for sequence-specific recognition of any segment of DNA. At present, the rules relating to zinc finger sequences/DNA-binding preferences and redesigning of DNA-binding specificities of zinc finger proteins are well understood only for the 5′-GNN-3′ triplets.

Although numerous structural studies of zinc finger protein complexes with DNA are available, the information is not yet sufficient for rational design of zinc finger domains that bind to any given triplet. The creation of zinc finger chimeric nucleases (ZFN) that recognize

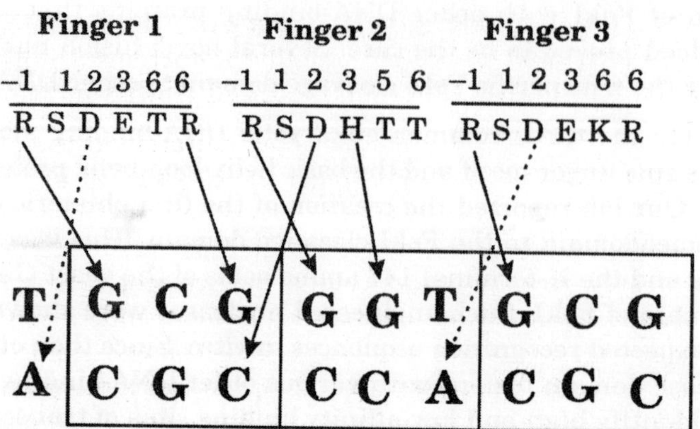

Fig. 3.2. Key base contacts deduced from the crystal structure of Zif 268-DNA complex. Each finger makes contact with its target 3-bp site. In addition, Asp² at position 2 in each finger makes contact with base outside the 3-bp site.

and cleave any target sequence depends on the reliable creation of zinc finger proteins (ZF) that can specifically recognize a target sequence. A phage display selection method could be used to identify zinc finger proteins that bind to a particular DNA site. This is the method of choice for designer zinc finger proteins. Briefly, a library of zinc fingers with randomized amino acids at specific positions (–1 to 6) within the α-helix is fused to a viral coat protein of a phage that is used to infect bacteria.

The virus is then allowed to replicate and reproduce in culture to make more copies. Since the fusion proteins are displayed on the surface of the virions, they can be enriched for highly specific and functional zinc finger proteins that bind a chosen target substrate with high affinity. Three different selection methods based on the phage display-parallel selection, sequential selection, and bipartite selection-have been reported in literature using Zif268-derived phage libraries for selection of designer zinc fingers. These libraries are constrained by the target site overlap problem because of the presence of Asp² at position 2 of the α-helix in each of the three-zinc finger domains of the Zif268. The Asp² of all three fingers make a cross-strand contact to a base outside the canonical triplet site. This precludes the presence of a simple general recognition code that will facilitate the rational design of zinc finger-based DNA-binding domains. This also makes selection of zinc finger proteins by using phage display libraries more difficult.

Three strategies have since been developed to overcome this problem : (1) zinc fingers 1 to 3 can be sequentially selected in the context of neighboring zinc fingers; (2) targets can be selected libraries with simultaneous randomization of amino acid residues in two adjacent fingers; and (3) bipartite selection which com-bines the sequential selection with the parallel selection technique. However, these approaches do not provide for the selection of zinc finger domains that function as independent, modular recognition motifs. They select for targets that are dependent on their inter-domain interaction for their sequence specificity. An alternate approach based on a bacterial two-hybrid system has been described for the isolation of a candidate zinc finger protein in an in vivo context from a randomized Zif268-derived mutant library of > 10^8

members in size. This method is analogous to the yeast two-hybrid system previously described by Hochschild and coworkers to study protein-protein interactions. In the bacterial two-hybrid system, zinc finger-DMA interactions are required for cell growth and survival.

However, this approach also suffers from the same limitation as the phage display methods in that it does not allow for selection of zinc finger motifs that act as inde-pendent, modular recognition domains since it also utilizes a Zif268-derived mutant library. In collaboration with Dr. Marc Ostermeier's lab, we are developing a double-reporter, one-hybrid system for rapidly selecting zinc finger proteins and improving their sequence specificities. This system will also allow for identification of zinc finger motifs that act as independent modular units. This will be done by using a mutant zinc finger library that is based on consensus back-bone framework for each and every finger within the protein; and by limiting the amino acid at position +2 of the α-helix of each finger to a glycine residue thus eliminating the cross-strand base contact that occurs outside the 3-bp site in the Zif268 derived libraries due to the presence of Asp^2. The one hybrid, double reporter system that is being developed in our lab is based on the one hybrid system originally described by Hochschild and coworkers that is used to detect protein-DNA interactions.

In their system, the gene for a DNA binding domain is fused to a subunit of *E. coli* RNA polymerase. The fusion is then used to activate transcription from a lac-derived promoter provided the binding site for the DNA-binding domain is present and centered at the –63 position. The reporter gene under the control of the lac-derived promoter is β-galactosidase. Thus, this method provides a way to screen and assay for interaction between the DNA-binding domain and various DNA-binding sites positioned at –63. Our system is designed to create one hybrid, double reporter system for evaluating and evolving DNA-binding specificity of zinc finger proteins.

The gene coding for the zinc finger is fused to a subunit of *E. coli* RNA polymerase. The fusion protein is then used to activate transcription of a reporter gene under the control of a lac-derived promoter provided the zinc finger binding site is placed at an appropriate distance upstream of the promoter. In our system, we incorporate two separate operons each containing one reporter gene under the control of a lac-derived promoter. The only difference between the two operons is the nature of the reporter gene and the target zinc finger binding sites, which are placed upstream of the promoter. In this way, binding of a zinc finger protein to two different sites can be evaluated simultaneously. We plan to employ two different reporter systems as well.

In the first, the antibiotic resistance genes coding for chloramphenicol and tetracycline, respectively, will be placed under the control of lac-derived promoters on separate operons. In the second, we have chosen two reporter genes namely the green fluorescent protein (GFP) and the red fluorescence protein (dsRED) to be placed under the control of lac-derived promoters on separate operons. The fluorescent system offers selection of zinc finger mutants with improved sequence specificity using quantitative flow cytome-try and fluorescence activated cell sorting. Recent advances suggest that a combination of design and selection is best suited to identify custom zinc finger DNA-binding proteins for known target sites. So far, only zinc finger modules that specifically recognize 5′-GNN-3′ and 5′-ANN-3′ triplets have been identified through design and selection strategies.

Studies are underway to determine the zinc finger recognition preferences for the other triplets, 5′-CNN-3′ and 5′-TNN-3′, respectively. In many instances, the current knowledge of zinc finger preferences for 5′-GNN-3′ and 5′-ANN-3′ is more than sufficient for designing and/or selecting a zinc finger protein to target a specific site within a gene of the human genome.

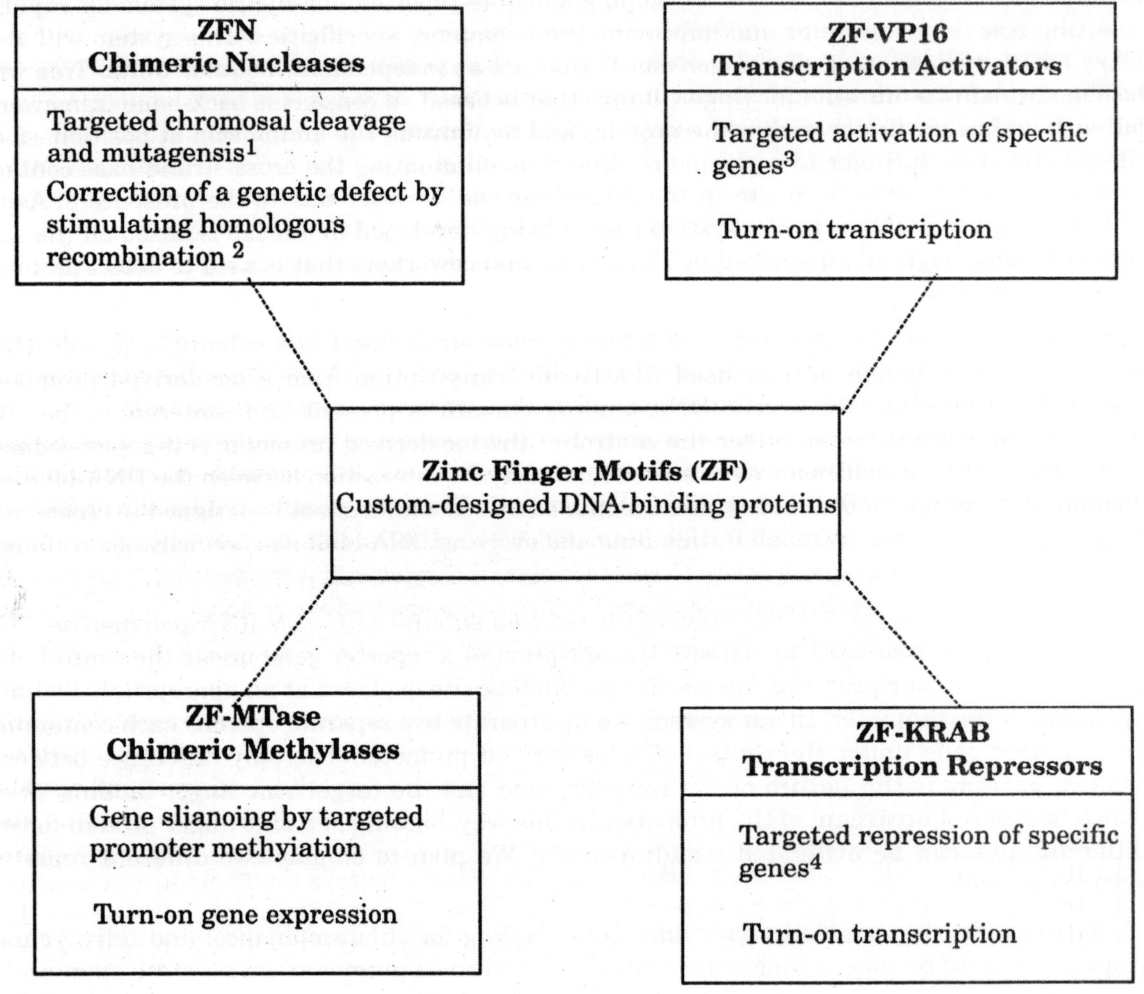

Fig. 3.3. Zinc finger protein platform technology. Additional functional domains like the FokI cleavage domain (N), activator domain (VP16), represser domain (KRAB) and methylases (Mtase) can be fused to the designed zinc finger motifs (ZF) to form, chimeric proteins that act as chimeric nucleases (ZFN), transcriptional activators (ZF-VP16), transcriptional repressers (ZF-KRAB) and targeted methylases (ZF-Mtase), respectively.

The ability to design or select zinc fingers with desired specificity implies that DNA-binding proteins containing zinc fingers will be made to order. Therefore, we reasoned that one could design "artificial" nucleases (ZFN) that will cut DNA at any preferred site by making fusions of zinc finger proteins (ZF) to the cleavage domain (N) of FokI endonuclease. We have been successful in engineering several novel ZFN by fusing three-zinc finger proteins to the cleavage domain of FokI. In these fusions, the three Cys_2His_2 zinc finger DNA-binding domain is at the N-terminus and the FokI nuclease domain is at the C-terminus; they are connected by a flexible 15 amino acid $(Gly_4Ser)_3$ linker to achieve efficient DSB at the predicted 9-base pair recognition site.

We have shown that the fusion of FokI cleavage domain to the zinc finger motif does not alter the sequence specificity of the zinc finger protein and it does not change its binding affinity significantly. It must be made clear that sequence specificity of the engineered ZFN is only as good as the zinc finger motif that was used to make it. If a zinc finger shows affinity for degenerate sites, then, the engineered ZFN will cut at those degenerate binding sites albeit with lower affinity. Although, our lab was the first to fuse another functional moiety (FokI cleavage domain) to zinc finger proteins, our long-term focus and efforts have remained on the development of chimeric nucleases.

Other functional domains, like activator domains and represser domains have been fused to the designed zinc finger motifs to form hybrid proteins that act as transcription activators and transcription repressers within cells. More recently, engineered transcription factors have been shown to stimulate blood vessel growth and wound healing by the induction of the gene coding for VEGF-A (vascular endothelial growth factor A) in mice. Cytosine methylation has been targeted preferentially to pre-determined sequences by attaching CpG-specific DNA methyltransferase to Zif 268 three-zinc finger proteins. It appears that important applications in medicine and biological research arising from zinc finger platform technology are almost certain to follow.

Mechanism of Cleavage by Zinc Finger Chimeric Nucleases (ZFN)

A three-zinc finger protein recognizes nine base pairs. An issue relevant to gene targeting is the ability of the cleavage reagent to distinguish the chosen target from all other sequences in a complex genome. In theory, any sequence shorter than 16-bp would be present in multiple copies in the human genome. In particular, any given 9-bp sequence is predicted to occur 10^4 times within the human genome. Therefore, the three-zinc finger nucleases are expected to make about 10,000 DSB within the human genome. How then does one achieve a unique, targeted chromosomal DSB within a genome? Our study on the mechanism of cleavage by the three-zinc finger nucleases indicates that the cleavage domain must dimerize in order to cut DNA.

Therefore, efficient cleavage requires two zinc finger-binding sites situated in close proximity in an inverted orientation. Such a configuration of binding sites orients the nuclease domains towards each other promoting dimerization. As expected, the spacer length between the inverted repeats is critical for the dimerization of the nuclease domains to occur. This in turn is determined by the $(Gly_4Ser)_3$ linker, which connects the zinc fingers to the FokI cleavage domain. Thus, ZFN prefer inverted repeats as their substrate to make a DSB. However, at high enzyme concentrations, single binding sites are cut; one molecule binds at its cognate site, while the second probably binds non-specifically nearby on the DNA facilitating the dimerization of the FokI catalytic domain.

The first clue for FokI dimerization came from the crystal structure of the FokI endonuclease, which revealed the interacting interface between the catalytic domains. In addition, kinetic studies on the mechanism of cleavage by FokI further supported that dimerization of FokI enzyme was required to produce a DSB. Since ZFN requires two copies of the 9-bp recognition sites in a tail-to-tail orientation in order to dimerize and produce a DSB, it effectively has an 18-bp recognition site. In principle, the binding sites need not be identical provided the ZFN that bind both sites are present. We have shown that two ZFN with different sequence specificities collaborate to produce a DSB when their binding sites are appropriately placed and oriented with respect to each other. Since the recognition specificity of zinc fingers can be manipulated experimentally, ZFN could be engineered so as to target a unique site within a genome.

Application of Chimeric Nucleases

The modular structure of FokI endonuclease and zinc finger proteins has made it possible to create artificial nucleases that will cut DNA at a pre-determined site. This approach opens the way to generate many novel ZFN with tailor-made sequence specificities desirable for various applications in biotechnology and medicine.

Recombinogenic Repair

How might these ZFN be used in the correction of a genetic defect? One approach would be to recruit the pre-existing cellular machinery to achieve the goal. Cells of many different organisms use recombination to repair DNA damage, especially DSB. If left un-repaired such DSB in a cellular chromosome would be lethal. Carroll and coworkers (1996) have shown that recombination in frog oocytes proceeds by exonuclease resection and then annealing of complementary strands. This is called single-strand annealing and this same mechanism is thought to operate in mammalian cells. One could utilize this process for gene-targeting experiments.

The essence of this approach is to target the chromosomal DNA for sequence-specific cleavage at or near the site where one wants the recombination to occur. Exonuclease would then resect the DNA at that chromosomal site, generating single-strand tails (Lin et al. 1984). Recombination would proceed by a single-strand annealing mechanism with the exogenous DNA that is present in the cell. The homologous foreign DNA would thus be incorporated at this chromosomal site. If the chromosomal target site is not cleaved, then it would not be accessible for exonucleolytic resection. In this case, the exogenous DNA will be degraded while the target remains unchanged. Thus, making a targeted DSB would greatly stimulate homologous recombination between the exogenous DNA and a chromosomal sequence.

Such experiments have been performed using Group I intron-encoded homing endonucleases in yeast, cultured mammalian and plant cells. Although the target sites for these enzymes range from 15-40-bp in length, they exhibit a broad range of sequence specificities and cutting frequencies because of their binding to degenerate sites. In addition, their sequence specificities are not amenable to easy manipulation. As a result they have only a limited use in gene-targeting experiments. Recently, we have shown in collaboration with Dana Carroll's lab that the three-zinc finger chimeric nuclease, Zif-QQR-F_N finds and cleaves its target in vivo. This was tested by microinjection of DNA substrates and the enzyme into frog eggs. Double-strand cleavage required an inverted repeat of the 9-bp target.

When the appropriate sites were placed in the recombination substrate, this DNA was cleaved in frog oocytes by the injected enzyme and homologous recombination ensued. These microinjection experiments have shown that greater than 90% of the substrate was cleaved in vivo and almost all cleaved substrates underwent homologous recombination. Although, in this proof-of-principle experiment, an extrachromosomal target was used to test the idea, two important conclusions can be drawn: (1) chimeric nucleases find and cleave their targets within cells; and (2) the targeted DSB stimulates homologous recombination at the site of cleavage, provided an undamaged template of the homolog is available for recombinogenic repair within cells. This is what normally occurs in cells : when one copy of a gene on one chromosome is damaged, it is repaired using the other undamaged allele on the paired chromosome as a template.

Chromosomal Cleavage and Mutagenesis

The obvious next step was to show specific cleavage of a chromosomal target within cells by chimeric nucleases. Recently, Dana Carroll's lab reported specific chromosomal cleavage and mutagenesis in *Drosopila melanogaster* (Bibikova et al. 2002). They designed a pair of ZFN that targeted the yellow (Y) gene on the X-chromosome of *Drosophila.* The nucleases were cloned under the control of a heat shock promoter and then introduced into *Drosophila* genome via P element-mediated transformation. Only when both ZFN were expressed in the developing larvae did somatic mutations occur specifically in the Y gene (yellow gene of *Drosophila* X-chromosome).

The yellow mosaics were observed in 46% of males expressing both ZFN. No yellow mosaics were observed in control larvae that expressed a single ZFN or without heat shock. Germline Y mutations were recovered from 5.7% of males but none from the females tested. This was as expected in female fruit flies because the targeted DSB would be repaired by recombination with the uncut Y homolog. In males, only simple ligation via nonhomologous end joining (NHEJ) is available to repair the chromosomal damage. DNA sequence analysis of the mutants revealed that all mutations were small deletions and/or insertions localized to the targeted site, which was typical of repair by NHEJ.

Many of these mutations led to frame shifts or deletion of essential codons of the Y gene. As a result, the male larvae emerging from the heat shock experiment showed yellow patches in the otherwise dark posterior abdomen. In subsequent experiments, Carroll's lab has recovered *y* mutations in female fruit flies, but always at a much lower frequency than that in male flies. This suggests that if the cleavage efficiency of ZFN was high enough to cut both homologues of the Y gene simultaneously, one can produce female gennline mutations. Since accurate repair from the homolog appears not to be fully efficient, it might be possible to induce germline mutations in autosomal recessive genes using ZFN (Bibikova et al. 2002). Furthermore, since DSB stimulates mutagenic repair that essentially operates in all organisms, targeted cleavage by ZFN could facilitate generation of directed mutations in a variety of cells and organisms.

In a corollary experiment, Bibikova et al. (2003) have shown in *Drosophila,* that gene targeting is enhanced using designed zinc finger nucleases. The ZFN-induced gene-targeting frequencies are about ten fold greater than those observed without target cleavage. Several important conclusions can thus be drawn from this study : (1) chimeric nucleases can be used to target and cleave a specific chromosomal site within cells; (2) organisms that express ZFN, like the *Drosophila* embryos in this case, are viable and they go on to mature into adults; (3) chimeric nucleases can be used to generate directed mutations in organisms and cells, including targeted germline mutations in autosomal recessive genes—provided the ZFN sequence-specificity and cleavage is high enough to target both copies of the alleles within the genome; and (4) gene targeting—the process of gene replacement by homologous recombination—is enhanced through targeted DSB using designed ZFN. Similar experiments are underway in our lab using mammalian cells.

We plan to correct a tyrosinase gene mutation in albino mouse melanocytes using ZFN. The long-term goal is to correct a tyrosinase gene mutation in albino mouse embryonic stem cells. Melanocytes are melanin pigment-producing cells of the skin. Tyrosinase is a key enzyme for melanin synthesis and pigmentation and it catalyzes the conversion of L-tyrosine to melanin. Melanocytes from albino mice contain a homozygous point mutation (TGT to TCT) in the tyrosinase gene. As a result, the amino acid cysteine is changed to a serine in the tyrosinase enzyme. Correction of this point mutation should restore the tyrosinase enzyme activity; and hence confer pigmentation to the albino melanocytes, resulting in dark cells. We have engineered ZFN to target a specific site within the tyrosinase gene for cleavage.

In vitro studies indicate that as expected ZFN bind and cleave the target site contained in a plasmid. These constructs are currently being tested in cell culture studies. Continued expression of ZFN is toxic to the melanocytes, which results in cell death. This warrants controlled expression of ZFN in these cells as in the case of experiments done using *Drosophila* larvae. After completion of this study, we plan to perform similar experiments using mouse embryonic stem cells with the eventual goal of generating dark progeny from albino mouse parents.

Gene Targeting in Human Cells Using ZFN

Gene targeting is a powerful technique to incorporate genetic changes in mammalian genomes. However, this process is very inefficient even in murine embryonic stem cells where it has been successfully applied to produce transgenie mouse "knockouts". Only one cell among 10^6 cells treated with donor DNA has the desired mutant phenotype. Powerful selection techniques are then needed to enrich for the cells carrying the directed mutation. Recently, Porteus and Baltimore (2003) have described a system based on the correction of a mutated GFP gene to study gene targeting in somatic cells. They have shown (hat gene targeting using ZFN is stimulated over 2000-fold by targeted chromosomal cleavage. This corresponds to a targeted correction rate of 3-5% of the cells. This is very encouraging indeed since these targeting rates are in the range of being therapeutically useful.

Future Experiments

The next logical step is to consider gene targeting using chimeric nucleases as a form of gene therapy for the correction of genetic defects that give rise to human diseases, particularly, in the treatment of monogenic diseases, which arise from deleterious mutations in a single gene. It would entail the following steps:

1. Identify a target site within the gene of interest. Inverted sequences of the form $(NNC)_3$...$(GNN)_3$ separated anywhere between 6 to 16 bases make for excellent targets. The target sequence could be within several hundred base pairs from the mutation site for gene conversion.
2. Design and/or select zinc finger proteins (ZF) that recognize the target.
3. Convert the engineered ZF to zinc finger chimeric nucleases (ZFN).
4. Deliver the ZFN and normal gene directly into the nucleus either by microinjection or through viral vectors; ZFN direct targeted chromosomal DSB and stimulate homologous recombination (through recombinogenic repair) with the exogenously provided normal gene. To achieve targeted chromosomal cleavage and mutagenesis by NHEJ (through mutagenic repair) only ZFN need to be delivered into the cells.
5. Monitor for homologous recombination at the target site.

Direct and rigorous experimentation is required to unequivocally establish the induction of homologous recombination at the targeted chromosomal site using human cell lines. Furthermore, it is important to show that unwanted recombination or insertions of the therapeutic gene do not occur at nonhomologous sites at other parts of the genome due to residual cleavage by ZFN. The genomic integrity and stability of treated cells need to be intact at all other loci except at the targeted chromosomal DSB after the ZFN treatment. These studies should be followed by an application to correct a genetic defect in a transgenic animal model.

The focus of our lab is now on two such research projects. In the first, the goal is to achieve targeted chromosomal cleavage and mutagenesis of the CCR5 (P chemokine receptor 5) gene in human cell lines. In the second, our aim is to attempt targeted correction of a gene mutation in the CFTR (cystic fibrosis transmembrane conductance regulator) gene associated with cystic fibrosis in humans. Both of these projects are described briefly below.

Targeted Chromosomal Cleavage and Mutagenesis of the CCR5 Gene in a Human Cell Line

A 32-nucleotide deletion (Δ32) within the β-chemokine receptor 5 (CCR5) gene has been identified in individuals who remain uninfected despite extensive exposure to HIV-1. CCR5 is the major co-receptor on $CD4^+$ cells through which HIV gains entry into the cells. The virus is unable to infect $CD4^+$ lymphocytes and macrophages that are homozygous for the A32 mutant CCR5 allele. The goal here is to induce targeted mutagenesis of the CCR5 gene in primate or human cell lines by using engineered ZFN. We have identified two zinc finger target sites near the A32 locus of the CCR5 gene. Wr have engineered ZFN to target and cleave one of these sites.

In vitro studies indicate that the engineered ZFN bind and cleave the target site encoded in a plasmid as expected. Experiments are underway to test this con-struct in cell culture studies. Our long-term goal is to induce targeted A32 deletion at the chromosomal locus encoding the CCR5 gene in hematopoietic stem cells (CD34+ cells) of individuals who are at high risk for HIV infection. The HIV-1 resistant autologous cells will then be amplified and expanded in cell culture. These cells will then be used to reperfuse the bone marrow of these individuals, thereby making their CD4+ lymphocytes and macrophages resistant to HIV-1 infection.

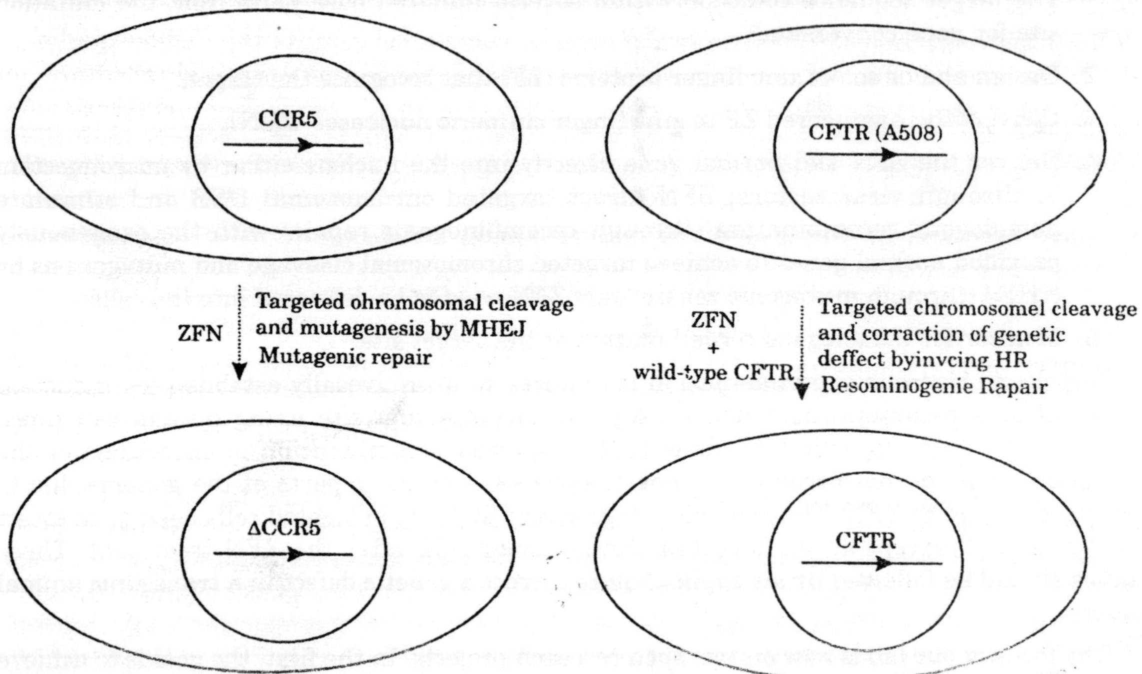

Fig. 3.4. Gene target ing in a human cell line using ZFN. A Targeted chromosomal cleavage and mutagenesis by nonhomologous end joining (NHEJ) through mutagenic repair using engineered chimeric nucleases. In this experiment, only ZFN are introduced into cells. B Targeted chromosomal cleavage and correction of a genetic defect by inducing homologous recombination *(HR)* through recombinogenic repair. In this experiment, both ZFN and the normal gene are introduced into cells.

Targeted Correction of a CFTR Genetic Defect in a Human Cell Line Using ZFN

Cystic fibrosis (CF) is a lethal autosomal recessive hereditary disorder characterized by chronic obstruction and infection of the airway, pancreatic insufficiency, and abnormally high electrolyte concentration in sweat. Although CF is a disease that affects multiple organs, respiratory disease is the major cause of morbidity and mortality: 95% of CF patients die of respiratory failure. Normal efflux of chloride ions occurs across respiratory epithelial cell membranes in response to elevated cAMP (cyclic adenosine monophosphate). This is lacking in cells derived from CF individuals, which in turn limits water secretion from epithelial cells.

As a result, dehydration and excessively thick airway mucus ultimately results in mucous obstruction, infection and inflammation, followed by damage to the airway epithelium. CF is caused by mutations of the cystic fibrosis transmembrane conductance regulator gene (CFTR gene). The most common mutation is a 3-bp deletion leading to the loss of phenylalanine at position 508 in the CFTR protein, known as AF508 mutation. Our goal is to attempt correction of AF508 mutation using ZFN in a human cell line. We have identified target sites within the CFTR gene near the AF508 mutation. We have designed ZFN to bind and cleave these targets. These are being tested in vitro for DNA binding and sequence-specific cleavage of the target.

Targeted correction of AF508 mutation involves somatic cells unlike the CCR5 experiment. The latter is to be done using hematopoietic stem cells that can be expanded in culture and reperfused into patients. The delivery of the CFTR gene to the target site and correction of the gene defect needs to occur in a sufficient number of cells in order to be helpful to the patients. One might have to use lentiviral vectors for efficient delivery of the therapeutic gene.

Potential Limitations of the Chimeric Nuclease Technology

Several things need to be considered when applying this technology in gene therapy.

These include the following : (1) ZF modules that recognize specifically each and ever) one of all possible triplets have not yet been identified. Only ZF modules for 5′-GNN-3′ and 5′-ANN-3′ triplets are known. In principle, this is likely to limit the number of human genes that could be targeted by ZFN approach. However, because the target sequence could be within several hundred bp from the mutation site for gene conversion, this may not pose a problem. (2) Targeting a single site within a gene of the human genome may not be very efficient. The identified target, if present in multiple copies within the gene of interest, will be an ideal locus for efficient cleavage. (3) Controlled and regulated expression of ZFN is necessary and highly preferable since continued expression of ZFN within cells is toxic and cell death by apoptosis ensues.

Ideally, one would like to express ZFN for a short while within cells, where hopefully they will find their target and induce a DSB, and then disappear via protein degradation pathways. Therefore, methods are needed for controlled and regulated expression of ZFN within mammalian cells. (4) In most cases, both alleles of autosomal recessive genes need to be targeted and mutated to observe a phenotypic effect. This is true of CCR5 gene targeting experiments as well. Heterozygous cells for CCR5A32 mutant allele confer only partial protection for about 1-2 years against HIV-1 infection before all cells eventually get infected.

Homozygous cells for CCR5A32 mutant allele are resistant to HIV 1 infection. While it may be feasible to target both copies of the CCR5 gene within cells using ZFN, it may not be a very robust process and may prove to be laborious and cumbersome because multiple rounds of selection may be needed. As an alternate route, one could envision targeted insertion of an inhibitor sequence like for a small interfering RNA (siRNA) within one allele of CCR5, and thereby, control the gene expression from the other homolog.

Incidentally, one can use a similar strategy to simultaneously control HIV-1 infection by both M-tropic and T-tropic viruses, which use CCR5 and CXCR4 (α-chemokine receptor 4) co-receptors, respectively. Furthermore, it appears that both copies of CCR5 gene can be mutated without any adverse side effects to individuals. The simplest explanation for this is that there are alternate mechanisms available within cells to compensate for the loss of CCR5 function. This makes CCR5 an ideal locus within the human genome for targeted insertion of other genes for controlled and regulated expression within cells.

Future Outlook

Gene targeting using chimeric nucleases is an emerging new technology. Recent advances now make it possible to design and/or select zinc finger proteins capable of recognizing virtually any 18-bp target sequence. This is long enough to specify a unique address within mammalian genomes. In principle, chimeric nucleases (ZFN) that combine the nonspecific cleavage domain of FokI endonuclease with zinc finger proteins (ZF) offer a general way to deliver site-specific DSb to the genome, and stimulate homologous recombination at that site. In the use of chimeric nucleases, binding of two ZFN each recognizing a 9-bp inverted site is necessary, since dimerization of the FokI cleavage domain is requited to produce a DSB. Therefore, ZFN effectively have an 18-bp recognition Site. Recent experiments from several labs indicate that ZFN find and cleave their chromosomal target; and as expected, they induce local homologous recombination at the site of cleavage.

These studies provide a glimpse of potential future therapeutic applications of ZFN in modifying and rewiring the human genome itself. This is a great time to be a biologist. There is an excited anticipation that this novel emerging technology will make targeted correction of a genetic defect feasible, especially in treating monogenic diseases. Several areas of reseaiv a appear to be converging and coalescing to make gene therapy a reality. These include the availability of the first draft sequence of the human genome. The *complete nudeotide sequence of the genome* is soon to follow this technical feat. Also, great strides and progress are being made in *stem cell* research. The first applications of chimeric nuclease technology using ZFN to treat human diseases are likely to occur in *ex-vivo gene therapy using stem celh.* Here, the desired cells can be identified through selection, expanded in cuiiurc and replenished into patients. Thus, the availability of chimeric nucleases, a new type of engineered molecular scissors that target a unique site within the human genome, will contribute and greatly aid the feasibility of genome engineering that is targeted rewiring of the genome.

Ethical issues aside, it is not unreasonable to expect that in a decade or two, all the technical problems associated with gene delivery will be overcome; and that gene therapy will be routinely used in a clinical setting, finally signifying a paradigm shift in the treatment of human diseases. Only then, the full impact of the Human Genome Project will be felt, its full potential and goals realized, and its promises fulfilled.

CHAPTER

4 Biological Catalysts

Enzymes occupy an important place in analytical biochemistry and many investigations require their are the biological catalysts and detection and quantitation. Studies of the enzyme content of blood plasma are particularly usefully in clinical biochemistry both in the monitoring of normal metabolic processes and in the detection of abnormal levels of enzyme production or release. Enzyme assays also provide convenient methods for assessing the quality of foodstuffs and checking the efficiency of sterilization and pasteurization processes. Enzymes are valuable analytical tools and offer sensitive and specific methods of quantitation for many substances.

The increasing availability of highly purified enzyme preparations, both in solution and in immobilized forms, permits the development of a wide range of methods. Because of the difficulties in measuring the amount of enzyme in the conventional units of mass or molar concentration, the accepted unit of enzyme activity is defined in terms of reaction rate. The *International Unit (IU)* is defined as that amount of enzyme which will result in the conversion of 1 μ mol of substrate to product in 1 minute under specified conditions. The SI unit of activity, which is becoming more acceptable, is the *katal* and is defined as that amount of enzyme which will result in the conversion of 1 mol of substrate to product in 1 second. A convenient sub-unit is the nanokatal, which is equal to 0.06 International Units.The *specific activity* of an enzyme preparation is expressed as the catalytic activity per milligram of protein (units mg^{-1} protein) and is a convenient way of comparing the purity of enzyme preparations.

The turnover number expresses the catalytic activity in terms of units per mole of pure enzyme rather than milligrams of protein. This allows direct comparison of the catalytic activity of different enzymes, *i.e.* to say that one enzyme is X times more active than another. It can only be used if the relative molecular mass of the enzyme and protein content of the sample are known and the sample is known to be a pure preparation. Enzymes are classified under six main headings, which relate to the chemical reactions involved. Each class is then subdivided in order to classify the other features of the reaction. The main headings are listed below but further details may be found in the Appendix.

1. **Oxidoreductases.** Enzymes that catalyse the transfer of hydrogen or oxygen atoms or electrons.

2. **Transferases.** Enzymes that catalyse the transfer of specific groupings.

3. **Hydrolases.** Enzymes that catalyse hydrolytic reactions.

4. **Lyases.** Enzymes that catalyse the cleaving of bonds by reactions other than hydrolysis.

5. **Isomerases.** Enzymes that catalyse intramolecular rearrangements.

6. **Ligases.** Enzymes that catalyse the formation of bonds and require ATP.

Each enzyme has been given a four-digit number by the Enzyme Commission of the International Union of Biochemistry. The first three digits relate to the reaction catalysed by the enzyme and the final one is required if several enzymes with different protein structures catalyse the same reaction.

Name of Enzyme (EC W.X.Y.Z)

EC — Enzyme Commission number system
W — indicates the reaction catalysed (1-6)
X — indicates the general substrate or group involved
Y — indicates the specific substrate or coenzyme
Z — the serial number of the enzyme.

THE NATURE OF ENZYMES

All enzymes are proteins although many are conjugated proteins and are associated with non-protein groups. Their catalytic activity depends on the maintenance of their native structure and slight variations may result in significant changes in this activity. A common feature of enzymes is the presence of a cleft or depression in the structure which is lined with mainly hydrophobic amino acid residues and into which the substrate fits.

Certain amino acid residues which are concerned with either the orientation of the substrate, and hence the specificity of the enzyme, or are involved in the catalysis of the reaction, are located in this cleft. Those amino acid residues that are associated with the latter role form the active site of the enzyme and are often located towards the base of this cleft. In most cases they are ionic or reactive and include histidine, lysine, cysteine and serine as well as glutamic. The nature of enzymes and aspartic acids. In addition, the binding of ions from the solution, particularly cations, may also aid either the location of the substrate or the catalysis of the reaction.

Factors Affecting Enzyme Activity

Temperature

An increase in temperature increases the rate of all chemical reactions, including those catalysed by enzymes, but it also increases the rate of denaturation of enzyme protein. Denaturation also occurs more readily with pure solutions of enzymes than with impure. It is sometimes suggested that enzymes show an optimum operating temperature but the most suitable temperature for a particular reaction is a compromise between maximal activity for a short period of time and a falling activity due to denaturation for a longer period of time.

Many assays are performed at 37 °C either because it is mistakenly assumed that the human body temperature is the optimum temperature for an enzyme or, more realistically, because above this temperature the rate of inactivation of the enzyme becomes far more significant.

The International Union of Biochemistry originally recommended that 25 °C should be regarded as a standard temperature but subsequently raised it to 30 °C because of difficulties in keeping to the lower temperature in hot climates. It is possible to convert enzyme activities quoted for one temperature to equivalent activities at another temperature by using predetermined conversion factors, but the validity of such methods is questionable. Currently no standard temperature is specified but it is recommended that the assay temperature should be quoted in all references to enzyme activities.

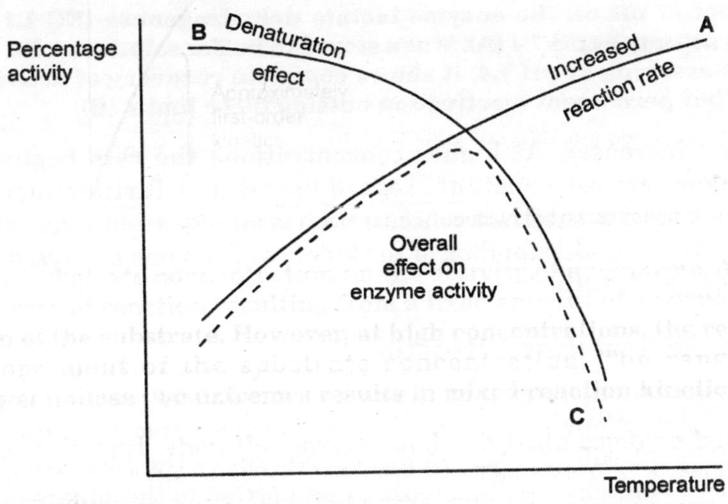

Fig. 4.1. The effect of temperature on enzyme-catalysed reactions. The velocity of a chemical reaction increases with increasing temperature (A) but because of the increasing denaturation of the protein, the proportion of active enzyme falls (B). These two processes result in the characteristic temperature profile of an enzyme (C).

pH

Enzymes are very sensitive to changes in pH and function best over a very limited range with a definite pH optimum. The effects of pH are due to changes in the ionic state both of the amino acid residues of the enzyme and of the substrate molecules. These alterations in charge will affect substrate binding and the resulting rate of reaction. Over a narrow pH range, these effects will be reversible but extremes of acidity or alkalinity often cause serious distortion of protein structure and result in permanent denaturation. It is important to appreciate that the pH optimum of the enzyme reaction may differ for different substrates. Thus a quoted value may not necessarily hold true for every assay method and this should always be determined when designing enzyme assays.

Substrate Concentration

Experimental studies on the effect of substrate concentration on the activity of an enzyme show consistent results. At low concentrations of substrate the rate of reaction increases as

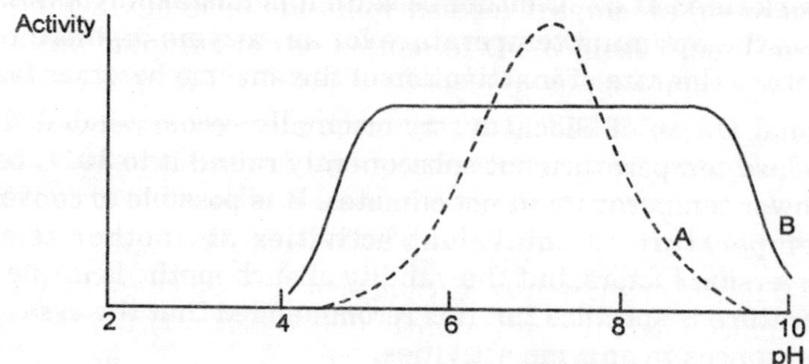

Fig. 4.2. The effect of pH on the enzyme lactate dehydrogenase (EC 1.1.1.27). The enzyme shows maximum activity at pH 7.4 (A). When stored in buffer solutions with differing pH values for 1 h before re-assaying at pH 7.4, it shows complete recovery of activity from pH values between 5 and 9 but permanent inactivation outside these limits (B).

the concentration increases. At higher concentrations the rate begins to level out and eventually becomes almost constant, regardless of any further increase in substrate concentration. The choice of substrate concentration is an important consideration in the design of enzyme assays and an understanding of the kinetics of enzyme-catalysed reactions is needed in order to develop valid methods.

Kinetics of Enzyme-catalysed Reactions

The law of mass action states that the rate of a chemical reaction is proportional to the product of the concentrations of the reactants. This means that the rate of a reaction which has a single component will increase in direct relation to the increasing concentration but for a two-component reaction it will increase in proportion to the square of the concentration. These relationships may be expressed by the following equations :

Rate = k_1 (concentration) (single reactant)

Rate = k_2 (concentration) × (concentration) (two reactants)

where k_1 and k_2 are the reaction velocity constants or rate constants for the reactions. The reactions are said to show first- and second-order kinetics respectively. Occasionally situations arise where increases in the concentration of a reactant do not result in an increase in the reaction rate. Such reactions are said to show zero-order kinetics.

The effect of increasing the concentration of the substrate can be explained most satisfactorily by the formation of an enzyme-substrate complex as a key stage in the reaction. It is the breakdown of this single component, the ES complex, which results in the formation of the products as illustrated by equation and hence first-order kinetics apply.

$$E + S \rightleftharpoons ES \rightleftharpoons E + P \quad ...(4.1)$$

Michaelis-Menten Equation

The concept of an enzyme-substrate complex is fundamental to the appreciation of enzyme reactions and was initially developed in 1913 by Michaelis and Menten, who derived an equation that is crucial to enzyme studies. Subsequent to Michaelis and Menten several other workers approached the problem from different viewpoints and although their work is particularly useful in advanced kinetic and mechanistic studies, they confirmed the basic concepts of Michaelis and Menten.

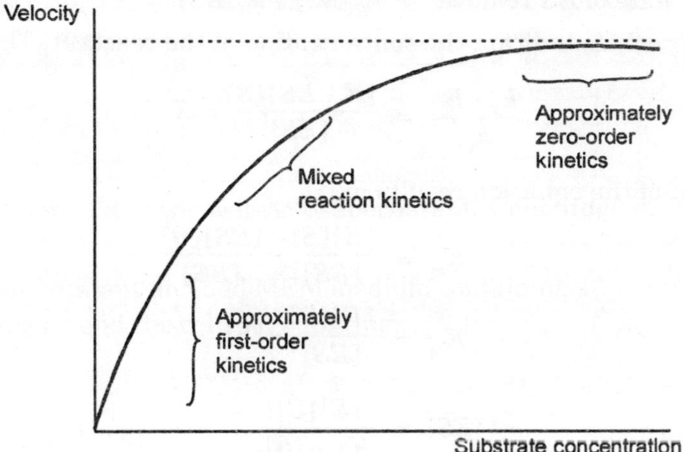

Fig. 4.3. The effect of substrate concentration on the activity of an enzyme. At low substrate concentrations, the rate of reaction resulting from a fixed amount of enzyme is proportional to the concentration of the substrate. However, at high concentrations, the reaction is almost constant and independent of the substrate concentration. The range of substrate concentrations between these two extremes results in mixed reaction kinetics.

The ES complex is formed when the enzyme and substrate combine but it may also be formed by the combination of the enzyme and product. The complex normally breaks down to form free enzyme and product but it may also revert to the free enzyme and substrate. These four reactions are summarized by equation:

$$E + S \underset{k_2}{\overset{k_1}{\rightleftharpoons}} ES \underset{k_4}{\overset{k_3}{\rightleftharpoons}} E + P \qquad ...(4.2)$$

The dissociation constant of the complicated equilibrium involving the enzyme-substrate complex is known as the Michaelis constant, K_m, and involves the rate constants for each of the four reactions involving the ES complex :

$$K_m = \frac{k_2 + k_3}{k_1 + k_4} \qquad ...(4.3)$$

When concentration of the complex is constant, the rate of formation of the complex is balanced by the rate of disappearance of the complex. If measurements of the reaction rate for any substrate concentration are made before the product is present in appreciable amounts the rate of the reverse reaction will be negligible and the rate constant k_4 can be ignored. Under these conditions:

$$K_m = \frac{k_2 + k_3}{k_1} \qquad ...(4.4)$$

It is extremely difficult to determine the relative amounts of free and complexed enzyme but it is possible to measure the total activity of the enzyme. The proportion of free enzyme can be represented as the difference between the total enzyme (E) and that complexed with substrate (ES). Hence :

$$\text{Rate of } ES \text{ formation} = k_1 [E - ES] [S]$$

$$\text{Rate of } ES \text{ removal} = k_2 [ES] + k_3 [ES]$$

where the square brackets indicate the concentration of the reactant. Therefore :

$$\frac{k_2 + k_3}{k_1} = \frac{[E - ES] [S]}{[ES]} = K_m$$

Rearrangement of the equation results in :

$$K_m = \frac{[E] [S]}{[ES]} - \frac{[ES] [S]}{[ES]}$$

$$K_m = \frac{[E] [S]}{[ES]} - [S]$$

$$[ES] = \frac{[E] [S]}{K_m + [S]} \qquad ...(4.5)$$

By definition, the observable rate of product formation (ν) is proportional to the concentration of the enzyme substrate complex, ES :

$$\nu = k_3 [ES]$$

Therefore substituting in equation :

$$\nu = \frac{k_3 [E] [S]}{K_m + [S]} \qquad ...(4.6)$$

The true concentration of enzyme is difficult to measure especially in terms of molar concentration but if the substrate concentration is large compared with that of the enzyme, all of the enzyme will be present as the ES complex and the reaction will proceed at maximum velocity. Under these conditions of excess substrate and maximum velocity (V_{max}) :

$$V_{max} = k_3 [E]$$

Substituting in equation results in the common form of the Michaelis equation :

$$\nu = \frac{V_{max} \times [S]}{K_m [S]} \qquad ...(4.7)$$

The equation gives a measure of the Michaelis constant (K_m) in terms of the measured velocity of the reaction (ν) which results from a substrate concentration ([S]) and the maximum velocity (V_{max}) which can be achieved using very high concentrations of substrate.

The value for the maximum velocity is related to the amount of enzyme used but the Michaelis constant is peculiar to the enzyme and is a measure of the activity of the enzyme. Enzymes with large values for K_m show a reluctance to dissociate from the substrate and hence are often less active than enzymes with low K_m values. The substrate concentration required for a particular enzyme assay is related to K_m and when developing an assay, the value for K_m should be determined.

Determination of the Michaelis Constant

The reaction velocity produced by a fixed amount of enzyme with varying concentrations of substrate is determined and a plot of the two variables shows a characteristic shape. The substrate concentration that results in a velocity that is half the maximum velocity is numerically equal to the Michaelis constant :

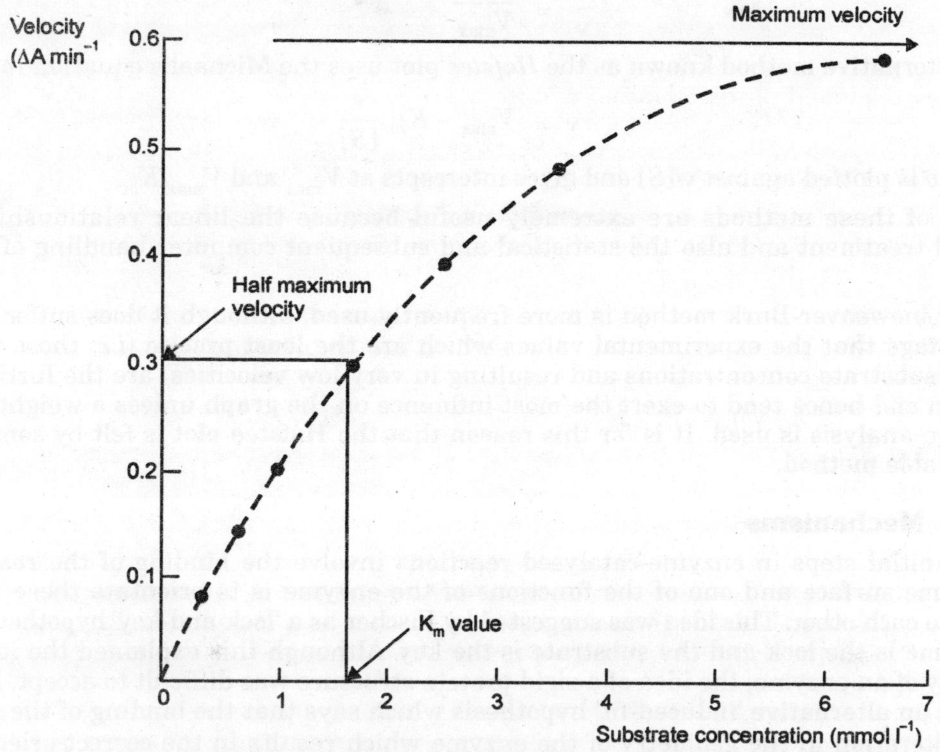

Fig. 4.4. Determination of the Michaelis constant (K_m).

$$\frac{V_{max}}{2} = \frac{V_{max} \times [S]}{K_m + [S]}$$

$$[S] = K_m$$

While this method is extremely simple it is also experimentally inaccurate. Because of the hyperbolic nature of the relationship, the curve approaches maximum velocity asymptotically making the deduction of a value for V_{max} difficult. Any error in assessing this value will be reflected in the value ascribed to K_m.

Lineweaver and Burk (1934) described a method for the determination of K_m which uses the reciprocal form of the Michaelis equation converting it to a linear relationship (Procedure 4.1) :

$$\frac{1}{v} = \frac{K_m}{V_{max}} \times \frac{1}{[S]} + \frac{1}{V_{max}} \qquad ...(4.8)$$

A plot of the reciprocal of velocity against reciprocal of the substrate concentration gives a straight line graph with intercepts of

$$\frac{1}{[S]} = -\frac{1}{K_m}$$

and

$$\frac{1}{\nu} = \frac{1}{V_{max}}$$

An alternative method known as the *Hofstee* plot uses the Michaelis equation in the form

$$\nu = V_{max} - K_m \frac{\nu}{[S]}$$

in which v is plotted against $\nu/[S]$ and gives intercepts at V_{max} and V_{max}/K_m.

Both of these methods are extremely useful because the linear relationships make graphical treatment and also the statistical and subsequent computer handling of the data simple.

The Lineweaver-Burk method is more frequently used, although it does suffer from the disadvantage that the experimental values which are the least precise (*i.e.* those involving very low substrate concentrations and resulting in very low velocities) are the furthest from the origin and hence tend to exert the most influence on the graph unless a weighted linear regression analysis is used. It is for this reason that the Hofstee plot is felt by some to be a more reliable method.

Enzyme Mechanisms

The initial steps in enzyme-catalysed reactions involve the binding of the reactants to the enzyme surface and one of the functions of the enzyme is to orientate these reactants relative to each other. This idea was suggested by Fischer as a 'lock-and-key' hypothesis, where the enzyme is the lock and the substrate is the key. Although this explained the idea of the specificity of an enzyme, the idea of a rigid protein structure was difficult to accept. Koshland proposed an alternative 'induced-fit' hypothesis which says that the binding of the substrate causes alteration in the geometry of the enzyme which results in the correct orientation of the appropriate group in both the substrate and the enzyme.

Procedure 4.1: Determination of the Michaelis constant (K_m) for the enzyme D-amino acid oxidase (EC 1.4.3.3) using the Lineweaver-Burk method

Data

Final substrate concentration ($m\ mol\ l^{-1}$)	*Rate of reaction* ($\Delta A_{340}\ min^{-1}$)
0.33	0.08
0.66	0.14
1.00	0.20
1.66	0.30
2.50	0.39
3.33	0.46
6.66	0.58

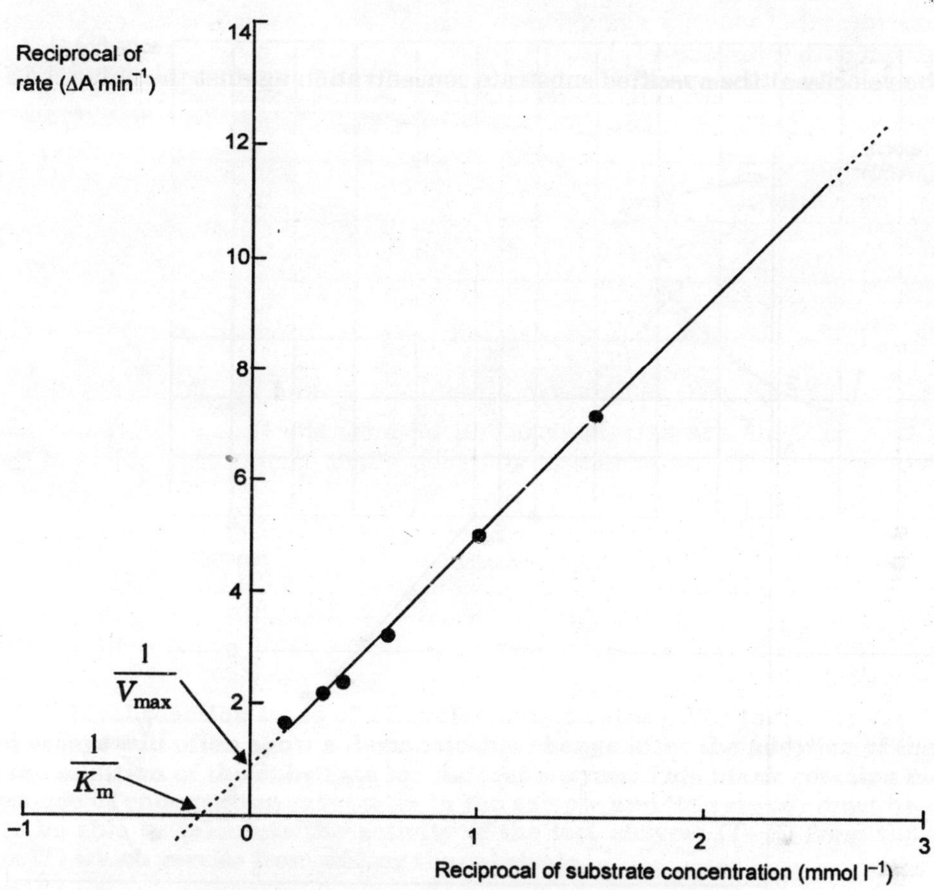

Fig. 4.5. Showing the Enzyme Mechanism.

Method

Plot the reciprocal of the reaction velocity against the reciprocal of the substrate concentration.

Calculation

The values for the Michaelis constant (K_m) and the maximum velocity are calculated from the intercepts on each of the axis which give the reciprocals of the K_m and the V_{max}.

Results

Michaelis constant (K_m) is 4.4 m mol l^{-1}

Maximum velocity is 1.1 absorbance units per minute

Procedure 4.2 : Determination of the Michaelis constant for the enzyme D-amino acid oxidase (EC 1.4.3.3) using the Hofstee method

Data

The same as for Procedure 4.1

Method

Plot the velocity at the specified substrate concentration against the value, $V/[S]$.

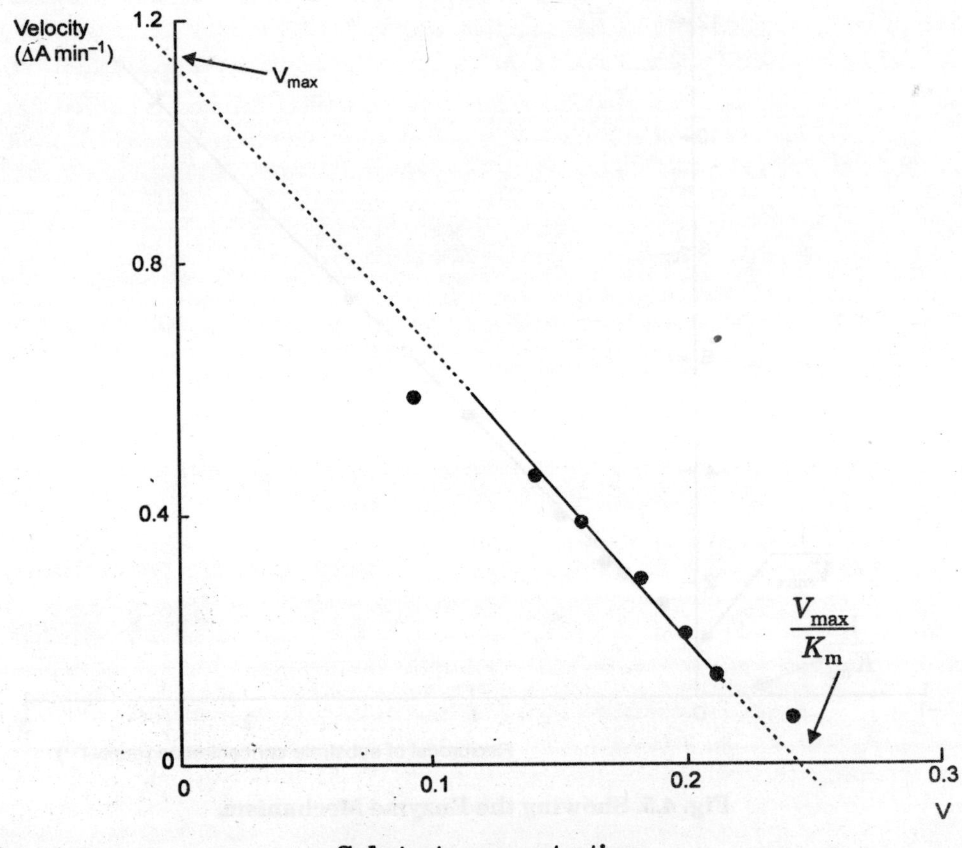

Substrate concentration

Fig. 4.6. Relationship between substrate concentration & rate of reaction.

Calculation

The values for the Michaelis constant (K_m) and the maximum velocity are calculated from the intercepts on each axis, which give the value for V_{max} and the fraction V_{max}/K_m.

Results

Michaelis constant (K_m) is 4.7 m mol l^{-1}

Maximum velocity as 1.14 absorbance units per minute.

Another hypothesis suggests that the binding of a substrate to an enzyme causes a strain or deformation of some of the bonds in the substrate molecule, which are subsequently broken. The effectiveness of this mechanism depends upon the strength of the binding force and does not necessarily involve any movement of the protein but suggests the idea of a flexible enzyme. Many enzymic reactions are a consequence of either nucleophilic or electrophilic attack on the substrate, the former being the more common. If such a species approaches another group it will tend to push electrons away from the 'positive centre'. If the reaction to be catalysed is

$$A—X \rightarrow A^+ + X^-$$

then an approach by a nucleophile (C^-, the catalyst) will make the reaction more feasible :

$$C^- \; A—X \rightarrow C—A + X^-$$

When a bond is broken the nature of the leaving group is important in determining the energy of activation. The basicity of the group is related to its nucleophilicity and if the basic nature of the leaving group is diminished so also will be the tendency to reform the broken bond:

$$H^+ + A—X \rightarrow A—X—H^+$$

$$C^- \; A—X—H^+ \rightarrow C—A + H^+X$$

This is an example of acid catalysis and the effect is to pull electrons away from the leaving group. Often both acid catalysis and nucleophilic attack are involved in enzyme-catalysed reactions in what are known as 'push-pull' mechanisms.

Activation of Enzymes

Many enzymes require additional substances in order to function effectively. Conjugated enzymes require a prosthetic group before they are catalytically active, such groups being covalently or ionically linked to the protein molecule and remaining unaltered at the end of the reaction. Catalase (EC 1.11.1.6), for instance, contains a haem group while ascorbate oxidase (EC 1.10.3.3) contains a copper atom. The purification of some enzymes inactivates them because substances essential for their activity but not classed as a prosthetic group are removed. These are frequently inorganic ions which are not explicit participants in the reaction.

Anionic activation seems to be non-specific and different anions are often effective. Amylase (EC 3.2.1.1), for example, is activated by a variety of anions, notably chloride. Cationic activation is more specific, *e.g.* magnesium is particularly important in reactions involving ATP and ADP as substrates. In cationic activation it seems very likely that the cation binds initially to the substrate rather than to the enzyme. In addition to these activators there is a group of substances which are not simple activators but are often essential co-substrates for many different enzymes; the most common examples are given.

Table 4.1. Cofactors as transporting systems

Cofactor	*Group transported*	*Type of reaction*
NAD^+, $NADP^+$	Redox carrier	Oxidation
Flavins	Redox carrier	Oxidation
Cytochromes	Redox carrier	Oxidation
AMP, ADP	Phosphate	Energy transfer
Pyridoxal phosphate	Amino	Transamination
UDP	Glycosyl	Carbohydrate metabolism
Coenzyme A	Acyl	Lipid metabolism

Inhibition of Enzymes

A substance that decreases the rate of an enzyme-catalysed reaction is known as an inhibitor and its effects may be permanent or transient. The inhibition of some reactions by substances which may be products of either that reaction or a subsequent reaction provides a control mechanism for cellular metabolism, while the selective inhibition of enzymes is the basis of many aspects of pharmacology and chemotherapy. Some inhibitors structurally resemble the substrate and are bound by the enzymes but cannot be converted to the products. Because the formation of the complex between the enzyme and the inhibitor is a reversible reaction, the inhibitor can be displaced by high concentrations of the normal substrate.

The inhibition causes an increase in the K_m value for the enzyme but the maximum velocity for the reaction remains unaltered. Such substances are known as *competitive inhibitors* and their effects can be reduced and even eliminated by high concentrations of substrate. The classical example of competitive inhibition is the action of the sulphonamide antibiotic drugs due to their structural similarity to the natural substrate *p*-aminobenzoic acid.

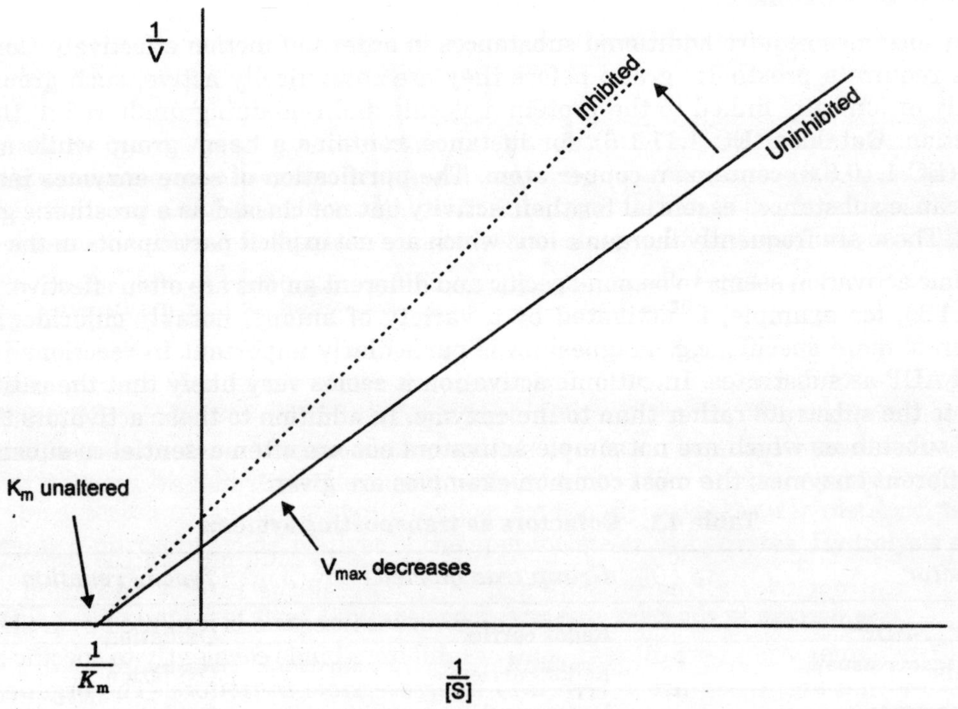

Fig. 4.7. The kinetic effects of a competitive inhibitor. The effects of a competitive inhibitor can be reversed by high concentrations of substrate and the increased amount required to achieve maximum velocity results in an increased value for the Michaelis constant.

Another type of inhibitor combines with the enzyme at a site which is often different from the substrate-binding site and as a result will inhibit the formation of the product by the

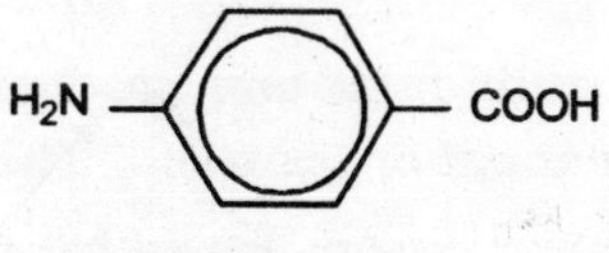

H_2N — (benzene ring) — COOH

p-Aminobenzoic acid

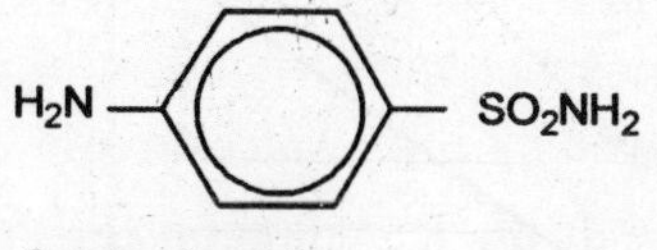

H_2N — (benzene ring) — SO_2NH_2

Sulphanilamide

H_2N — (benzene ring) — $\underset{O_2}{S}$ — $\overset{H}{N}$ — (pyrimidine ring, N, N)

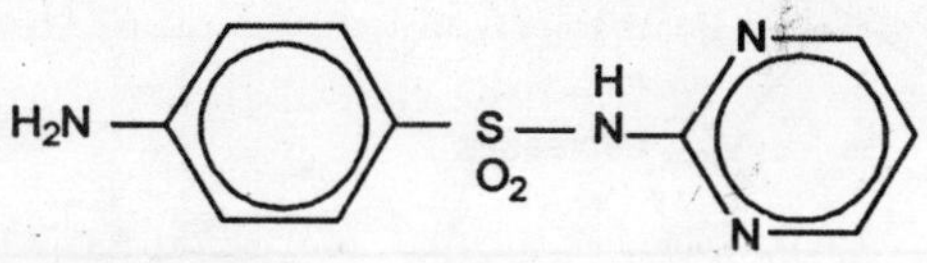

Sulphadiazine

Fig. 4.8. Competitive inhibition. Sulphonamide drugs compete with p-amino-benzoic acid, which is an essential growth factor for many bacteria.

breakdown of the normal enzyme-substrate complex. Such *non-competitive inhibition* is not reversed by the addition of excess substrate and generally the inhibitor shows no structural similarity to the substrate. Kinetic studies reveal a reduced value for the maximum activity of the enzyme but an unaltered value for the Michaelis constant. There are many examples of non-competitive inhibitors, many of which are regarded as poisons because of the crucial role of the inhibited enzyme.

Cyanide ions, for instance, inhibit any enzyme in which either an iron or copper ion is part of the active site or prosthetic group, *e.g.* cytochrome *c* oxidase (EC 1.9.3.1). Not all inhibitors fall into either of these two classes but some show much more complex effects. An *uncompetitive inhibitor* is defined as one that results in a parallel decrease in the maximum velocity and the K_m value.

The basic mode of action of such an inhibitor is to bind only to the enzyme-substrate complex and not to the free enzyme and so it reduces the rate of formation of products. Alkaline phosphatase (EC 3.1.3.1) extracted from rat intestine is inhibited by L-phenylalanine in such a manner. Another different class of inhibitors binds covalently to specific amino acids in the enzyme and these are referred to as *irreversible inhibitors.* The organophosphorus compounds, of which nerve gases are examples, inactivate enzymes which rely on the hydroxyl group of serine residues for their activity, *e.g.* cholinesterase (EC 3.1.1.8).

Allostery

Allosteric enzymes show various activation and inhibition effects which are competitive in nature and related to conformational changes in the structure of the enzyme. Such allosteric

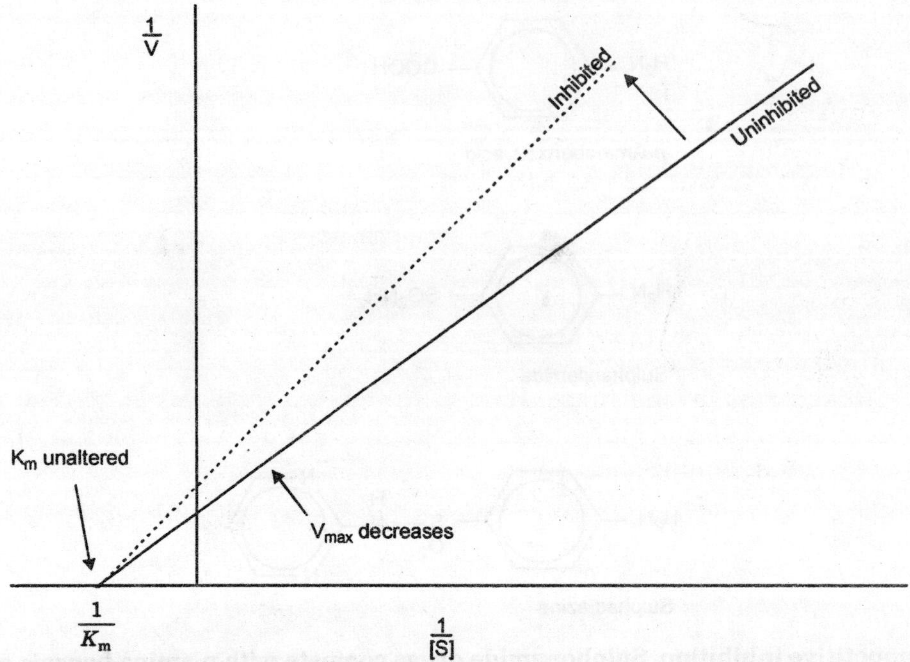

Fig. 4.9. The kinetic effects of a non-competitive inhibitor. The effect of a non-competitive inhibitor is not reversed by high concentrations of substrate and the enzyme reaction shows a reduced value for the maximum velocity. The enzyme remaining is unaltered and gives the same value for the Michaelis constant as originally shown by the uninhibited enzyme.

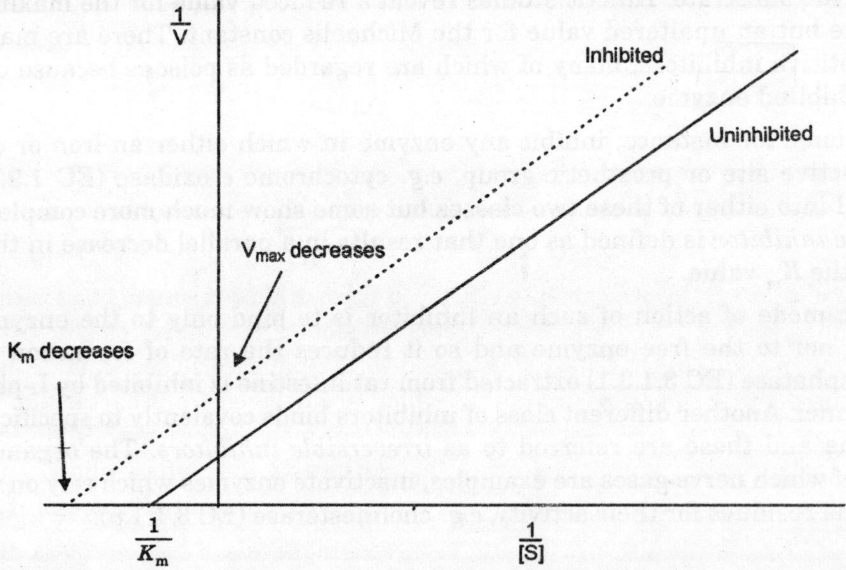

Fig. 4.10. The kinetic effects of an uncompetitive inhibitor. The effects of an uncompetitive inhibitor are very complex and the reaction usually shows a parallel decrease in both the maximum velocity and the Michaelis constant.

enzymes are often crucial enzymes in metabolic pathways and exert control over the whole sequence of reactions. The name allostery refers to the fact that inhibition of the enzyme is by substances that are not similar in shape to the substrate. The basic concept of allosteric enzymes initially proposed by Monod is an extension of Koshland's induced-fit theory.

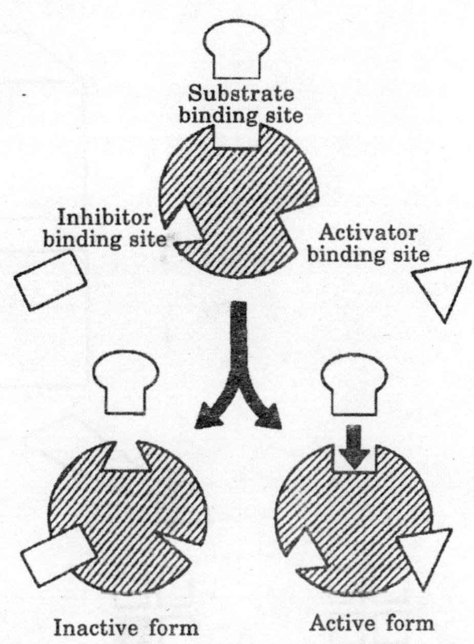

Fig. 4.11. Allosteric enzymes. The binding of an activator stabilizes the enzyme in an active form while the binding of an inhibitor distorts the active site, causing a loss of activity.

An allosteric enzyme in order to be catalytically active must be in a conformation that allows the binding of the substrate. Its structure is normally flexible and may only be stabilized by the binding of other molecules. Hence the binding of an activator at the activator site locks the enzyme in a form in which the substrate-binding site is available. The binding of an inhibitor at a separate inhibitor site causes distortion of both the substrate binding site and the activator site. 6-Phosphofructokinase (EC 2.7.1.11), a key enzyme in the glycolytic pathway, is inhibited by ATP and citrate but is activated by AMP, ADP, inorganic phosphate and the normal substrate, fructose-6-phosphate. These effects mean that its activity is reduced when the concentration of high energy compounds is high or there is an adequate source of energy from the citric acid cycle. Conversely, the enzyme is activated by compounds whose presence in increased amounts suggests a lack of high energy compounds in the cell.

Iso-enzymes

Enzymes that perform the same catalytic function are known as homologous enzymes and fall into two classes. Heteroenzymes are derived from different sources and although they catalyse the same reaction they show different physical and kinetic characteristics. The hydrolytic enzyme α-amylase (EC 3.2.1.1) is found in the pancreatic secretion in man and is different from the enzymes of the same name which are derived from bacteria or malt. Iso-enzymes, some-times referred to as isozymes, are different molecular forms of the same enzyme and are found in the same animal or organism although they often show a pattern of distribution between tissues. Many iso-enzymes are hybrids of a limited number of sub-units.

Some enzymes show multiple forms owing to increasing levels of polymerization of a single sub-unit; these should not really be called iso-enzymes, because they do not have any genetic difference between the different forms. Cholinesterase, for instance, shows five forms consisting of a single sub-unit existing as monomers, dimers, trimers, tetramers and pentamers.

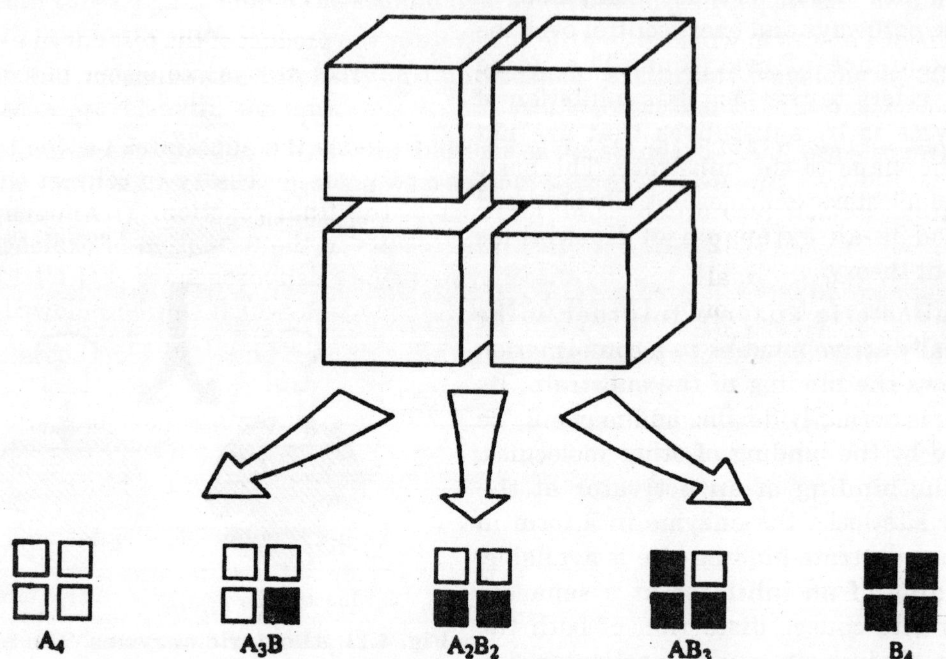

Fig. 4.12. The quaternary structure of proteins. The enzyme lactate dehydrogenase (EC 1.1.1.27) has a relative molecular mass of approximately 140000 and occurs as a tetramer produced by the association of two different globular proteins (A and B), a characteristic that results in five different hybrid forms of the active enzyme. The A and B peptides are enzymically inactive and are often indicated by M (muscle) and H (heart). The A_4 tetramer predominates in skeletal muscle while the B_4 form predominates in heart muscle but all tissues show most types in varying amounts.

Lactate dehydrogenase (EC 1.1.1.27) shows a classical iso-enzyme pattern. In man it is composed of two different protein sub-units known as A and B arranged into tetramers giving five different hybrid forms of the two basic units. The A and B sub-units are enzymically inactive but all the tetramers are active and because of their different composition can be separated from each other by electrophoresis or ion-exchange chromatography. They also show different catalytic properties, *e.g.* substrate specificity and inhibition kinetics, and these properties often provide the basis for relatively specific iso-enzyme assays.

Enzymes with iso-enzyme variants often show a characteristic distribution of the iso-enzymes in different tissues and the B_4 form of lactate dehydrogenase is found mainly in heart muscle and the A_4 mainly in skeletal muscle. While iso-enzymes show regular variations in their molecular structure, it is the difference in their catalytic activity that is most significant. They often show different inhibition and activation effects, permitting one iso-enzyme to function under conditions that would reduce the activity of another and it is likely that such variations aid the control of the same reaction under different cellular or tissue conditions.

ENZYME ASSAY METHODS

Enzyme assays in which either the substrate or the product of the test enzymes is measured are known as direct assays. For many enzymes direct assays are either not convenient or not feasible, and coupled assays are used.

The reaction mixture for a coupled assay includes the substrates for the initial or test enzyme and also the additional enzymes and reagents necessary to convert the product of the first reaction into a detectable product of the final reaction. The enzyme aspartate aminotransferase (EC 2.6.1.1), for instance, results in the formation of oxaloacetate, which can be converted to malic acid by the enzyme malate dehydrogenase (EC 1.1.1.37) with the simultaneous conversion of NADH to NAD^+, a reaction which can be followed spectrophotometrically at 340 nm :

$$\text{asparate + oxoglutarate} \xrightarrow[\text{aminotransferase}]{\text{asparate}} \text{glutamate + oxaloacetate}$$

$$\text{oxaloacetate + NADH} \xrightarrow{\text{malate dehydrogenase}} \text{malate + NAD}^+$$

In this example, malate dehydrogenase is known as the indicator enzyme.

Many assays have been described in which the initial product forms the substrate of an intermediary reaction involving auxiliary enzymes. The assay of creatine kinase (EC 2.7.3.2), for example, involves hexokinase (EC 2.7.1.1) as the auxiliary enzyme and glucose-6-phosphate dehydrogenase (EC 1.1.1.49) as the indicator enzyme:

$$\text{creating phosphate + ADP} \xrightarrow{\text{creatine kinase}} \text{creatine + ATP}$$

$$\text{ATP + glucose} \xrightarrow{\text{hexokinase}} \text{ADP + glucose-6-phosphate}$$

$$\text{glucose-6-phosphate + NADP}^+ \xrightarrow{\text{glucose-6-phosphate dehydrogenase}} \text{6-phosphogluconate + NADPH}$$

Optimized Assays

Prior to any experimental work in developing an assay, information about the enzymes should be obtained from the literature, such as pH optima, activators, inhibitors and K_m values. It is important that all such information is confirmed experimentally using the selected method. The pH optimum of an enzyme will often vary from one substrate to another and must be determined for each substrate. The buffer system used will often affect the overall activity of an enzyme and may alter its pH optimum. In general, the amino buffers such as glycylglycine and tricine, etc. Result in a greater enzyme activity than do the simple inorganic buffers such as phosphate and carbonate.

Buffers are most effective over a narrow pH range (approximately two units) which centres on their pK_a value. Those buffers with pK_a values similar to the known optimum pH of the enzyme should be tested for their effect on the activity of the enzyme over a limited pH range. It is difficult to determine the optimum pH of each enzyme when designing a coupled assay because of the presence of other enzymes and it is necessary to use a direct assay technique for each one. The effect of buffers, activators and inhibitors must also be checked using the same direct method.

Table 4.2. Buffers

Common name	*Base compound*	pK_a
ACES	Aminosulphonic acid	6.8
ADA	Iminoacetic acid	6.6
BES	Aminosulphonic acid	7.1
BICINE	Hydroxyglocine	8.3
CAPS	Aminosulphonic acid	10.4
EPPS	Piperazinesulphonic acid	8.0
GG	Glycylglycine	8.2
HEPES	Piperazinesulphonic acid	7.5
MES	Morpholinosulphonic acid	6.1
MOPS	Morpholinosulphonic acid	7.2
TAPS	Aminosulphonic acid	8.4
TES	Aminosulphonic acid	7.5
TRICINE	Hydroxyglycine	8.1
Tris	Hydroxyaminomethane	8.1

The selection of the overall assay pH may not be easy because of the differences in optima between the various enzymes. In general, it is advisable to select the pH optimum of the test enzyme in order to achieve maximum sensitivity and the proportionate loss of activity for each of the additional enzymes can be determined from their pH profiles compensated for by adding more of the enzyme to the assay mixture. It is essential to identify the need for any activators and to determine their optimal concentrations.

In many coupled assays it is not uncommon to find that an activator for one enzyme can inhibit another. The selected optimum concentration of all activators must be checked against all the other enzymes for any possible inhibition effects. Problems can often be solved by careful control of concentrations and by increasing the amounts of any inhibited supplementary enzymes. To achieve maximum velocity, a substrate concentration which is at least ten times greater than the K_m value for the enzyme should be used.

Although maximum velocity is only theoretically achieved at an infinite substrate concentration, it is possible using the Michaelis-Menten equation to calculate the percentage of maximum velocity given by any concentration of substrate. For a substrate concentration of ten times greater than the K_m value the velocity (ν) achieved will be :

$$\nu = \frac{V_{max} \times 10}{1+10}$$

$$\nu = \frac{10}{11} = 91\% \text{ maximum velocity}$$

While it is desirable to use as high a concentration as possible, some enzymes are subject to substrate inhibition and give a characteristic Lineweaver-Burk plot. In such cases the chosen substrate concentration must give the highest reaction velocity possible. It is important, when describing any enzyme assay, to report the percentage maximum velocity which the method will give. The only rate-limiting factor in a coupled assay should be the concentration of the initial and linking products and all other reagents should be in excess.

The role of the auxiliary and indicator enzymes is essentially that of a substrate assay system and under optimum assay conditions the rate of the indicator reaction should be equal to the rate of formation of the initial product. The indicator reaction must be capable of matching the different test reaction rates and its velocity can be defined by the Michaelis-Menten equation in the usual way:

$$\text{test velocity} = \text{indicator velocity} = \frac{V_{max} \times [P]}{K_m + [P]}$$

where [P] is the concentration of the test product.

The product concentration is normally very low and almost constant due to a steady state being established between its formation and use in the indicator reaction. Hence :

$$\text{Velocity} = \frac{V_{max}[P]}{K_m}$$

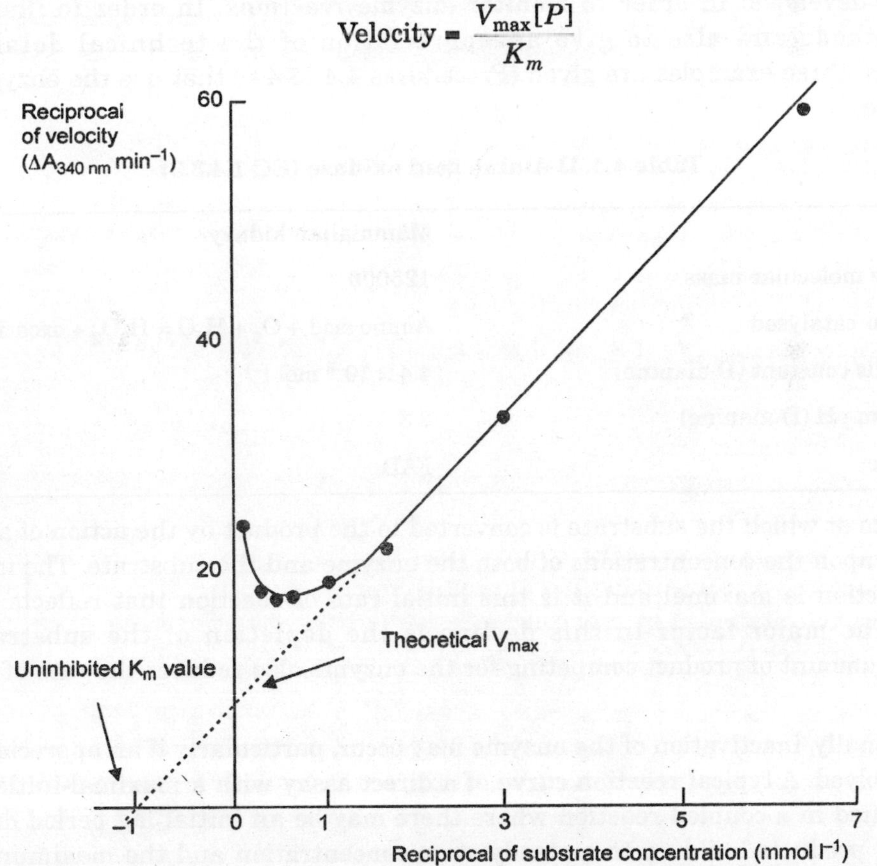

Fig. 4.13. Substrate inhibition. The enzyme L-amino acid oxidase (EC 1.4.3.2) suffers substrate inhibition at concentrations of L-leucine above 3.0 m mol l^{-1}. A Lineweaver-Burk plot shows the characteristic bend to the usual straight line.

In order for the indicator reaction to parallel the test reaction the ratio, V_{max}/K_m, for the indicator enzyme should be significantly greater than the same ratio for the test enzyme. The value for V_{max} is the only variable in the equation and is determined by the amount of enzyme present, usually quoted in units per assay volume. An increase by a factor, of 100 is usually recommended for each stage and will permit concentrations of approximately 1/100 of the K_m value to give a rate comparable to the test reaction rate. It may be necessary, if pH or inhibition effects reduce the activity of the enzyme by a known proportion, to increase further the amount of enzyme added to the assay. Some examples of coupled and direct assays for a range of enzymes.

Measurement of Enzyme Activity

In order to follow the progress of an enzyme-catalysed reaction it is necessary to measure either the depletion of the substrate or the accumulation of the product. This demands that either the substrate or the product show some measurable characteristic which is proportional to its concentration. This is not always the case and a variety of techniques have been developed in order to monitor enzyme reactions. In order to illustrate some of the methods and also to give an appreciation of the technical details and the calculations, three examples are given (Procedures 4.4 to 4.6) that use the enzyme D-amino acid oxidase.

Table 4.4. D-Amino acid oxidase (EC 1.4.3.3)

Source	Mammalian kidney
Relative molecular mass	125000
Reaction catalysed	Amino acid + O_2 + H_2O = H_2O_2 + oxoacid + NH_3
Michaelis constant (D-alanine)	4.4×10^{-3} mol l^{-1}
Optimum pH (D-alanine)	8.3
Cofactor	FAD

The rate at which the substrate is converted to the product by the action of an enzyme is dependent upon the concentrations of both the enzyme and the substrate. The initial rate of such a reaction is maximal and it is this initial rate of reaction that reflects the enzyme activity. The major factor in this decline is the depletion of the substrate but the increasing amount of product competing for the enzyme also reduces the rate of the forward reaction.

Additionally, inactivation of the enzyme may occur, particularly if an appreciable reaction time is involved. A typical reaction curve of a direct assay with a maximal-initial velocity is often modified in a coupled reaction where there may be an initial lag period during which the linking products build up to a steady-state concentration and the maximum velocity is determined from the slope of the steepest section of the curve.

Table 4.3. Examples of kinetic spectrophotometric methods of enzyme assay

Enzyme	*EC number*	*Direct or coupled*	*Substrate used*	*Product detected*	*Linking substrate*	*Linking enzymes*
Lactate dehydrogenase	1.1.1.27	D	Pyruvate	NADH		
D-Amino acid oxidase	1.4.3.3	C	D-Alanine	NADH	Ammonia	Glutamate dehydrogenase
γ-Glutamyl transpeptidase	2.3.2.2	D	γ-Glutamyl-*p*-nitroanilide	*p*-Nitroaniline		
Aspartate aminotransferase	2.6.1.1	C	L-Aspartate	NADH	Oxaloacetate	Malate dehydrogenase
Alkaline phosphatase	3.1.3.1	D	*p*-Nitrophenyl phosphate	*p*-Nitrophenol		
5′-Nucleotidase	3.1.3.5	C	5′-AMP	NADH	Adenosine	Adenosine deaminase
					Ammonia	Glutamate dehydrogenase
Argininosuccinate lyase	4.3.2.1	D	Argininosuccinate	Fumarate		
Aldolase	4.1.2.3	C	Fructose-1,6-diphosphate	NADH	Glyceraldehyde-3-phosphate	Glycerophosphate dehydrogenase
Glucose phosphate isomerase	5.3.1.9	C	Fructose-6-phosphate	$NADP^+$	Glucose-6-phosphate	Glucose-6-phosphate dehydrogenase
Glutamine synthetase	6.3.1.2	C	L-Glutamate	NADH	ADP Pyruvate	Pyruvate kinase Lactate dehydrogenase

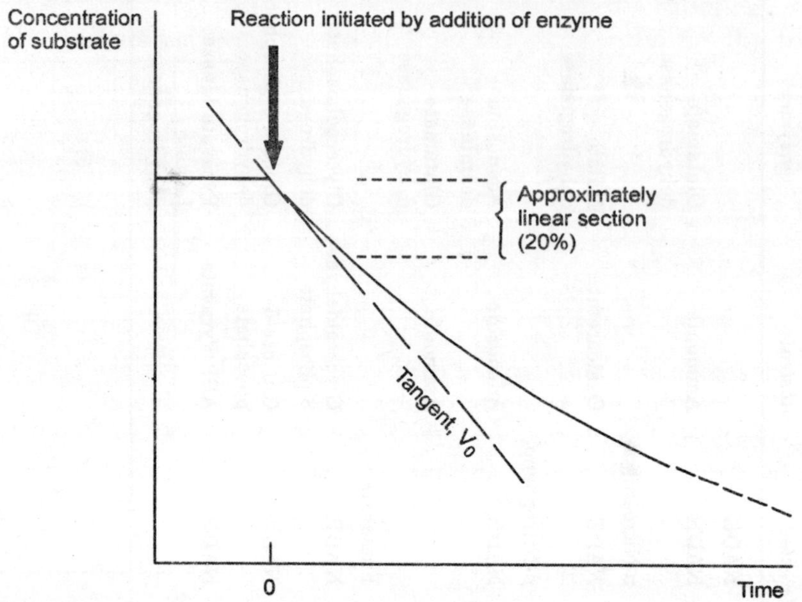

Fig. 4.14. Characteristic progress curve of an enzyme-catalysed reaction. The rate of reaction declines from a maximum initial velocity, V_0, which can be represented by the tangent to the curve at zero time. The rate falls least during the first 15-20% of the total reaction change.

The measurement of reaction velocity is most frequently done in one of three ways. The velocity of the reaction in terms of substrate depletion or product accumulation may be monitored by recording the concentration change and plotting the values. Such a technique is known as a *kinetic* or *continuous assay*. The most valid measure of the rate of the reaction is the initial velocity, V_0, and this may be determined either by drawing a tangent to the curve at zero time or with the use of electronic equipment to measure the change in the first few seconds. Hence :

enzyme concentration a velocity (V_0).

It is not always possible to measure initial velocities in this way owing to difficulties in detecting either substrate or product, in which case an alternative method can be used in which the amount of substrate used over a relatively long but specified period is measured.

While this may be prone to error owing to the declining reaction rate, it can be a valid analytical method because the first section of a reaction progress curve is relatively linear. This linear section usually comprises the first 10-20% of the total change possible and provided that the assay time has been demonstrated to be within these limits the enzyme activity can be related to the amount of substrate or product converted. Such methods are known as *fixed time assays.* Hence :

$$\text{enzyme concentration} \propto \frac{1}{\text{amount of substrate used}}$$

$$\alpha \text{ amount of product formed}$$

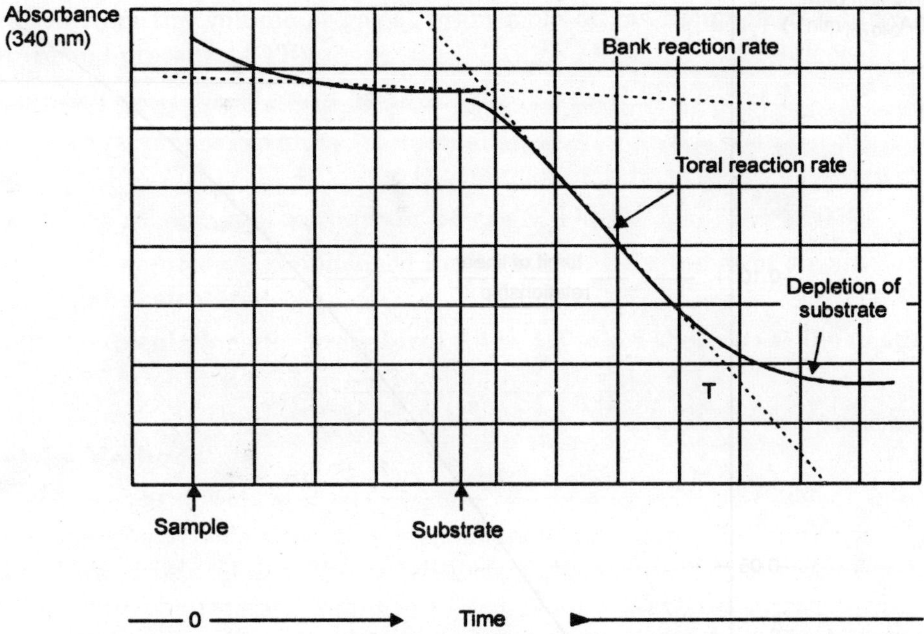

Fig. 4.15. Typical reaction trace of a coupled enzyme assay. The indicator reaction in many coupled assays will often show a demonstrable change after the addition of the sample but before the addition of the substrate for the test enzyme. This blank reaction may be due to the presence of endogenous substrates in the sample and its rate *(B)* must be measured in order to be able to calculate the activity of the test enzyme *(T– B)* from the total rate of reaction *(T)* which results from adding the substrate.

A technique that depends upon the same basic assumption as fixed time assays but relates the enzyme activity to the time taken for a fixed amount of product to be formed is called a *fixed change assay.* It is particularly useful for reactions which result in a pH change and can be monitored potentiometrically.

Hence :

$$\text{enzyme concentration} \propto \frac{1}{\text{time}}.$$

Quantitation of Enzyme Activities

It is normally not possible to use standard preparations of enzymes and although pre-assayed samples of enzymes are commercially available, they are usually used in quality assurance programmes and not as standards and, as a consequence, all test values have to be calculated from measured reaction rates or product concentrations. Because such calculations assume that there is a direct relationship between the rate of reaction and the concentration of the enzyme it is essential, when setting up an assay, to determine the effective range of activity for which this relationship is linear. This is most conveniently done by assaying carefully prepared dilutions of a sample containing a high enzyme activity and plotting the resulting reaction rates against the percentage dilution of the sample.

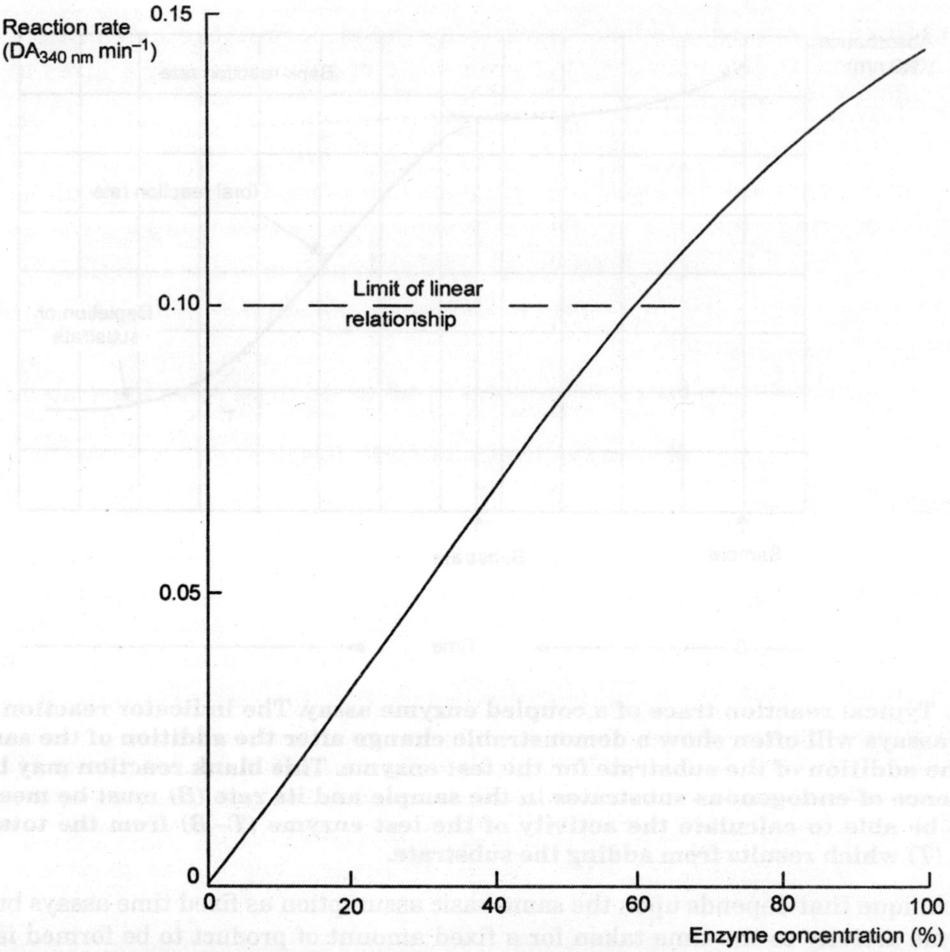

Fig. 4.16. The effective analytical range of an enzyme assay. The assay of D-amino acid oxidase (EC 1.4.3.3), using the method detailed in Procedure 4.5, shows a valid analytical range up to a maximum reaction rate of 0.10 absorbance change per minute.

The enzyme activity in katals is the number of moles of substrate converted by the enzyme in 1 second. Considerable difficulty is often experienced in calculating the activity from first principles and it is important to appreciate the necessary stages in the calculation. Almost invariably all of these steps can be condensed into a single conversion factor which may be applied to the measured reaction rate or instrument reading in order to calculate the enzyme activity in a sample. Procedures 4.4 to 4.6 give details of the calculations.

PROCEDURE 4.3: SEQUENCE OF STEPS FOR CALCULATING ENZYME ACTIVITY

1. Calculate the concentration of product formed in 1 second, expressed in moles per litre

For fixed time assays this most frequently involves the use of standards and a calibration graph. Some methods, *e.g.* the use of the molar absorbance coefficient in spectrophotometry, do not require standards and gasometric methods permit the calculation of molar concentration from the volume of gas (1 gram mole of gas occupies 22.4 litres at standard temperature and pressure, STP).

2. *Calculate the actual amount of product formed in 1 second*

Knowing the total assay volume the actual amount of product can be calculated from the concentration value, giving the value for the number of units of enzymes activity in the assay.

3. *Calculate the enzyme activity in a standard volume of the sample*

Knowing the volume of sample used in the assay, the number of enzyme units per litre or millilitre can be calculated.

Consult the calculations for Procedures 4.4 to 4.6 to see these steps illustrated.

MONITORING TECHNIQUES

Gasometric Methods

Some enzyme-catalysed reactions result in the production or uptake of a gas and some of the earliest assay methods used this phenomenon as a basis for monitoring the reaction. Classically, Krebs, in elucidating the metabolism of glucose, used such methods. They measure either the pressure changes that result in a fixed volume (Warburg manometry) or the volume changes required to keep a constant pressure (Gilson respirometry). The latter is potentially more useful because the molar concentration of the gas can be more easily calculated from volume than from pressure changes but both types of method are extremely tedious and difficult to perform and, as a result, are rarely used nowadays. They have been largely superseded by electrochemical gas measuring devices. The uptake of oxygen by oxidative enzymes and the generation of carbon dioxide by decarboxylating enzymes are the most frequent gaseous reactions used.

Some reactions involved both gases and it is possible to measure the uptake of oxygen in the presence of carbon dioxide by including a compartment in the reaction vessel containing a sodium hydroxide solution to absorb the carbon dioxide. If ammonia is generated it is normally retained in aqueous solution provided that the pH is not too high. It may be released subsequently by the addition of a concentrated solution of sodium hydroxide.

Table 4.5. Examples of gasometric methods of enzyme assay

Enzyme	*EC number*	*Substrate*	*Product*	*Gas exchanged*
D-Amino acid oxidase	1.4.3.3	D-Amino acids/oxygen	Ammonia	Oxygen uptake
Malate dehydrogenase	1.1.1.38	Pyruvate/carbon dioxide	Malate	Carbon dioxide uptake
Catalase	1.11.1.6	Hydrogen peroxide	Oxygen	Oxygen release
Cholinesterase	3.1.1.8	Acetylcholine	Acetic acid	Carbon dioxide release from bicarbonate buffer
Histidine decarboxylase	4.1.1.22	L-Histidine	Carbon dioxide	Carbon dioxide release
Pyruvate carboxylase	6.4.1.1	Oxaloacetate/ADP	Carbon dioxide	Carbon dioxide release

The actual volume of gases exchanged during such reactions is very small and therefore the measuring devices must be very sensitive. Because slight variations in temperature significantly affect either the volume or pressure of a gas, strict temperature control is necessary. Reaction vessels are usually the characteristically shaped Warburg flasks with varying numbers of side arms.

The majority of the reactants are contained in the main vessel while those that will initiate the reaction are held in the side arms. The centre well is used to hold a small filter paper wick soaked in a solution of sodium hydroxide to absorb any carbon dioxide generated during the reaction. The reaction is started by tilting the flask to mix the contents of the side arms and the main vessel. With all gasometric methods it is vital that all the connections and stoppers are gas-tight and usually these glass-to-glass connections are lubricated with lanolin grease. However, it requires very careful technique to ensure that the system is leak-proof and it is desirable to test for leaks once a system has been set up and prior to initiating the reaction.

Warburg Manometry

The basic design of a Warburg manometer is illustrated and consists of a U-tube manometer connected by a side arm to the reaction flask, which is immersed in a constant temperature bath. The manometer limb is calibrated from a mid-point of zero over a 300 mm range and the composition of the fluid is such that the pressure exerted by 10000 mm is equal to 1 atm. The readings obtained reflect the rate of the reaction and in many cases are sufficient although they may not give the quantitative information regarding the amount of gas evolved.

For quantitative work, it is necessary to determine the flask constant (*k*), which takes into account the total gas volume of the system, the solubility of the gas in water and the absolute temperature of the reaction. This factor can then be used to convert the pressure readings into gas volume.

Gilson Respirometry

The Gilson respirometer uses a constant pressure system, and variations in the gas volume are compensated for by altering the volume of the flask by means of a calibrated syringe. The pressure generated by the test flask and its contents is balanced against a reference volume of gas using a small indicating manometer. As gaseous exchange occurs, the manometer fluid is kept at the same level by altering the syringe piston (volumeter), which is calibrated in microlitres. The reading of the volumeter gives a direct measure of the volume of gas exchanged but it is still necessary to correct the value for standard temperature and pressure if critical results are required :

$$\text{Gas volume} \times \frac{T_0}{T} \times \frac{P}{P_0}$$

where T is the absolute temperature of the reaction,

T_0 is absolute zero temperature (–273°C),

P is the test atmospheric pressure,

P_0 is the standard atmospheric pressure (760 mm mercury).

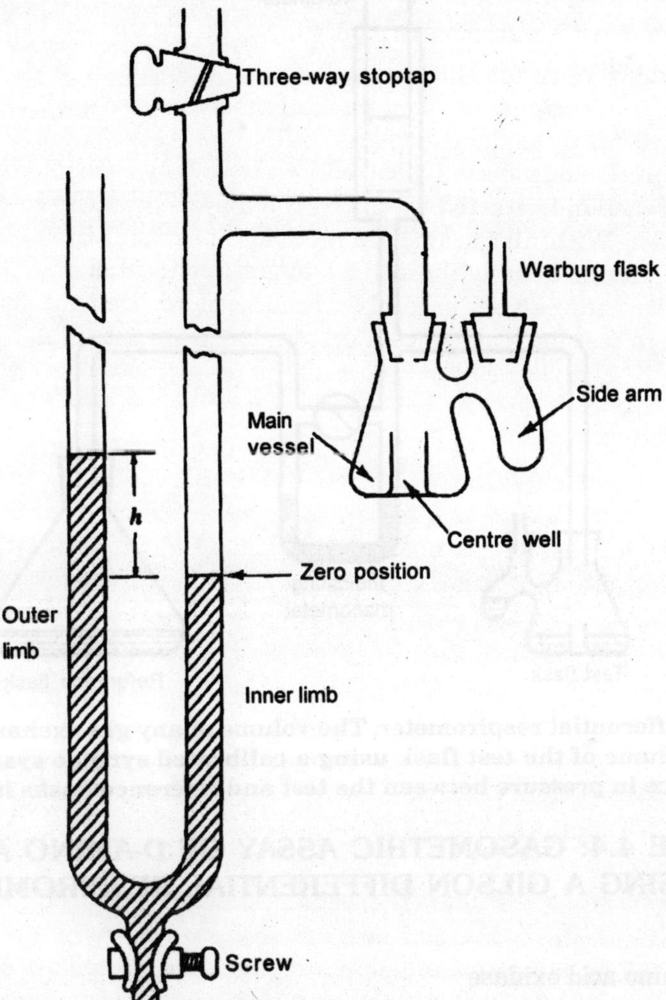

Fig. 4.17. Warburg manometer. Prior to any readings being taken, the manometer fluid in the inner limb is adjusted to the zero position by means of the pressure screw at the base of the manometer. Changes in the height *(h)* will then reflect the pressure changes in the flask which result from any gaseous exchange during the reaction.

The volume of gas can be converted to moles of gas using the fact that 1 mole of gas at standard temperature and pressure occupies 22.4 litres.

The method described for the assay of D-amino acid oxidase in Procedure 4.4 illustrates the use of the Gilson respirometer in a gasometric assay.

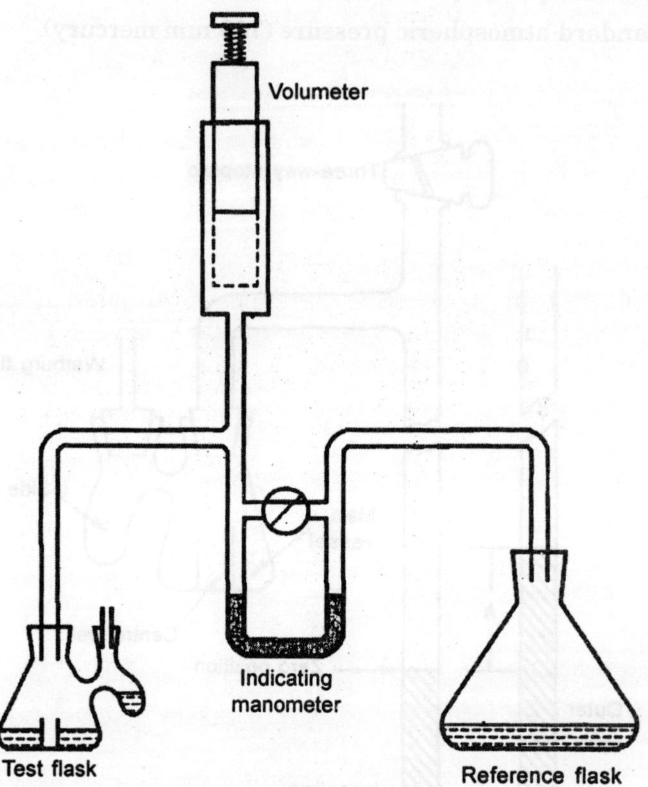

Fig. 4.18. Gilson differential respirometer. The volume of any gas exchanged can be measured by altering the volume of the test flask using a calibrated syringe system (volumeter) until the original balance in pressure between the test and reference flasks is restored.

PROCEDURE 4.4: GASOMETRIC ASSAY OF D-AMINO ACID OXIDASE USING A GILSON DIFFERENTIAL RESPIROMETER

Reaction

Catalysed by D-amino acid oxidase

$CH_3CHNH_2COOH + O_2 \rightarrow CH_2COCOOH + H_2O_2 + NH_3$

Reagents

Glycylglycinse buffer	0.1 mol l^{-1} adjusted to pH 2:3
FAD	0.3 m mol l^{-1}
Sodium hydroxide	100 g l^{-1}
D-Alanine	0.5 mol l^{-1}

Method

1. Into a Warburg flask pipette :

   Main compartment 1.2 ml buffer solution

	0.1 ml D-alanine
	0.1 ml FAD
Side arm	0.1 ml sample
Well	0.1 ml sodium hydroxide with a filter paper wick.

2. Seal the joints of the flasks with lanolin and equilibrate at 37°C for 15 min with the system open to the air.
3. Close the system and monitor the reading for a further 15 min.
4. Initiate the reaction by mixing the contents of the main vessel and the side arm. Monitor the change in volume due to the uptake of oxygen for 30 min and record the results as microlitres of gas per second.

Calculation

The enzyme activity in katals per litre can be calculated from the volume change after correcting for temperature and pressure :

$$\text{Recorded volume} \times \underbrace{\frac{273}{310} \times \frac{P}{760}}_{A} \times \underbrace{\frac{1}{224 \times 10^6}}_{B} \times \underbrace{\frac{1000}{0.1}}_{C}$$

The various steps in the calculation are.

A. Correction for temperature and pressure (P mm Hg).

B. Converting microlitres of gas to moles.

C. Correction for volume of sample used.

Spectrophotometric Methods

The use of spectrophotometry to monitor enzyme-catalysed reactions is a very convenient and popular method owing to the simplicity of the technique and the precision that is possible. The technique lends itself readily not only to temperature control using water-jacketed or electrically heated cell holders but also to the measurement of initial velocities by continuous monitoring and recording techniques or by automated analysis systems. Although only a small number of substrates or products show absorption maxima in the visible region of the spectrum a considerable number show maxima in the ultraviolet.

However, in order to be able to measure the rate of substrate depletion or product formation, it is necessary that these two components of the reaction show different absorption characteristics to permit the measurement of one without interference from the other.

The natural substrates which are most frequently used are the nucleotide coenzymes NAD^+ and $NADP^+$, which are reversibly reduced by many enzymes:

$$\underset{\text{(oxidized)}}{NAD(P)^+} + H_2 \rightleftharpoons \underset{\text{(reduced)}}{AD(P)H} + H^+$$

The two forms of these coenzymes (oxidized and reduced) have different spectral characteristics, which permit the reduced form to be detected at 340 nm in the presence of the oxidized form.

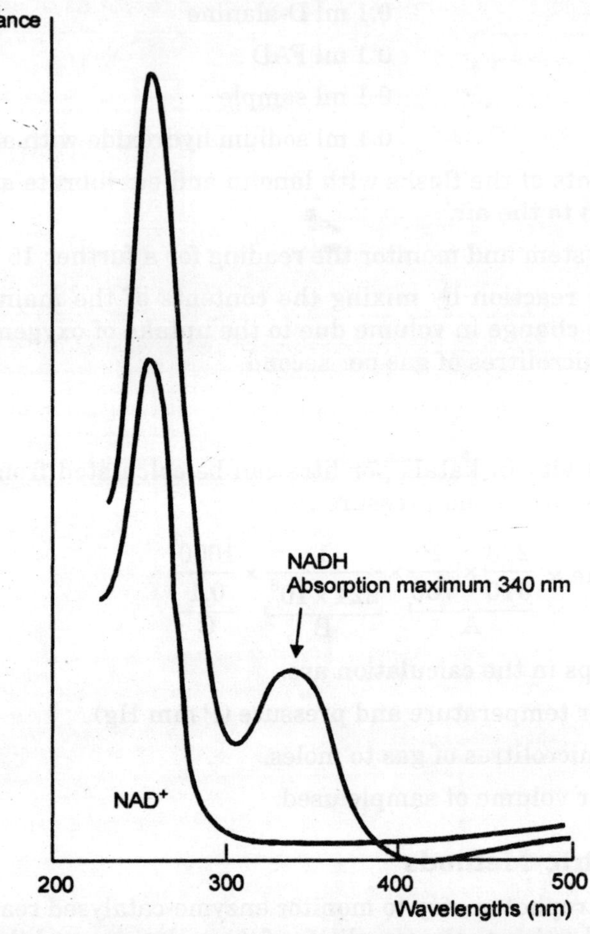

Fig. 4.19. Absorption spectra of NAD^+ and NADH.

The natural substrates for many enzymes do not show any significant spectral properties and alternative artificial substrates (analogues) may be used. Many hydrolytic enzymes can most easily be detected by using substrate analogues; the glycosidases, for instance, may be measured using *p*-nitro-phenyl derivatives of the appropriate carbohydrates. Hydrolysis results in the production of *p*-nitrophenol, which can be measured at 404 nm.

Table 4.6. Examples of fixed time spectrophotometric methods of enzyme assay

Enzyme	*EC number*	*Substrate*	*Product*	*Measurement*
Glucose oxidase	1.1.3.4	Glucose	Hydrogen peroxide	Oxidation of a chromogen
Aspartate aminotransferase	2.6.1.1.	L-Aspartate	Oxaloacetate	Formation of a hydrazone
Cholinesterase	3.1.1.8	Acetylcholine	Acetate	Conversion to hydroxamate
Aldolase	4.1.2.3	Fructose-1,6-diphosphate	Dihydroxyacetone phosphate	Formation of a hydrazone
Glucose phosphate isomerase	5.3.1.9	Glucose-6-phosphate	Fructose-6-phosphate	Reaction with resorcinol
L-Glutamate synthetase	6.3.1.2	L-Glutamate/ATP	Phosphate	Determination of phosphate

Table 4.7. Examples of fluorimetric methods of enzyme assay

Enzyme	*EC number*	*Substrate*	*Product*	*Excitation wavelength (nm)*	*Emission wavelength (nm)*
Monoamine oxidase	1.4.3.4	Kynurenine	**4-Hydroxyquinoline**	315	380
UDP-glucuronyl-transferase	2.4.1.17	**4-Methylum-belliferone**	4-Methylglucuronide	340	440
Cholinesterase	3.1.1.8	Indoxylacetate	**Indoxyl**	395	470
Lipase	3.1.1.3	Dibutyrylfluorescein	**Fluorescein**	490	520

Note: All NAD^+- and $NADP^+$– linked enzyme assays are also capable of being monitored fluorimetrically. The fluorescent compounds in each assay are shown in bold type.

Table 4.8. Examples of electrochemical methods of enzyme assay

Enzyme	*EC number*	*Substrate*	*Product*	*Device*
D-Amino acid oxidase	1.4.3.3	D-Amino acids/oxygen	Ammonia	Ammonia-selective electrode
Urate oxidase	1.7.3.3	Uric acid/oxygen	Allantion	Oxygen electrode
Cholinesterase	3.1.1.8	Acetylcholine	Acetic acid	pH electrode
Lipase	3.1.1.3	Triacylglycerols	Fatty acids	pH electrode
Carbonic anhydrase	4.2.1.1	Carbon dioxide	Bicarbonate	pH electrode
Glutamate decarboxylase	4.1.1.15	L-Glutamate	4-Aminobutyrate/CO_2	Carbon dioxide electrode
Arginase	3.5.3.1	L-Arginine	L-Ornithine/urea	Urea-selective electrode

Many methods have been developed in which a product of the reaction is chemically modified to produce a substance with a particular spectral property. The inorganic phosphate released by the hydrolysis of phosphate esters may be measured by simple chemical methods (Fiske and Subbarow) after the enzyme reaction has been stopped.

Such techniques are often convenient but do not lend themselves to the measurement of initial velocity. Coupled enzyme assays provide a good alternative to chemical modification and permit a kinetic technique to be employed. In a coupled assay, the rate at which the product is formed is measured by using the product of the reaction as a substrate for a second enzyme reaction which can be monitored more easily. Coupled assays offer great flexibility in enzyme methodology while still retaining all the advantages of continuous monitoring techniques and are illustrated by Procedure 4.5.

PROCEDURE 4.5: SPECTROPHOTOMETRIC ASSAY OF D-AMINO ACID OXIDASE (KINETIC)

Test reaction

Catalysed by D-amino acid oxidase

$$CH_3CHNH_2COOH + O_2 \rightarrow CH_3COCOOH + H_2O_2 + NH_3$$

Indicator reaction – catalysed by glutamate dehydrogenase

$$NH_3 + HOOCCH_2CH_2COCOOH + NADH \rightarrow HOOCCH_2CH_2CHNH_2COOH + NAD$$

Assay reagent

Glycylglycine buffer pH 2.3	(0.1 mol l^{-1})	2.05 ml
2-Oxoglutarate	(0.2 mol l^{-1})	0.20 ml
NADH	(2.25 m mol l^{-1})	0.20 ml
FAD	(0.3 m mol l^{-1})	0.10 ml
Glutamate dehydrogenase	(20 m katal or 1200 units)	0.05 ml
ADP	(15.0 m mol l^{-1})	0.10 ml

Method

1. Pipette the assay reagent (2.7 ml) into a clean dry cuvette and allow to attain a temperature of 37°C in a thermostated cell holder.
2. Monitor the absorbance of the solution at 340 nm continuously.
3. Add 0.2 ml sample and mix carefully.
4. When the absorbance shows either no further fall or a slow but steady fall (the blank reaction) initiate the reaction by adding 0.1 ml 0.5 mol l^{-1} D-alanine.
5. Monitor the reaction for at least 5 min or until a linear section of the reaction trace can be clearly defined.
6. The rate of the blank reaction and the total reaction are determined graphically and reported as absorbance change per second. Under conditions described the reaction gives 84% maximum velocity for any enzyme preparation.

Calculation

The enzyme activity in katals per litre may be calculated from the molar absorption coefficient for NADH at 340 nm (6.22×10^3 l mol^{-1} cm^{-1}).

$$\underbrace{\frac{(\Delta A_{340}\ \text{Test} - \Delta A_{340}\ \text{Blanks})}{6.22 \times 10^3}}_{A} \times \underbrace{\frac{3.0}{1000}}_{B} \times \underbrace{\frac{1000}{0.2}}_{C}$$

The various steps in the calculation are :

A. The use of the molar absorption coefficient for NADH to calculate the concentration change (mol l^{-1} s^{-1}).

B. Calculation of the amount of NADH oxidized in the 3.0 ml reaction volume in 1 second.

C. Correction for the sample volume used.

Fluorimetric Methods

The use of fluorescent substrates gr products permits sensitive kinetic measurement of enzyme reactions to be undertaken and although there are relatively few natural fluorescent substrates, analogues can sometimes be used. Some products can be converted to fluorescent compounds and can be used in fixed time assays.

Fluorimetric methods are useful for monitoring reactions involving the nucleotide coenzymes. The natural fluorescence of the reduced forms in the region of 460 nm can be used in kinetic assays. However, this fluorescence is destroyed at pH values below 2.0, whereas any oxidized forms of the co-enzymes present are stable. If the pH of the solution is then raised above 10.5 and heated, the oxidized forms are themselves converted to fluorescent derivatives. This latter procedure lends itself to fixed time assays such as is illustrated in Procedure 2.6. Fluorimetric methods, while being inherently sensitive, have significant disadvantages which relate to the technical difficulties in fluorescence measurements.

Many fluorescent compounds are unstable particularly in ultraviolet light, although it is sometimes possible to increase the stability by including additional reagents, *e.g.* imidazole. The presence of other fluorescent substances in either the sample or the reagents will give a background fluorescence which will reduce the sensitivity of the measurements. This is a problem particularly when analysing tissue extracts, etc., in which aromatic amino acids in the proteins will contribute to the fluorescence. A major problem in fluorimetric measurements is caused by the fact that many substances, often present in small amounts as impurities in reagents and solvents, can quench the radiation emitted by the fluorescent substance. For this reason it is necessary to use only reagents of the highest purity and to ensure that all equipment is scrupulously clean.

PROCEDURE 4.6: FLUORIMETRIC ASSAY OF D-AMINO ACID OXIDASE (FIXED TIME)

Test reaction – catalysed by D-amino acid oxidase

$$CH_3CHNH_2COOH + O_2 \rightarrow CH_3COCOOH + H_2O_2 + NH_3$$

Indicator reaction – catalysed by glutamate dehydrogenase

$$NH_3 + HOOCCH_2CH_2COCOOH + NADH \rightarrow HOOCCH_2CH_2CHNH_2COOH + NAD$$

Reagents

Assay reagent	as for Procedure 5
Sodium hydroxide	10 mol l^{-1} containing imidazole (10 m mol l^{-1})
Standard solution of NAD^+	1.5 m mol l^{-1}
D-Alanine	0.5 mol l^{-1}

Method

1. Pipette the assay reagent (2.7 ml) into each of the two test-tubes labelled TEST and BLANK. To each add 0.2 ml of sample.
2. After incubating at 37°C for about 5 min, initiate the reaction in the TEST by adding 0.1 ml of D-alanine.
3. Incubate both tubes for a further 60 min. Add 0.1 ml D-alanine to the BLANK tube and immediately transfer 1.0 ml from each tube to appropriately labelled tubes containing 0.1 ml HCl (2.0 mol l^{-1}). Allow to stand at room temperature for 15 min to destroy the remaining NADH.

4. To both tubes, add 2.0 ml of the sodium hydroxide/imidazole reagent and heat at 56°C for 15 min in the dark.
5. Cool and measure the intensity of fluorescence (excitation wavelength, 365 nm; emission wavelength, 455 nm).
6. The fluorescence produced by 1.0 ml dilutions of the standard NAD^+ are also measured in the same manner after heating with sodium hydroxide reagent.

Calculation

From the calibration curve, the concentration of NAD^+ in both the test and blank can be determined. From the difference in the results the amount of NAD^+ generated in 1 second can be calculated (mol l^{-1} s^{-1}) and from this the enzyme activity in katals per litre of sample.

$$\text{Concentration change} \times \underbrace{\frac{3}{1000}}_{A} \times \underbrace{\frac{1000}{0.2}}_{B}$$

The various steps in the calculation are:

A. Calculation of the amount of NAD^+ produced in the 3.0 ml reaction volume in 1 second.

B. Correction for the volume of sample used.

Luminescence Methods

Bioluminescence provides the basis for sensitive enzymic assay methods both for substrate assays and coupled enzyme assays. Firefly luciferase (EC 1.13.12.5) catalyses the production of light (540-600 nm) by the oxidation of luciferin (D-LH_2).

$$\text{D-LH}_2 + \text{ATP} + \text{O}_2 \rightarrow \text{AMP} + \text{PP} + \text{D-L} + \text{CO}_2 + \text{light}$$

The enzyme is activated by magnesium ions, has an optimum pH of about 7.5 and offers a very sensitive assay method for ATP but has much wider applications when used in coupled assays with ATP-converting enzymes.

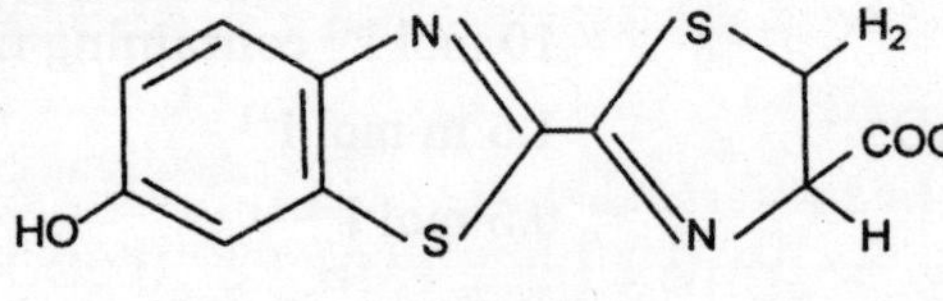

D-Luciferin

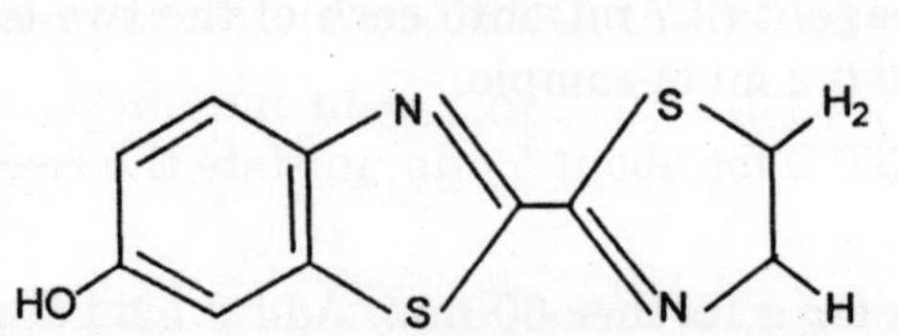

Oxyluciferin

Fig. 4.20. Luciferin, a substrate for bioluminescence.

Luciferase from bacterial sources catalyses the oxidation of long chain aliphatic aldehydes and requires the coenzyme FMN. The wavelength of the emitted radiation in this reaction is approximately 490 nm:

$$FMNH_2 + RCHO + O_2 \rightarrow FMN + RCOOH + H_2O + light$$

The reaction can be coupled to the oxidation of either NADH or NADPH by the presence of a flavin dehydrogenase, which is usually present complexed with the luciferase:

$$NAD(P)H + FMN \rightarrow NAD(P)^+ + FMNH_2$$

A wide range of coupled assays involving ATP, NAD(P)H and FMN can be developed using these two enzymes and provide increased levels of sensitivity over other coupled assays (Procedure 4.7).

Electrochemical Methods

The measurement of pH using the glass electrode provides a very convenient method for monitoring reactions in which either an acid or a base is produced, notably certain hydrolytic enzymes. The change in pH resulting from the reaction has an effect on the activity of the enzyme and hence on the rate of the reaction, and to maintain the reaction at the optimum pH and yet measure the amount of acid or alkali formed a pH-stat system can be used. In such an instrument, slight changes from an initial pre-selected pH activate a solenoid valve to allow the addition of either an acid or an alkali to maintain the pH at the selected value.

The reaction is monitored not as a change in pH but as the volume of titrant required to maintain the pH. Such a method also avoids the difficulties of making direct measurements of pH change in the presence of the buffers in the reaction mixture. The use of ion-selective electrodes opens up the wider use of electro-chemical methods for monitoring the formation of particular products, *e.g.* ammonia, while simple platinum electrodes can be used to monitor redox reactions catalysed by the oxidoreductases or hydrolytic enzymes using substrate analogues which cause redox changes. Oxygen exchange may be monitored with an oxygen electrode as an alternative to manometry, and other electrodes, *e.g.* carbon dioxide, are also available.

PROCEDURE 4.7: BIOLUMINESCENCE ASSAY OF FMN USING HACTERIAL LUCIFERASE (EC 1.14.14.3)

Reaction—catalysed by luciferase

$$FMNH_2 + C_{10}H_{23}CHO + O_2 \rightleftharpoons FMN + C_{10}H_{23}COOH + H_2O + light$$

Assay reagent

Glycylglycine buffer pH 7.0	($0.1\ mol\ l^{-1}$)	1.0 ml
Luciferase	(see below)	0.1 ml
Decylaldehyde	(0.1% in buffer)	0.1 ml
NADH solution	($10\ mg\ ml^{-1}$ in buffer)	

Luciferase (*Vibrio fischeri*) 1×10^4 light units per ml in buffer containing $10\ mg\ l^{-1}$ albumin and $6\ mol\ l^{-1}$ dithiothreitol.

Method

1. Pipette the reagent into a tube appropriate for the model of luminometer being used.
2. Add 0.1 ml of the FMN standard or sample solution.
3. Mix and measure any background luminescence.
4. Add 0.1 ml NADH solution.
5. Quickly mix and monitor the luminescence for 5 min, recording the maximum level of emission.

Calculation

Using a series of standard solutions of FMN, a calibration graph can be drawn and used to determine the concentration of FMN in the test samples.

Microcalorimetric Methods

Almost all enzyme-catalysed reactions are exothermic and the enthalpy change (ΔH) involves the release of energy as heat. The resulting increase in temperature, although very small, provides a non-specific method of monitoring the reaction. Most calorimeters used for this purpose measure the increase in temperature in an insulated container (adiabatic calorimeters) and although the temperature change may be as small as 1×10^{-3} °C, it can be detected and measured using thermistors. The amount of heat released during a reaction is proportional to the amount of substance involved but the relationship is complicated in enzyme studies by secondary reactions.

Although the use of entropy constants means that calorimetry theoretically does not require standardization, in many instances this will be necessary. The initial energy change can often be enhanced, giving an increase in the sensitivity of the method. Hydrogen ions released during a reaction, for instance, will protonate a buffer with an evolution of more heat.

$$\text{Glucose} + \text{ATP} = \text{glucose-6-phosphate} + \text{ADP} + H^+ \qquad \Delta H = -27 \text{ kJ mol}^{-1}$$

$$\text{Tris} + H^+ = \text{Tris } H^+ \qquad \Delta H = -47 \text{ kJ mol}^{-1}$$

$$\text{Total } \Delta H = -74 \text{ kJ mol}^{-1}$$

Similarly, the use of coupled reactions can provide further energy release, *e.g.* the coupled determination of glucose using glucose oxidase (EC 1.1.3.4) and catalase (EC 1.11.1.6).

$$\text{Glucose} + O_2 = \text{gluconic acid} + H_2O_2 \qquad \Delta H = -83 \text{ kJ mol}^{-1}$$

$$H_2O_2 = H_2O + \frac{1}{2}O_2 \qquad \Delta H = -125 \text{ kJ mol}^{-1}$$

$$\text{Total } \Delta H = -208 \text{ kJ mol}^{-1}$$

Because the energy changes are so small, microcalorimetry lends itself to substrate assays in the presence of excess enzyme rather than to enzyme assays and is particularly useful when immobilized enzymes are used.

TREATMENT OF SAMPLES FOR ENXYME ASSAY

Studies of the enzyme content of cells frequently involve the use of coarse tissue samples of either animal or plant origin. In such cases some preliminary dissection of the tissue may be necessary to isolate the relevant tissue components and remove unwanted structural material such as collagen, cellulose, etc., before moving on to the more critical disruption of the cells. Sometimes it is possible to use the technique of tissue culture to provide pure cell preparations for subsequent studies.

Tissue Culture

Tissue culture, more frequently used as cell culture, enables animal and plant cells to be cultured in large numbers by techniques comparable to those used in microbiology but, because of the fragile nature of the cells, does require special cultural conditions. The culture media used must supply all the essential factors for growth, such as a wide range of amino acids, nucleotides, enzyme co-factors as well as indeterminate factors that can only be supplied in special products, *e.g.* foetal bovine serum.

The environmental conditions must be carefully controlled, particularly pH, and this is frequently maintained by culturing in a bicarbonate buffer system and a carbon dioxide saturated atmosphere. The build-up of waste products inhibits growth and it is necessary to change the culture medium at regular intervals, usually every few days. As with all cell culture techniques, asepsis is essential but very difficult to maintain. Antibiotics may therefore be incorporated into the culture media to assist in maintaining sterility.

Storage of Samples for Enzyme Assay

All samples must be stored appropriately to minimize the loss of activity due to protein denaturation, lack of stabilizers or presence of inhibitors. Optimal storage conditions will vary for different enzymes and the nature of the sample, blood, tissue, etc. Such information would be sought from specialist text-books.

In general, storage at low temperature, with protecting molecules present, is often to be recommended, *e.g.* storage at –20°C in bovine serum albumin or glycerol. A 50% (v/v) solution of glycerol is particularly useful because it is still fluid at –20°C and, being non-protein, can subsequently be removed very easily. For enzymes that are prone to oxidative effects, the presence of mild reducing substances such as the amino acid cysteine or similar substances may be beneficial.

Treatment of Tissue and Cell Samples

Very few enzymes are secreted by cells and it is often necessary to disrupt the cells in order to demonstrate enzyme activity. When the total enzyme content of cells is required, it is necessary to disrupt them completely and to remove all paniculate material by centrifugation, leaving the dispersed enzyme protein in the supernatant fluid. Many enzymes are membrane bound and are only released from fragmented membraues if a suitable detergent is incorporated into the medium. The detergents used include Triton X100 and the bile salts, *e.g.* sodium deoxycholate-Enzymes that fail to exhibit maximum activity without the releasing effects of detergents are said to show 'latency'.

Many cellular studies, however, require the careful disruption of the cell and the subsequent separation of the various cellular organelles such as mitochondria, nuclei, etc., by a process known as cell fractionation. Gross homogenization of tissues may be achieved

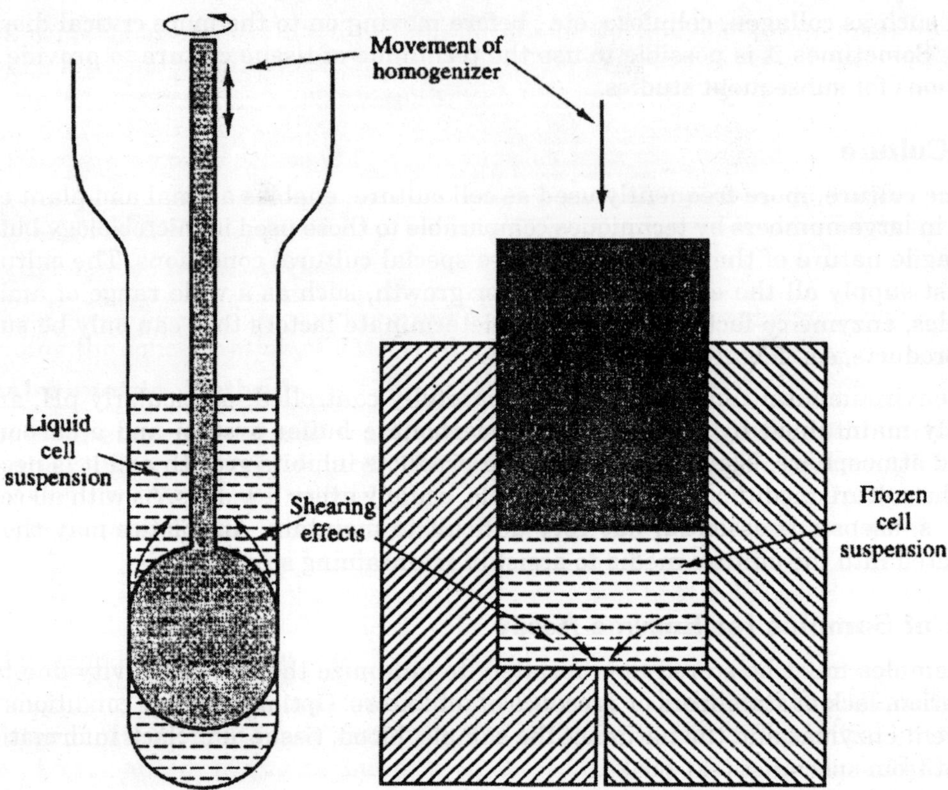

Fig. 4.21. Cell homogenizers. The cell membranes are disrupted by the shearing effects of being forced through narrow apertures. In the Potter homogenizer (*a*) the gap between the plunger and the walls of the tubes is critical in determining the extent of the disruptive effect and in the cell press systems, (*b*) disruption is caused by rapidly forcing a frozen cell suspension through a small hole.

with one of the high speed bladed instruments, usually known as blenders, and while they are generally used to produce cellular dispersions from chopped tissues, they also cause considerable damage to the cells when used at high speeds. They are therefore mainly used in preliminary homogenization of tissue samples prior to their complete cellular disruption using an alternative technique.

For smaller samples of tissue, there are various cylindrical rotating blade instruments which, while causing the general disruption of tissues, can be controlled to produce more specific cellular damage. As well as showing the cutting effects of the rotating blade, such instruments often generate ultrasonic effects due to the castellated design of the blades. Many cells, especially bacteria, yeast, etc., require very drastic measures to disrupt them and various cell disintegrators are available for this purpose.

Ultrasonic vibrations cause the formation of minute bubbles, a phenomenon known as cavitation, which is caused by the extreme variations in pressure generated by the sound waves, although the generation of heat may cause problems unless the samples are cooled frequently during the treatment. An additional hazard lies in the potential damage to the hearing and it is necessary to use some sound-proofing around the instrument. Shearing effects are often used to disrupt cells and probably the most frequently used is the Potter homogenizer, which consists of a glass tube with a close-fitting PTFE pestle which may be either used manually or driven by a variable speed motor.

The clearance between the pestle and the walls of the tube is critical in order to achieve the degree of fragmentation required. A gap of approximately 0.25 mm is frequently used for the preparation of mitochondrial suspensions. Alternative disruptive shearing techniques include forcing a frozen sample through a very small orifice under high pressure or shaking with small glass beads.

Many methods for the fractionation of cellular organelles have been described and it is probably not possible to have a method that is equally effective at preserving each individual component. For each type of cell and organelles the most appropriate method must be selected and evaluated.

Sub-cellular Enzymes

It is often necessary to assess the efficiency of cell fractionation procedures. Electron microscopy of the prepared fractions is very informative but gives no quantitative indication of the purity of the fraction. It is often easier to measure the relative concentrations of marker enzymes in each fraction. These marker enzymes are assumed to be associated with only one of the cellular organelles and hence their relative concentrations in each fraction will give some indication of the degree of cross-contamination in the preparations.

There is considerable debate about the validity of this assumption and more precise information may be gained from specialist texts.

Table 4.9. Organelle markers

Structure	*Enzyme*	*EC number*
Cytoplasm	Lactate dehydrogenase	1.1.1.27
Membrane	Adenylate cyclase	4.6.1.1
Endoplasmic reticulum	Glucose-6-phosphatase	3.1.3.2
Lysosomes	Acid phosphatase	3.1.3.2
Peroxisomes	Catalase	1.11.1.6
Mitochondria		
Inner membrane	Cytochrome *c* oxidase	1.9.3.1
Outer membrane	Amine oxidase	1.4.3.4
Matrix	Glutamate dehydrogenase	1.4.1.3
Nucleus	NMN adenyltransferase	2.7.7.1

SUBSTRATE ASSAY METHODS

Enzymes are extremely useful analytical tools, primarily because of their specificity. Currently a very large number of enzymes are available commercially in various forms of purity and, although their cost is often high, only small amounts are usually required. Procedure 4.8 describes a fixed time method for the quantitation of D-alanine using D-amino acid oxidase.

PROCEDURE 4.8: SPECTROPHOTOMETRIC DETERMINATION OF D-ALANINE USING D-AMINO ACID OXIDASE

Test reaction – catalysed by D-amino acid oxidase

$$CH_3CHNH_2COOH + O_2 \rightarrow CH_3COCOOH + H_2O_2 + NH_3$$

Indicator reaction – catalysed by glutamate dehydrogenase

$$NH_3 + HOOCCH_2CH_2COCOOH + NaDH \rightarrow HOOCCH_2CH_2CHNH_2COOH + NAD$$

Reagetns

Glycylglycine buffer pH 2.3	0.1 mol l^{-1}
2-Oxoglutarate	0.2 mol l^{-1}
NADH	2.5 m mol l^{-1}
Glutamate dehydrogenase	1.5 μkatal or 100 units ml^{-1} in ADP solution (0.5 m mol ml^{-1})
D-amino acid oxidase	1.5 μkatal or 100 units ml^{-1}
Standard solution D-alanine	5.0 m mol l^{-1}

Method

1. Prepare a bulk assay mixture as follows :
   19.0 ml buffer solution
   1.0 ml NADH solution
   1.0 ml 2-oxoglutarate solution
   1.0 ml glutamate dehydrogenase preparation
   1.0 ml D-amino acid oxidase.
2. Into the cuvettes pipette 2.9 ml of this mixture and monitor the absorbance for several minutes to ensure a steady reading or at least a very slow fall in absorbance.
3. To cuvette (BLANK) add 0.1 ml distilled water and note absorbance B_1. To the other cuvette (TEST) add 0.1 ml amino acid solution and again record absorbance T_1.
4. Incubate both cuvettes for 15 min at 37°C and again record absorbance values B_2 and T_2.
5. The amount of NADH oxidized and hence the amount of amino acid is proportional to the absorbance change.
   $(T_1 - T_2) - (B_1 - B_2)$

Calculation

The concentration of D-alanine in the sample (mol l^{-1}) can be calculated from the molar absorption coefficient for NADH :

$$\underbrace{\frac{(T_1 - T_2) - (B_1 - B_2)}{6.22 \times 10^3}}_{A} \times \underbrace{30}_{A}$$

The steps in the calculation are :

A. The use of the molar absorption coefficient for NADH to calculate the concentration change (mol l^{-1}).

B. Corrections for the dilution of the original sample.

End-point Methods

In the most frequently used method of substrate quantitation, the total amount of change resulting from the presence of the substrate is measured and used to calculate the original amount of substrate present. There must be an adequate enzyme activity to ensure that the conversion of substrate to product takes place in a reasonable period of time, but towards the end of a reaction the concentration of the substrate will be very low and the time taken to reach the equilibrium position will be relatively long. The ratio V_{max}/K_m is useful in assessing the amount of enzyme that is required and numerical values of about 1.0 are recommended.

Provided that the concentration of the substrate is less than the K_m value, such conditions will result in a reaction time of just over 3 min for approximately 99% conversion of the substrate. Reactions do not necessarily go to completion and regardless of the amount of enzyme used, the equilibrium position of the reaction will not change. It is important for quantitative measurements that the reaction goes as near to completion as possible and this may be achieved by a variety of methods. The equilibrium position may be altered by changing the pH away from the optimum for the enzyme. For example, the equilibrium position for the reaction in which pyruvate is converted to lactate by lactate dehydrogenase (EC 1.1.1.27) lies very much towards pyruvate at the normal pH of 7.6 but at pH 9.0 the equilibrium is altered towards lactate. In bi-substrate and tri-substrate reactions the use of high concentrations of the second and third substrates will displace the equilibrium of the test reaction towards the formation of the product and the use of a second reaction which results in the regeneration of one of the initial substrates from its product will similarly force the conversion of the test substrate.

The assay of glutamate, for instance, can be achieved by the enzyme glutamate dehydrogenase (EC 1.4.1.3), in which oxoglutarate is the product, and the reaction can be forced almost to completion by incorporating pyruvate and lactate dehydrogenase (EC 1.1.1.27) to regenerate the NAD^+ used in the test reaction. The amount of oxoglutarate formed can subsequently be determined and the initial amount of glutamate calculated. The ADP/ATP systems are also very amenable to this approach using the enzyme pyruvate kinase (EC

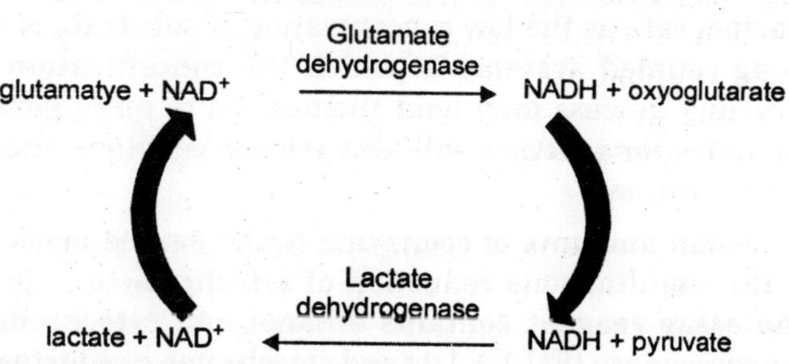

Fig. 4.22. The recycling of NAD^+ in an assay of L-glutamate.

2.7.1.40) and phosphoenolpyruvate to regenerate ATP. End-point assays can also use coupled enzyme systems in which the product of the test reaction provides the substrate for the subsequent auxiliary and indicator reactions.

The choice of pH is less critical than with the measurement of enzyme activity: frequently a compromise pH is used and increased amounts of enzyme used to compensate for any loss in activity. Alternatively, the reaction may be completed in two stages, changing the pH with the careful use of buffer systems for the second stage of the reaction. The quantitation of sucrose uses β-D-fructofuranosidase (invertase) (EC 3.2.1.26) at pH 4.6 and after a suitable period of time the pH is changed to 7.5 for the second stage of the reaction to be completed :

$$\text{sucrose} \xrightarrow{\beta\text{-D-fructofuranosidase}} \text{glucose + fructose} \quad \text{pH 4.6}$$

$$\text{glucose} \xrightarrow{\text{glucose oxidase}} \text{gluconic acid} + H_2O_2 \quad \text{pH 7.0}$$

$$H_2O_2 + \text{chromogen} \xrightarrow{\text{peroxidase}} \text{colour} \quad \text{pH 8.0}$$

Enzymic Cycling

Very low concentrations of substrates may be assayed by recycling the test substrate for an appreciable but definite period of time and measuring the amount of product formed. The coenzyme NADPH, for instance, may be assayed using the two enzymes glutamate dehydrogenase (EC 1.4.1.3) and glucose-6-phosphate dehydrogenase (EC 1.1.1.49):

$$\text{NADPH + oxoglutarate} + NH_3 = \text{NADP}^+ + \text{L-glutamate}$$

$$\text{NADP}^+ + \text{glucose-6-phosphate} = \text{NADPH + 6-phosphogluconate}$$

The reaction mixture contains oxoglutarate, ammonia, glucose-6-phosphate and the two enzymes. The reaction is started by adding the test substrate (NADPH) and allowed to continue for a defined time. The reaction is stopped and the enzymes are inactivated, usually by heat. The amount of 6-phospho-gluconate which has accumulated is determined using an appropriate method and this relates to the amount of NADPH originally present. In this case the product may be determined using glucose-6-phosphate dehydrogenase and monitoring the oxidation of excess NADPH. Such methods do require calibration with known amounts of the test substrate.

Kinetic Methods

Direct kinetic methods comparable to those used for enzyme assays are generally feasible only with automated instrumentation because of the difficulty in measuring the rapidly falling reaction rate as the low concentration of substrate is further depleted. However, kinetic assays using coupled systems in which the concentration of the substrate is held constant by a recycling process may lend themselves to more general use. In such a situation the rate of the indicator reaction will also remain constant and can be related to the original substrate concentration.

Small amounts of coenzyme NAD^+ can be measured by coupling its reduction to NADH to the simultaneous reduction of cytochrome *c*, a process that can be monitored at 550 nm. The assay reagent contains ethanol and cytochrome *c*, together with the enzymes alcohol dehydrogenase (EC 1.1.1.1) and cytochrome c reductase (EC 1.6.99.3). On initiating the reaction with the sample containing NAD^+, a steady state rapidly develops in which the rate of reduction of cytochrome *c* reflects the concentration of NAD^+. Direct kinetic assays are the only valid methods for the measurement of activators and inhibitors and calibration plots of the percentage activation or inhibition by known amounts of the substance can be made.

Examples of inhibition assays include the quantitation of organophosphorus pesticides using the inhibition of cholinesterase (EC 3.1.1.7) while manganese can be measured in amounts as low as 1×10^{-12} mol using its activating effect on isocitrate dehydrogenase (EC 1.1.1.41).

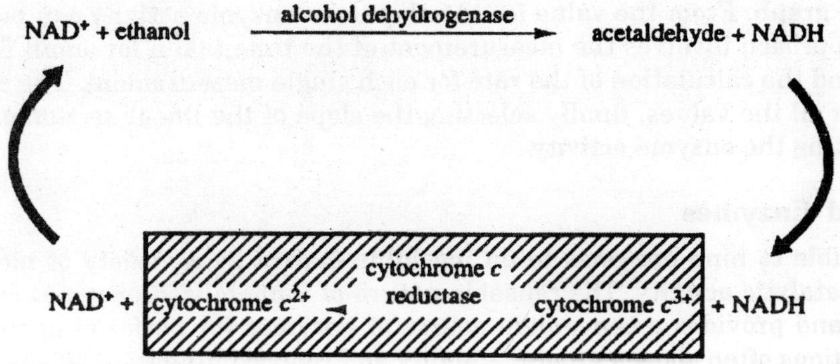

Fig. 4.23. A kinetic assay of NAD^+. The rate of increase in absorbance at 550 nm as cytochrome *c* is reduced is a measure of the steady-state concentration of NAD^+.

Automated Analysis

There are many instruments designed for either enzyme assays or substrate assays using enzymes. Information on the analytical capabilities of these instruments will be supplied by the manufacturers. This will often include protocols for specified assays using kits of commercially available pre-prepared reagents. These may be in liquid or dry form and may, for substrate assays, include immobilized enzymes. The facility to be able to develop additional automated methods on a particular instrument will depend upon its design and some instruments are dedicated solely to specified analyses. Many automated instruments measure enzyme activity using fixed time colorimetric methods.

Some, however, can be classed as reaction rate analysers, *e.g.* the centrifugal analysers, and these instruments determine the reaction rate from either the initial slope of the reaction curve or from repeat measurements at fixed intervals. In both methods the slope of the line is taken to represent the activity of the enzyme. The main technical function of a reaction rate analyser is the automated addition and mixing of the reagents and sample under carefully controlled temperature conditions and the monitoring of the reaction immediately after initiation. The data generated are presented in various forms but all involve some degree of computerization.

The most versatile types of instrument are those using spectrophotometrie detection systems and using NADH-linked assays. Some instruments are dedicated to specific methodologies which often involve ion-selective electrodes or commercially prepared reagents and as a result may be restricted for general use. The presence of a lag period in many coupled assays and difficulties in determining the linear portion of a curve present the main problems in the calculation of enzyme activity using reaction rate analysers. In the simplest instruments the slope of the curve in the first few seconds of the reaction is extrapolated into a straight line or, if the reaction is known to show a lag period, the rate of reaction after a defined period of time can be measured.

The more sophisticated instruments use microcomputers to determine the linear portion of the curve and calculate the enzyme activity directly from the slope. The second derivative of the reaction progress curve (rate of change of the slope) can be monitored by the computer and when a value of zero is held for a period of time (10–15 seconds) this indicates a linear section of the graph. From the value for the slope, the enzyme activity can be calculated. An alternative approach involves the measurement of the time taken for small fixed changes in absorbance and the calculation of the rate for each single measurement. The computer stores and compares all the values, finally selecting the slope of the linear section of the curve and again calculating the enzyme activity.

Immobilized Enzymes

It is possible to bind enzymes to an insoluble matrix by a variety of methods and still retain their catalytic activity. The reusable nature of immobilized enzymes can significantly reduce costs and provides a convenient source of enzymes for performing substrate assays. Such preparations often show a greater stability and reduced inhibition effects than do soluble enzymes, although occasionally optimum pH values may be altered slightly. Covalent linking to a variety of polymers using non-essential amino acid residues is a convenient method of immobilization.

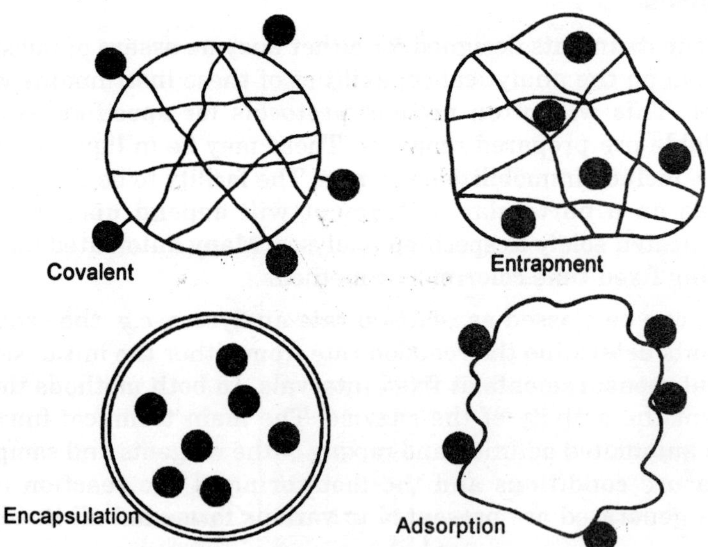

Fig. 4.24. General methods of immobilizing enzymes.

Enzyme inactivation during the process can be appreciably reduced by the presence of a competitive inhibitor of the enzyme. Physical immobilization techniques are less drastic than covalent bonding but some changes in the properties of the enzyme may occur, *e.g.* pH optima and activation effects. Enzymes may be adsorbed on a variety of polar materials such as charcoal, silica gel, alumina and ion-exchange resins and although this is a simple technique, a major problem lies in the ease with which enzymes may be desorbed as a result of variations in pH and buffer concentration, etc. The formation of a cross-linked polymer, often polyacrylamide, in the presence of the enzyme results in trapping of the enzyme molecules within the matrix of the polymer but still permits the access of small substrate molecules to the enzymes. Alternatively, enzymes may be encapsulated by the production of an emulsion of an aqueous solution of the enzyme in an organic solvent.

The droplets are stabilized by the addition of a suitable polymer which, after forming a membrane around the droplets, is allowed to harden and the capsules can then be washed free from the other reagents. Suitable semipermeable membranes can be produced with a variety of reagents including collodion and polystyrene.

Immobilized enzymes may be used in affinity chromatographic methods but their use as catalysts may be in either the production or removal of compounds in chemical processes or as analytical tools. Many substrate assays can be performed using enzymes immobilized on a variety of surfaces, *e.g.* glass beads, plastic or nylon tubing. Alternatively they may be incorporated into gel or microparticulate layers on dry strips or slides. Immobilized enzymes used in conjunction with ion-selective electrodes provide very convenient methods of analysis.

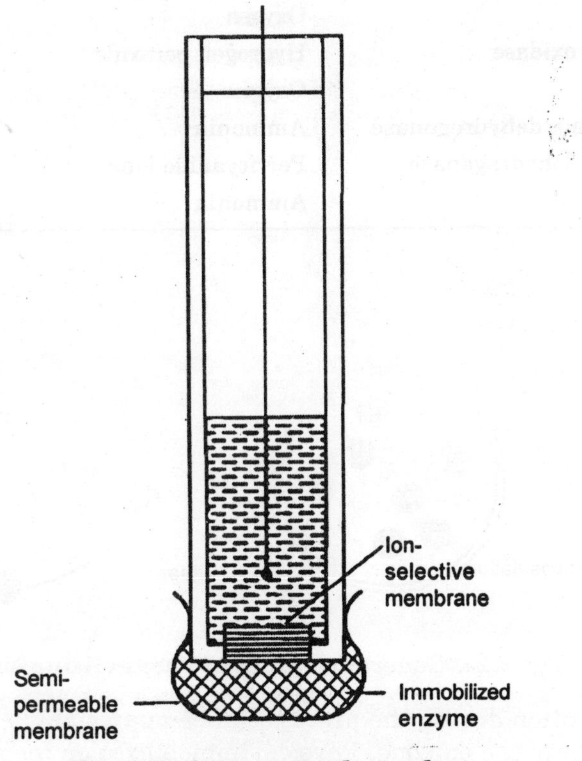

Fig. 4.25. An enzyme electrode.

The immobilized enzyme may be held in a gel or membrane around the electrode and the substance to be measured diffuses into the enzyme gel. Its conversion to the product alters the ionic equilibrium across the ion-selective membrane. It is important that the enzyme layer is thin, to minimize any problems caused by slow diffusion rates through the layer. The main advantage of enzyme electrodes lies in the simplicity of the method of analysis although the presence of interfering ions may present a problem, particularly when using cation-sensitive electrodes. Methods of enzyme electrode construction are discussed in the section on biosensors.

Substrate determinations, using enzyme electrodes must be performed under controlled conditions of temperature and pH and if standard solutions are used to calibrate the electrode they must be analysed at the same time. The response time of some enzyme electrodes may be several minutes and although for many the time is shorter this factor must be considered in the design of an assay method.

Table 4.10. Substrate assays using enzyme electrodes

Substrate	*Enzyme*	*Substrate or product measured*	*Electrode system*
Amino acids	D-and L-amino acid oxidases	Hydrogen peroxide	Platinum
		Ammonia	Cation
Ethanol	Alcohol oxidase	Hydrogen peroxide	Platinum
		Oxygen	Oxygen
Glucose	Glucose oxidase	Hydrogen peroxide	Platinum
		Oxygen	Oxygen
L-Glutamate	Glutamate dehydrogenase	Ammonia	Cation
Lactic acid	Lactate dehydrogenase	Ferricyanide ions	Platinum
Urea	Urease	Ammonia	Cation

CHAPTER

5 Ethyl Alcohol in Industry

Microorganisms are biocatalysts actively used in the fermentation process where product and biomass are obtained from fermentable sugars. The cells are harvested and reused. The immobilisation of cells has been practised in the production of vinegar. Enzymes and whole cells can be bound to solid support, fixed in the form of an active layer. When substrate passes over the surface, enzymatic reactions change the substrate to the desired product. Enzymes and suitable microorganisms are used in manufacturing food flavours, additives, medicines, and other goods are produced by variety of microbial metabolites. Immobilised cells may perform differently to an equivalent mass of freely suspended cells. They may produce extracellular polymers, which are usually lumped together with cells in dry biomass measurements, resulting in overestimation of the activity of immobilised biomass.

In addition, cell location in the form of biofilm can affect its quality and activity, because profiles of environmental conditions such as effect of substrate concentration often exit. However, in many studies results for freely suspended cultures have been used to model immobilised cell bioreactors, usually without a lot of attention in assessing the quality of the biomass in the system. Typically these models allow for the effect of difficult limitations in and around the biofilm, and assume that the immobilised cells have the same quality and activity is the suspended cell culture, regardless of the position in the biofilm. In fact significant substrate concentration gradients may exist for cells immobilised in biofilm.

Cells located close to the nutrient supply are likely to maintain higher quality and activity compared with cells located relatively further away, leading to differentiation in the quality or activity of the immobilised cell population. This differentiation is more pronounced if there are starvation regions. In practice, zero substrate concentration may exist inside the biofilm, because in these regions the cell physiology may be markedly different from that of the freely suspended cells. The cells activities have been described based on a multi-species biofilm model, and the microbial kinetics by a mathematical model. Using this model predicts that the biomass on the external surface of the biofilm has higher activity than the biomass near the solid support surface, and that condition may occur, after the biofilm has reached a critical dept or formed a thick multi-layers of the biofilm. The model was based on microbial growth kinetics determined by Wang *et al.*

However, no experimental results were presented in either of the immobilization studies to verify the model predictions, nor was the predicted immobilised biomass activity compared with that of freely suspended cells in a comprehensive way. Wang *et al., and* Najafpour *et al.*, worked with immobilised microbial cells of *Nitrobacer agilis, Saccharomyces cerevisiae* and *Pseudomonas aeruginosa* in gel beads, respectively. They found separately that the cells retained more than 90% of their activity after immobilisation by using specific oxygen uptake rate (SOUR) [mg $O_2 \cdot g^{-1}$ (dry biomass) h^{-1}] as the biomass activity indicator. Such differences in immobilised biomass and activity between free and immobilised biomass activities depend strongly on the particular characteristics of the microbial systems and their interaction with the support matrix.

IMMOBILISED MICROBIAL CELLS

Since 1978, several papers have examined the potential of using immobilised cells in fuel production. Microbial cells are used advantageously for industrial purposes, such as *Escherichia coli* for the continuous production of L-aspartic acid from ammonium furmarate. Enzymes from microorganisms are classified as extracellular and intracellular. If whole microbial cells can be immobilised directly, procedures for extraction and purification can be omitted and the loss of intracellular enzyme activity can be kept to a minimum.

Whole cells are used as a solid catalyst when they are immobilised onto a solid support. There are three general methods available for the immobilisation of whole microbial cells: carrier binding, entrapment and cross-linking.

Carrier Binding

As the carrier binding method is based on binding microbial cells directly to water-insoluble carriers. The binding is due to ionic forces between the microbial cells and the water-insoluble carriers. This technique has rarely been used, however, because of lyses during the enzyme reactions. Microbial cells may leak from the carrier, thereby disrupting the immobilisation. Therefore, this method has not been applied successfully.

Entrapping

In applying entrapment, microbial cells are directly entrapped into polymer matrices. This method has been applied to several microorganism by using gelatin, agar, polyacrylamide gel, calcium alginate, etc., as the entrapping agents. The following matrices are used extensively to immobilise microbial cells:

- Collagen
- Gelatin
- Agar
- Alginate
- Carrageenan
- Cellulose triacetate
- Polycarylamide
- Polystyrene

In this method, microbial cells are apparently lysed within the entrapping agent, but they retain the desired enzymatic activity. To prepare an efficient immobilization, the type and concentration of a bifunctional reagent for entrapment should be optimised. The advantage of this method is that cell leakage may act as a diffusion barrier, thus causing difficulty in transferring substrate and product through the matrix.

Cross-Linking

Microbial cells immobilised by cross-linking with bi- or multifunctional reagents such as glutaraldehyde have been reported to be successful, but active immobilisations have not been obtained with other reagents such as toluenediisocyanate. The microbial cell wall is composed of lipoproteins, with lipopolysaccharide extending from the cell membrane. Glutaraldehyde reacts with lysine within the protein in the lipid bilayer of the cell membrane. Furthermore, gelatin is physically absorbed on a solid support, which provides a covalent link between the microbial cells which is tightly bound to a solid support by absorption. Glutaraldehyde has been used as cross-linking reagent for ethanol production.

Maximum achieving rates was reported as high as 7.4 $g \cdot l^{-1} \cdot h^{-1}$ ethanol produced from glucose. The major advantage of the cross-linking method is that the immobilisation reagent does not act as diffusion barrier. Actually, a thin film of cells is provided in this system, making it ideal for many applications.

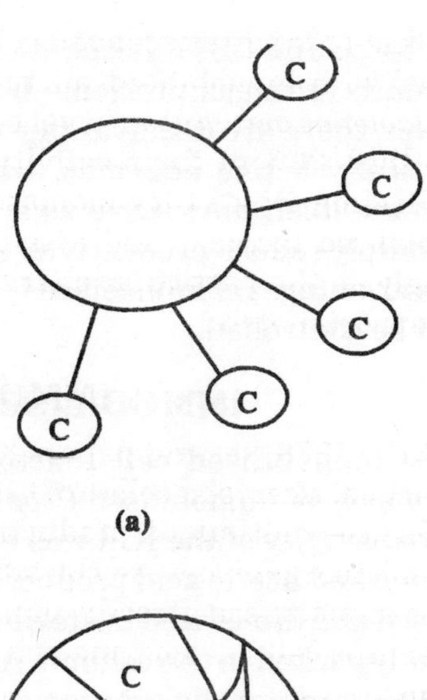

(a)

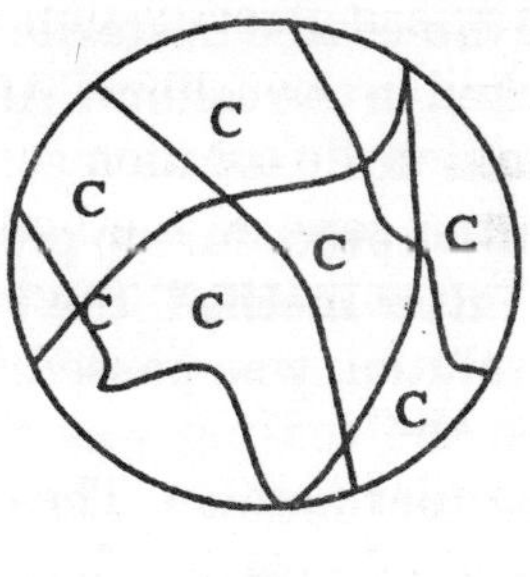

(b)

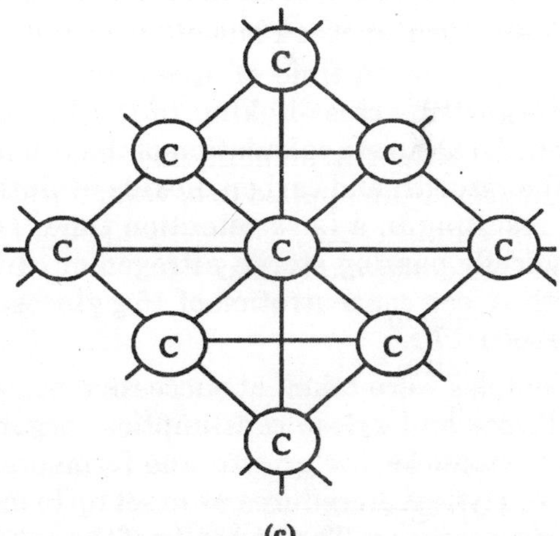

(c)

Fig. 5.1. Alternatives methods used to immobilize cells. (*a*) Carrir binding; (*b*) entrapment; (*c*) cross-linking.

Advantages and Disadvantages of Immobilised Cells

There are several advantages of an immobilised cell system over a batch, or CSTR system. The first and most obvious benefit is the capability of recycling or reusing the microorganisms since they are retained inside the reactor as the product leaves the reactor. Secondly, immobilisation can be easily used for a continuous process maintaining high cell density, thus providing a high productivity. Thirdly, nutrient depletion and any inhibitory compounds do not, in general, have a large effect on the immobilised cells because the cells are fixed by immobilisation.

In batch and CSTR fermentation, nutrient depletion, inhibition and accumulation of toxic by-products are major problems, but immobilised cells are usually unaffected by toxic by-products. However, there are disadvantages to using immobilised cells. The cell may contain numerous catalytically active enzymes, which may catalyse unwanted side reactions. Also, the cell membrane itself may serve as a diffusion barrier, and may reduce productivity. The matrix may sharply reduce productivity if the microorganism is sensitive to product inhibition. One of the disadvantages of immobilised cell reactors is that the physiological state of the microorganism cannot be controlled.

IMMOBILISED CELL REACTOR EXPERIMENTS

The immobilised cell reactor (ICR) experiments were undertaken to determine the performance of immobilised *Propionibacterium acidipropionici* in a plug-flow tubular reactor. The productivity of the ICR was evaluated by measuring glucose and xylose consumption, and propionic and acetic acid production, along the length of the reactor. The Rasching rings were sterilised and dip-coated in sterile 20% gelatin and 1.5% agar n solution. The coated rings were randomly packed in the column. After the coated rings had dried, the packing was sprayed with a 2.5% glutaraldehyde solution.

An alternative procedure to prepare the packing was also tested and proved to be satisfactory. In using the latter method, the reactor column was filled with clean Rasching rings. A hot gelatin agar solution was passed through the column, and allowed to wet the entire packing surface. When the coating was dried, the glutaraldehyde solution was passed through the column in a similar fashion. The column was allowed to stand for 24 hours and then washed with sterile, distilled water. The reactor was sterilised by passing ethylene oxide through the column. The ethylene oxide was allowed to stand in the column for eight hours before the system was purged with sterile nitrogen. The feed and product tubing were autoclaved by using steam at 15 psig. After sterilisation a 24-hour-old seed culture was pumped through the column.

An adaptation time of about 48 hours was allowed for the establishment of a film of microorganisms cross-linking to the Rashing rings. Feed media was then pumped into the reactor. An accurate calculation of the retention times in the reactor was difficult, owing to the fact that growth and carbon dioxide evolution may take a major fraction of void volume of the ICR, resulting in a false retention time. The microbial film thickness could be controlled by periodically passing sterile nitrogen or carbon dioxide through the column, to stop cell overgrowth. A feed concentration of 15 g glucose and 15 g xylose per litre was used over a feed rate of 20-200 ml/hr.

Samples were taken at successive points along the reactor length, and the usual analysis for glucose and xylose consumption, organic acid production and cell density were done. A kinetic model for the growth and fermentation of *P. acldipropionici* was obtained from these data. Analytical procedures were set up to measure cell density, sugar concentration and organic acid concentration. The turbidity of the culture (optical density) was determined by reading the absorbance with spectrophotometer and cell counts. Sugar concentration as a single substrate was measured by an industrial digital analyser. Total sugar was evaluated by a reducing agent such as dinitrosalicylic acid in alkaline solution. An orange colour was produced which was read on a colourimeter at 540 nm. Organic acids were determined by a gas chromatograph, with a flame ionisation detector.

ICR RATE MODEL

Presents the results of the ICR retention time studies, sugar concentration (dual substrate) studies and cell density. The kinetic model for ICR was derived on the basis of a first order reaction, plug flow and steady-state behaviour.

$$r_A = kC_A \quad ...(5.1)$$

Table 5.1. Continuous Fermentation of Dual Substrate (Glucose, Xylose) in ICR at 36° C.

Retention Time (h)	*Length of reactor (inches)*	*Substrate concentration (g l⁻¹)*		*Organic acid concentration (g l⁻¹)*		*Reaction rate of glucose ($g l^{-1} h^{-1}$)*	*Cell density (number of cells / ml)*
		Glucose	*Xylose*	*Propionic acid*	*Acetic acid*		
28	0	15	15	—	—		
	6	6.28	10	1.3	7.78	0.31	9×10^{11}
	14	4.52	8.6	1.98	8.45	0.37	9×10^{11}
	24	2.13	7.2	2.7	12.4	0.46	9×10^{11}
	34	1.36	5.4	3.25	17.2	0.49	9×10^{11}
	44	1.14	3	3.95	19.1	0.495	9×10^{11}
20	0	15	15	—	—		
	6	7.1	11	1.2	6.45	0.395	9×10^{11}
	14	4.72	9.8	1.46	8.1	0.514	9.5×10^{11}
	24	2.21	7.9	1.87	12	0.64	9.5×10^{11}
	34	1.5	7.15	2.12	16.4	0.675	9.5×10^{11}
	44	1.2	5.8	2.4	18	0.69	9.5×10^{11}
12	0	15	1.5	—	—		
	6	9.3	14.5	2.59	1.33	0.475	1×10^{10}
	14	6.02	14	3.26	2.37	0.75	2×10^{10}
	24	3.62	12	4.5	2.45	0.95	3×10^{11}
	34	2.65	13	5.07	2.5	1.03	3×10^{11}
	44	2.3	10	5.36	2.54	1.06	5×10^{11}
5	0	15	15	—	—		
	6	9.8	15	0.5	0.58	1.04	6×10^{10}
	14	7.27	14.5	0.6	0.76	1.55	1.8×10^{10}
	24	6.05	14	0.65	1.2	1.79	5×10^{10}
	34	5.84	14	0.7	1.36	1.83	5×10^{10}
	44	5.2	13.5	0.8	1.65	1.96	6×10^{10}

$$u\frac{dC_A}{dz} = r_A \quad ...(5.2)$$

where r_A is the reaction rate, k is the rate constant, C_A is the sugar concentration, z is the axial reactor length, and u is the bulk fluid velocity. Substituting (5.1) into (5.2) yields.

$$u\frac{dC_A}{dz} = kC_A \qquad ...(5.3)$$

Equation (5.3) is a linear first-order differential equation of concentration and reactor length. Using the separation of variables technique to integrate (5.3) yields.

$$\ln\left(\frac{C_A}{C_{A_o}}\right) = \frac{kz}{u} \qquad ...(5.4)$$

A plot of ln C_A/C_{A_o} as a function of dimensionless reactor length for the fermentation data obtained at a feed concentration of 15 g·l^{-1} glucose and 15 g·l^{-1} xylose at different retention times. A straight line is obtained at each retention time. The value of the slope increased with increasing retention time owing to decreasing velocity through the column Table 5.2 also indicates that the rate constant in the first-order reaction is fixed with different retention times.

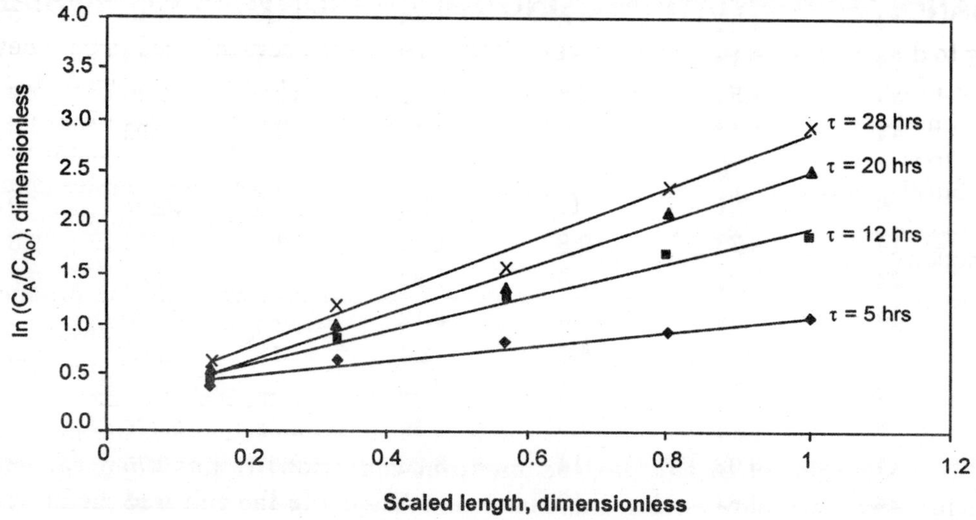

Fig. 5.2. Model test for ICR using *Propionibacterium acidipropionici.*

Table 5.2. ICR Kinetic Model for Immobilized Propionibacterium Acidipropionici.

Retention time τ, hour	*Flow rate ml / h*	*Rate constant h^{-1}*
5	214	0.70
8	135	0.69
12	90	0.68
20	55	0.68
28	38	0.51

The ICR flow rate was five to eight times faster than the CSTR. The overall conversion of sugars in the ICR at a 12 hour retention time was 60%, At this retention time, the ICR was eight times faster than CSTR, but in the CSTR an overall conversion rate of 89% was obtained. At the washout rate for the chemostat, the ICR resulted in a 38% conversion of total sugars. Also, the organic acid production rate in the ICR was about four times that of the CSTR. At a higher retention time of 28 hours, the conversion of glucose in the ICR and CSTR are about the same, but the conversion of xylose reached 75% in the ICR and 86% in the CSTR.

NOMENCLATURE

r_A Reaction rate, $g \cdot l^{-1} \cdot h^{-1}$
k Rate constant, $l \cdot h^{-1}$
C_A Sugar concentration, $g \cdot l^{-1}$
z Axial reactor length, cm
u Bulk fluid velocity, $cm \cdot h^{-1}$.

ETHANOL FERMENTATION USING *SACCHAROMYCES CEREVISIAE*

Owing to diminishing fossil fuel reserves, alternative energy sources need to be renewable, sustainable, efficient, cost-effective, convenient and safe. In recent decades, microbial production of ethanol has been considered as an alternative fuel for the future because fossil fuels are depleting. Several microorganisms, including *Clostridium* sp. and yeast, the well-known ethanol producers *Saccharomyces cerevisiae* and *Zymomonas mobilis,* are suitable candidates to produce ethanol. Microorganisms under anaerobic growth conditions have the ability to utilise glucose by the Embden–Mereyhof–Parnas pathway.

Carbohydrates are phosphorylated through the metabolic pathway; the end products are two moles of ethanol and carbon dioxide. During batch fermentation of *Saccharomyces cerevisiae,* other influential parameters can adversely influence the specific rate of growth, and inhibition can be caused either by product or substrate concentration. The viability of the *Saccharomyces cerevisiae* population, its specific rate of fermentation and the sugar uptake rate are directly related to the desired medium condition. It has been reported by Nagodawithana and Steinkraus, that the addition of ethanol to the cultured media was less toxic for *Saccharomyces cerevisiae* than ethanol produced by cell bodies. This indicates that there are other metabolic by-products that can cause inhibition, and may show impurity of ethanol produced in the fermentation system. Also, the death rates were lower with addition of pure ethanol than the similar condition with the endogenously produced ethanol concentration. Use of biofilm reactors for ethanol production has been investigated to improve the economics and performance of fermentation processes.

Immobilisation of microbial cells for fermentation has been developed to eliminate inhibition caused by high concentrations of substrate and product, also to enhance productivity and yield of ethanol. Recent work on ethanol production in an immobilised cell reactor (ICR) showed that production of ethanol using *Zymomonas mobilis* was doubled. The immobilised recombinant Z. *mobilis* was also successfully used with high concentrations of sugar (12% –15%). The potential use of immobilised cells in fermentation processes for fuel production has been described previously. If intact microbial cells are directly immobilised, the removal of microorganisms from downstream product can be omitted and the loss of intracellular enzyme activity can be kept to a minimum level. Recently, immobilised biomass activity has been given more attention, because it has been acknowledged to play a significant role in bioreactor performance.

Frequently, immobilised cells are subjected to limitations in the supply of nutrients. Thus, because of the presence of heterogeneous materials such as immobilised cells, there is no convective flow inside the beads and the cells can receive nutrients only by diffusion. Immobilisation of cells to a solid matrix is an alternative means of high biomass retention. The cells divide within and on the core of the matrix. Alginate is widely used in food, pharmaceutical, textile and paper products. Alginate is used in these products for thickening, stabilizing, gel and film forming.

Sodium alginate is a linear polysaccharide, normally isolated from many strains of marine brown seaweed and algae, hence the name 'alginate'. The copolymer consists of two uronic acids or polyuronic acid. It comprises primarily D-mannuronic acid (M) and L-glucuronic acid (G). Alginic acid can be either water-soluble or non-water soluble depending on the type of the associated salt. Interchange of sodium ions with calcium ions in the solution may follow solidification of sodium alginate in calcium chloride solution. The sodium salt, other alkaline metals and ammonia are soluble in water, whereas the polyvalent cations salts, *e.g.* calcium, are not water-soluble, except for magnesium ions. The purpose of this research was to obtain high ethanol production with high yields of productivity and to lower the high operating costs.

The ICR column fermenter was used-with *Saccharomyces cerevisiae* by an entrapment technique utilising alginate as a porous wall to retain the yeast cells. The effect of initial glucose concentration on the production of ethanol by *S. cerevisiae* was evaluated. The yield for large-scale ethanol production was compared with the ethanol produced in batch fermentation using the same microorganism.

MATERIALS AND METHODS

Experimental Reactor System

The immobilised cell reactor was a plug flow tubular column, constructed with a nominal diameter of 5 cm, internal diameter of 4.6 cm, Plexiglas of 3 mm wall thickness and 85 cm length. The medium was fed to the ICR column from a feed tank located above the column. A variable-speed Masterflex pump, model L/S Easyload (Cole-Parmer, Vernon Hills, IL, USA) was used to transfer feed medium from a 20 litre polypropylene autoclaveable Nalgene carboy, the carboy serving as reservoir. The effluent from the column was collected in a 20 litre polypropylene autoclaveable carboy serving as product reservoir. A flow breaker was installed between the column and feed pump, which prevented growth of microorganisms and contamination of feed line and feed tank. Samples from the ICR column were taken from the inlet and outlet compartments of the column. A 16 hour culture was harvested at exponential growth phase and mixed with 2% sodium alginate.

The slurry of yeast culture was converted to droplet form while it was dripped into a 6% bath of calcium chloride using a 50 ml syringe. Once the slurry was added to the bath, beads of calcium alginate with entrapped cells were formed. The bed consisted of uniformly packed 5mm beads. The solidified beads were transferred to the column. About 70% of the column was packed. The extra space was counted for bed expansion by the fresh media. The void volume was measured by volume of distilled water pumped through the bed. The packed ICR column was used in continuous mode for the duration of fermentation. The fresh feed was pumped an in upflow manner, while sugar and ethanol concentration was monitored during the course of continuous fermentation. The working volume of ICR after random packing was 740 ml. The bed volume was about 660 ml. The experimental set up of the ICR is shown in figure 5.3. There was no evidence of cell leakage from the beads to the surrounding media, the matrix was permeable to substrate and product, cell growth and glucose conversion were monitored in the ICR. Over-growth of beads after a few days of operation was controlled. Carbon dioxide was purged to eliminate over-growth of beads.

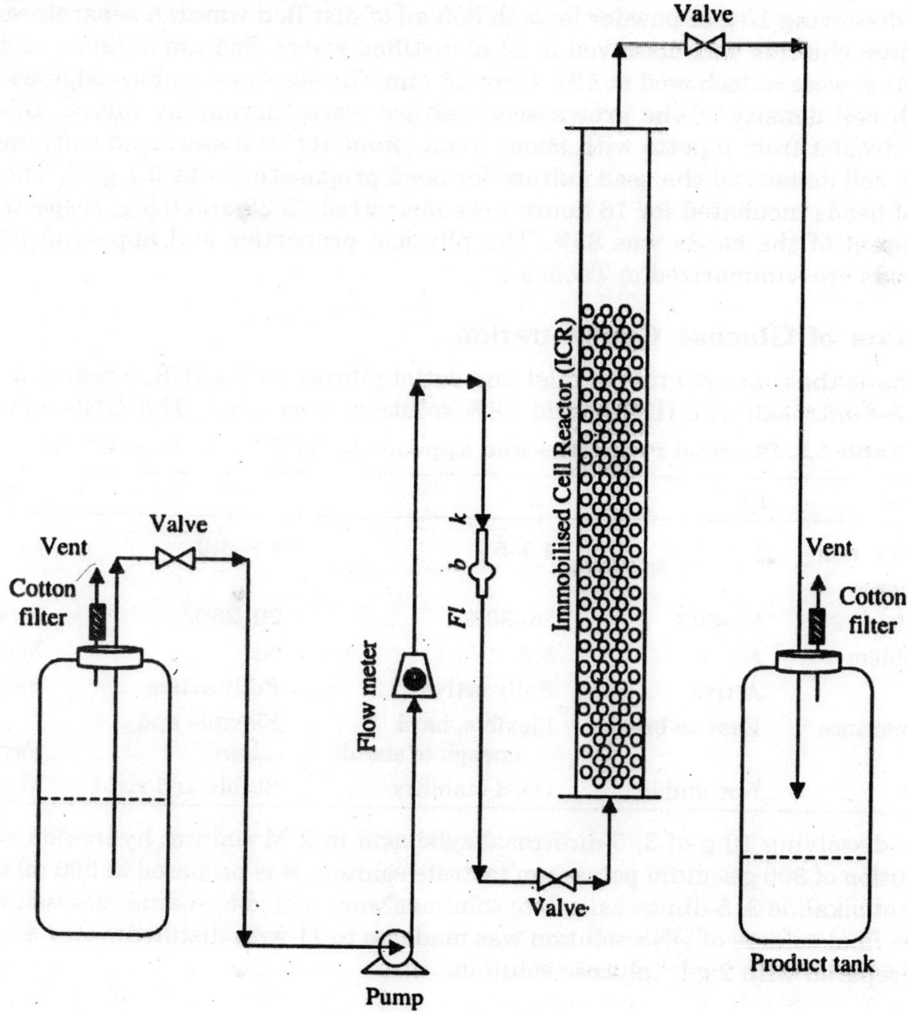

Fig. 5.3. Schematic diagram of ICR experimental setup.

The major over-growth occurred at the entrance region, in about the first 30 cm length of the column where the sugar concentration was very high. High sugar concentrations of 25, 35, 50 and 150 $g \cdot l^{-1}$ were used. Microbial over-growth was controlled with carbon dioxide passed through the bed. There was a maximum 30% increase in the beads' diameter at the lower part of column, where the glucose concentration was maximum. The void volume was measured by passing sterilized water. In addition to the carbon source, the feeding media consisted of 1 g·l^{-1} yeast extract pumped from the bottom of the reactor, while the flow rate was constant for a minimum duration of 24 hours. A seed culture of *S. cerevisiae* ATCC 24860 was grown in a media of 5 g glucose, and 0.5 g yeast extract, respectively, 1.5 g KH_2PO_4 and 2.25 g Na_2PO_4 phosphate buffer up to a total volume of distilled water, 500 ml.

The media was sterilized at 121 °C for 15 min. The stock culture of the microorganisms was transferred to the broth media for preparation of seed culture. Sodium alginate was prepared by dissolving 10 g of powder form in 500 ml of distilled water. A separate solution of 120 g of calcium chloride was dissolved in 21 of distilled water. Sodium alginate and calcium chloride solution were autochaved at 121 °C for 15 min. the sterilized sodium alginate solution and the high cell density of the grown seed culture were thoroughly mixed. Beads were prepared by droplet from pipette with about 5 mm diameter in a sterilized calcium chloride solution. The cell density of the seed culture for bead preparation was 3.1 g·l^{-1}. The wet and dry weight of beads incubated for 16 hours were measured : 3.28 and 0.5 g, respectively. The moisture content of the beads was 85%. The physical properties and appearance of the *S. cerevisiae* beads are summarized in Table 5.3.

Determinaiton of Glucose Concentration

To determine the concentration of inlet and outlet glucose in the ICR, a reducing chemical reagent, 3, 5-dinitrosalicylic (DNS) acid 98% solution, was used. The DNS solution was

Table 5.3. Physical Properties and Appearance of S. Cerevisiae Beads

Alginate, wt %	*1.5*	*2*	*3*	*6*
Beads diameter, mm	5	4.9–5	4.8–4.9	4.5
Growth, diameter expansion after 72 h	50–60%	25–30%	20–25%	No expansion
Diffusion problem	Nil	Nil	Nil	May exists
Cell activity	Active	Fully active	Fully active	Partly active
Physical appearance	Easy to break	Flexible, hard enough to stand	Flexible and hard	Very hard
Stability	Not stable	Good stability	Stable and rigid	Very rigid

prepared by dissolving 10 g of 3, 5-dinitrosalicylic acid in 2 M sodium hydroxide solution. A separate solution of 300 g sodium potassium tartrate solution was prepared in 300 ml of distilled water. The hot alkaline 3, 5-dinitrosalisylate solutions were added to sodium potassium tartrate solution. The final volume of DNS solution was made up to 11 with distilled water. A calibration curve was prepared with 2 g·l^{-1} glucose solution.

Detection of Ethanol

Ethanol production in the fermentation process was detected with gas chromatography, HP 5890 series II (Hewlett-Packard, Avondale, PA, USA) equipped with a flame ionisation detector (FID) and GC column Porapak QS (Alltech Associates Inc., Deerfield, IL, USA) 100/120 mesh. The oven and detector temperature were 175 and 185 °C, respectively. Nitrogen gas was used as a carrier. Isopropanol was used as an internal standard.

Yeast Cell Dry Weight and Optical Density

In batch fermentation, approximately 2ml sample was harvested every 2 hours. The absorbance of each sample during batch fermentation was measured at 620 nm using spectrophotometer, Cecil 1000 series (Cecil Instruments, Cambridge, England). The cells dry weight was obtained using a calibration curve. The cell dry weight was proportional to cell turbidity and absorbance at 620 nm. The cell concentration (dry weight) of 2.1 $g \cdot l^{-1}$ was obtained from the 24 hours culture broth, the free cell samples with absorbance of 1.6 at 620 nm.

Electronic Microscopic Scanning of Immobilised Cells

For electronic micrographs, samples were taken from fresh beads and 72 hour beads from the ICR column. The samples were dipped into liquid nitrogen for 10 minutes, then freeze dried for 7 hours (EMITECH, model IK750, Cambridge, UK). The sample was fixed on an aluminium stub and coated with gold-palladium by a Polaron machine model SD515 (EMITECH, Cambridge, UK) at 20 nm coating thickness. Finally the sample was examined under scanning electron microscope (SEM) by using a Stereoscan model S360 (SEM-Leica Cambridge, Cambridge, UK).

Statistical Analysis

The size of beads was uniform and consistent, the mean size of beads with 3% alginate and based on measurement of 20 samples; the mean value for the beads' diameter was 4.85 mm, with a standard deviation of 0.3 mm and calculated variance of 0.1 mm. The standard deviation was less than 5%. The data for the batch fermentation experiment with 50 $g \cdot l^{-1}$ glucose are presented they were replicated three times. The data for ICR with 25, 35, 50 and 150 $g \cdot l^{-1}$ glucose that were conducted at a wide range of flow rates are shown in figure 5.5 to 5.9, which were repeated in additional runs. The standard deviations of collected data for the batch experiments were approximately 5%, and the standard deviation for the ICR experimental data with a sugar concentration of 50 $g \cdot l^{-1}$ was approximately 10%. Data were analysed in a spreadsheet, Microsoft Excel 2000. The error analysis for the ICR data with a substrate concentration of 150 $g \cdot l^{-1}$ was slightly higher but it was in the range of 10-12%.

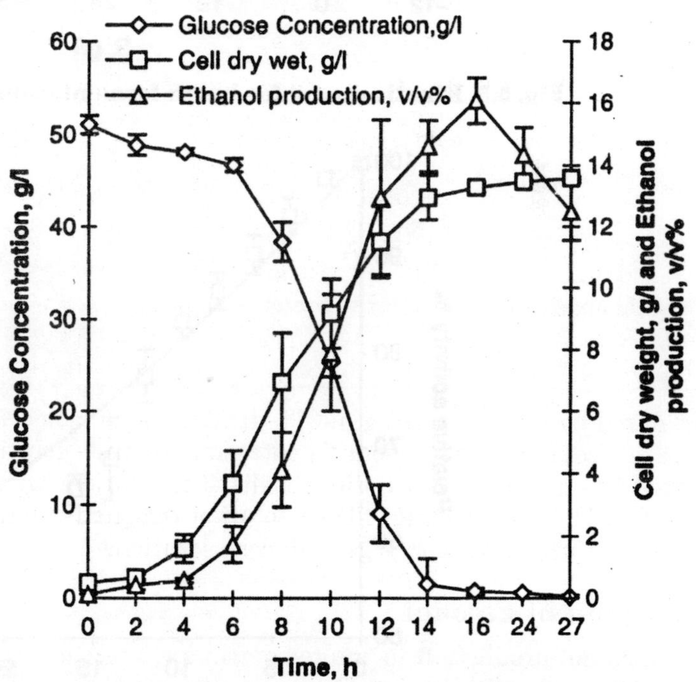

Fig. 5.4. Glucose concentration, cell density and production of ethanol in batch fermentation with intial concentration of 50 $g \cdot l^{-1}$ glucose versus time.

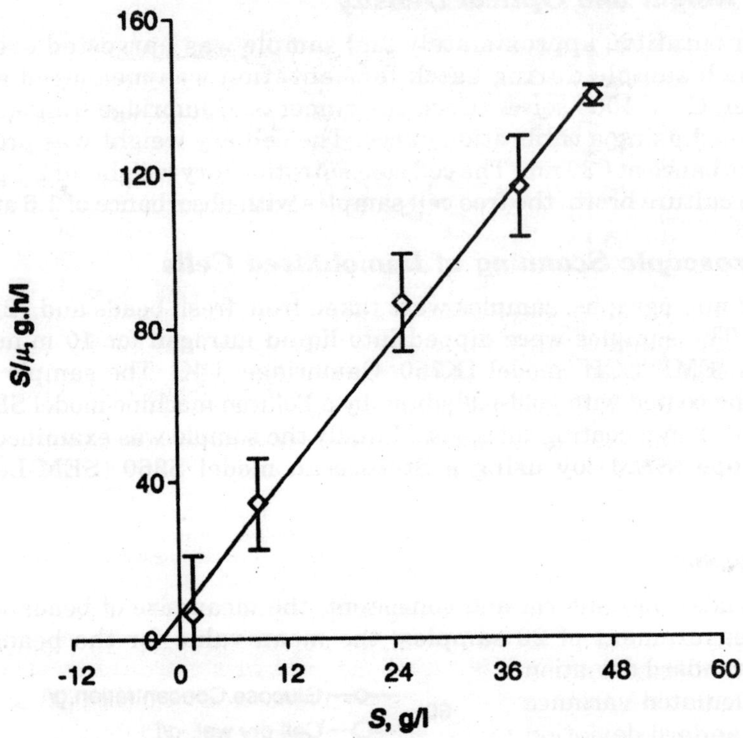

Fig. 5.5. Kinetic model for batch fermentation, Langmuir–Hanes plot.

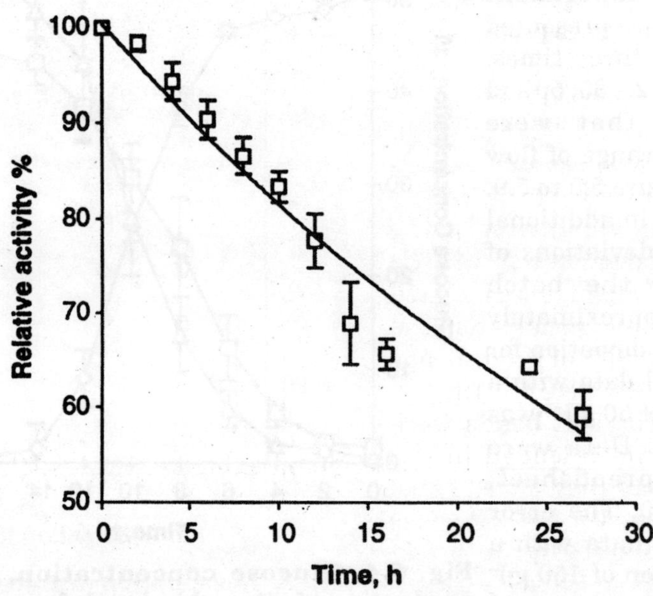

Fig. 5.6. Relative activities of *S. cerevisiae* in batch fermentation.

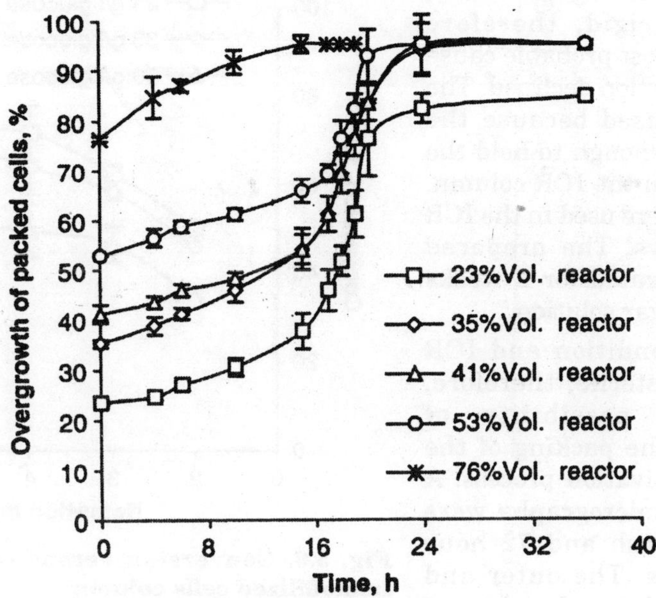

Fig. 5.7. Percentage, growth of immobilized cells with various initial beads loading.

RESULTS AND DISCUSSION

Evaluation of Immobilised Cells

In preparing immobilized cells, 1.5, 2, 3 and 6% alginate was used. By pressing them manually, the hardness and rigidity of the beads were tested. The physical criteria of the prepared beads summarized in Table 5.1. The suitable alginate concentration based on activity of the beads for ethanol production was 2%. The weight percentage of alginate was related to substrate and product penetration the beads and return to bulk of fluid. Beads with low alginate (1.5%) were too soft and easily breakable. The soft beads were pressed once they were loaded in the ICR column. Therefore they were unable to be used successfully; also the soft beads faced problems such as overgrowth and expansion of their diameter when grown in sugar solution. The high alginate

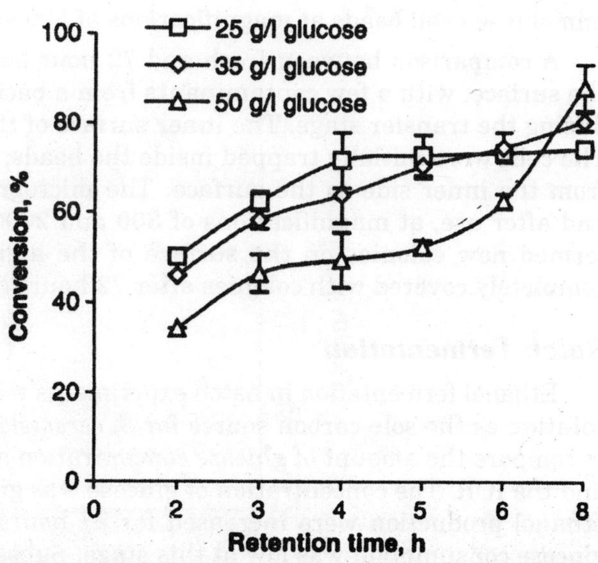

Fig. 5.8. Consumption of glucose in the immoblised cell column.

beads (6%) were very hard and almost unbreakable by pressing manually. They were very rigid, therefore diffusion was the most probable cause since ethanol production declined. The 2% alginate was used because the beads were strong enough to hold the weight of packing in the ICR column. The packed beads were used in the ICR column for 10 days. The prepared beads were refrigerated for 2 weeks and activated in sugar solution.

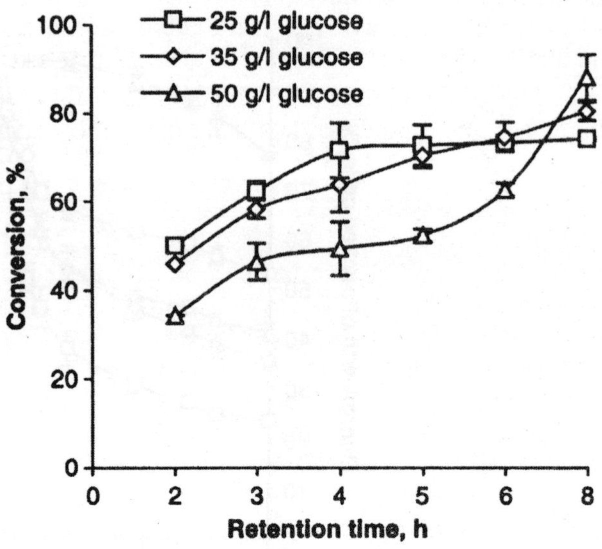

Fig. 5.9. Conversion versus retention time in the immobilized cells column.

The storage condition and ICR column were non-sterile; therefore, there was the possibility of contamination in the packing of the ICR and in the activation process. A series of electron micrographs were taken from the fresh and 72 hour immobilised beads. The outer and inner surfaces of the beads are respectively. These micrographs were used as a comparative indicator for yeast growth on the surface of the solid support (2% calcium alginate). There was no apparent leakage of cells from the beads into the bulk of the fluid. It was also apparent that by 72 hours in the ICR, the yeast was growing on the outer surface. Therefore the active sites were potentially available for ethanol production without diffusion problems. The outer surface of fresh and 72 hour immobilised cell beads at magnifications of 300 and 2000 can easily be seen.

A comparison between fresh and 72 hour beads indicates that the cells were present on the surface, with a few contaminants from a bacillus-type organism, This may have occurred during the transfer stage. The inner surface of the beads before and after use was compared. The cells were initially trapped inside the beads; after 72 hours the cells apparently migrated from the inner side to the surface. The micrographs of the inner sides of the beads before and after use, at magnifications of 300 and 2000. After 72 hours the cells appeared to have formed new colonies on the surface of the alginate layer. By contrast, the surfaces were completely covered with colonies after 72 hours of ethanol production in the ICR.

Batch Fermentation

Ethanol fermentation in batch experiments was carried out in triplicate with 50 $g{\cdot}l^{-1}$ glucose solution as the sole carbon source for *S. cerevisiae.* The purpose of the batch experiment was to compare the amount of glucose concentration and ethanol production in batch fermentation and the ICR. The concentration of glucose was gradually decreased while the cell density and ethanol production were increased for 27 hours. There was a lag phase of 4 –6 hours; the glucose consumption was low at this stage. Subsequently the concentration profile drastically decreased during batch fermentation after 8 – 16 hours. Since the cell density was initially low, sugar consumption was also low.

The resulting cell growth curve from the batch experiment was a typical sigmoidal (S-)shape. The maximum cell density of S. *cerevisiae* in batch fermentation was 13.7 g·l^{-1} with 50 g·l^{-1} glucose concentration. The average yield of biomass and product on substrates ($Y_{X/S}$ and $Y_{P/S}$) were 33 and 32%, respectively. The rate of ethanol productivity for 24 hours was 1.4 g·l^{-1} · h^{-1}.

A Langmuir-Hanes plot based on the Monod rate equation is presented in figure 5.5. The Monod kinetic model can be used for microbial cell biocatalyst and is described as follows :

$$\frac{S}{\mu} = \frac{K_s}{\mu_m} + \frac{S}{\mu_m} \qquad \text{...(5.5)}$$

The terms K_s and μ_m are defined as the Monod constant and maximum specific growth rate, respectively. The data generated in this study were linearly fitted with the model, as a function of concentration produced during the exponential phase versus time. From the plot, the maximum specific growth rate and Monod constant were determined to be 0.35 h^{-1} and 2.23 g·l^{-1}, respectively. The large value of the Monod constant may suggest that at high concentrations of substrate, more influence on the cell-substrate complex [C.S] dissociation occurred than [C.S] formation.

Relative Activity

The economics of an immobilised cell process depend on the lifetime of the microorganism and a continued level of clean product delivered by the fixed cells. It is important to eliminate the free cells from the downstream product without the use of any units such as centrifuge or filtration processes. Since the cells are retained in the ICR, the activity of intracellular enzymes may play a major role. It is assumed that the deactivation of the enzyme at constant temperature follows a first-order equation as shown below.

$$A_t = A_0 \exp(-k_d t) \qquad \text{...(5.6)}$$

where A_t and A_0 are activities of enzymes at time, t and initial time zero, respectively. Also, k_d is the dissociation constant. The plot of relative activity versus time in batch fermentation with free cells is shown in figure 5.6. The value of k_d was 0.36 h^{-1}. According to the first-order dissociation rate constant, the half-life of the *S. cerevisiae*, (τ_d = ln $2/k_d$), in batch fermentation suspended cells with 50 g·l^{-1} of glucose was 1.95 h. The free cells were apparently completely deactivated after 60 hours in batch operation. The data indicate that the free cells were deactivated rapidly compared with immobilised cells when used in a continuous mode for more than 7 days.

Reactor Set-up

The volume of reactor without beads was 1.41. The column was loaded with the solidified uniform beads of *S. cerevisiae*. The void volume of the reactor was 660ml when it was packed with immobilised beads. The growth of beads with different proportions of column packing is shown in figure 5.7. A fresh feed of 10 g·l^{-1} glucose solution was pumped from the bottom of the reactor. The optimum amount of packing obtained was 65-70% of the reactor volume. The trend of the collected data resembles the growth curve of yeast in suspended-cell culture. The diameters of the beads increased with the increase in *S. cerevisiae* cell density, which indicates that S. *cerevisiae* was growing within the solid support. Under these circumstances substrate would be easily consumed at the solid surface coated with immobilised yeast cells.

Thus, it was expected that substrate concentration at the surface would be less than the concentration of substrate in the bulk fluid. The main objective was to determine whether substrate penetrated into the beads.

Since the matrix of beads was quite porous, it was assumed that the concentration gradient was the major force that influenced the mass transfer process in immobilised cells of *S. cerevisiae*. Therefore, the immobilised yeast system was preferred compared with free cells in the solution. Moreover, economic aspects of immobilised yeast must be considered as it eliminates the need for the extra unit for free-cell removal from the product stream. The other advantages of the immobilisation system are that the substrate may not accumulate on the surface of the beads and that there was no evidence of cell leakage from the beads.

Effect of High Concentration of Substrate on Immobilised Cells

The fermentation was performed with various sugar concentrations to increase product concentrations. The initial sugar concentrations were 25, 35 and 50 g·l^{-1}. The sugar consumption profile in the ICR is presented. The sugar consumption trends of various glucose concentrations were similar, with a sharp reduction of substrate occurring within the first 3 hours. A range of 55 – 75% of the sugars was reduced within a 3 hour retention time. A 6 hour retention time indicated that this was the most suitable time to achieve high sugar consumption. A longer retention time was required for higher sugar concentrations of up to 150 g·l^{-1} in the ICR column (7 hours).

The amount of cell immobilized with calcium alginate was determined by the cell dry weight of immobilised cells and the assumption that 98% of the beads were considered to be loaded with active cells. Beads were measured before loading in the ICR column. Conversion of glucose versus dilution rate, used in the continuous fermentation process with immobilised *S. cerevisiae* is shown in figure 5.9.

As the sugar concentration increased the conversion may have decreased. At very high sugar concentrations, the conversion of sugar decreased. The maximum sugar conversion at 6 hours' retention time was obtained from 74.3, 80.5 and 88.2% for 25, 35 and 50 g·l^{-1} glucose concentrations, respectively. At high concentrations of sugar, conversion was highly influenced by dilution rate. The conversion decreased to less than 10% once the dilution rate approached wash out. The sugar conversion at low sugar concentrations appeared to be much

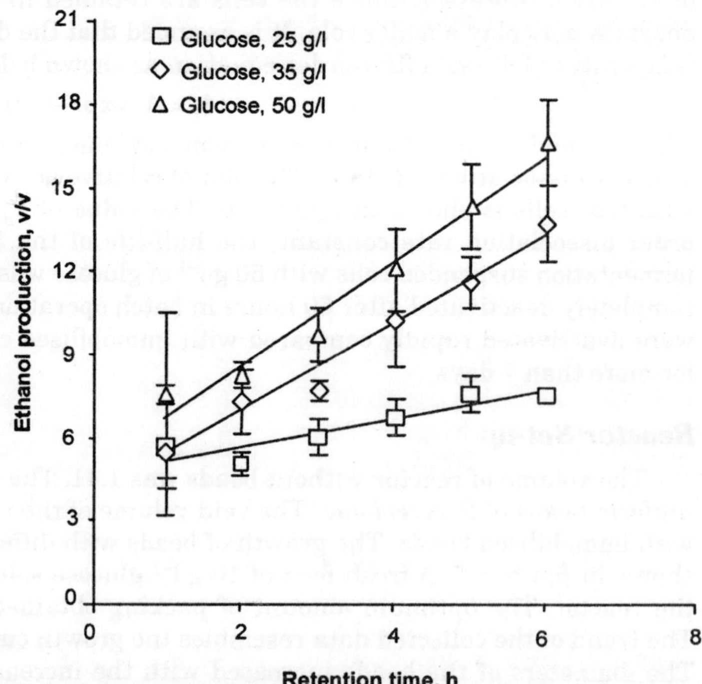

Fig. 5.10. Ethanol production versus retention time in the immobilised cell column.

less sensitive to feed dilution rate. The conversion for 25 $g \cdot l^{-1}$ glucose was constant at 74% for a retention time of greater than 3 hours. The trend of the data with 50 $g \cdot l^{-1}$ sugar showed that conversion increased with retention time.

At a higher sugar concentration, it required a longer retention time to achieve higher conversion. Reactor productivity was obtained by dividing final ethanol concentration with respect to sugar concentration at a fixed retention time. It was found that the rates of 1.3, 2.3 and 2.8 $g \cdot l^{-1} \cdot h^{-1}$ for 25, 35 and 50 $g \cdot l^{-1}$ glucose concentrations were optimal. Ethanol productivities with various substrate concentrations were linearly dependent on retention time. The proportionality factor may have increased while the substrate concentration increased. As the sugar concentration doubled, the slope of the line for ethanol productivity with 50 $g \cdot l^{-1}$ sugar increased fivefold. These results indicate that the ICR column has a high capacity to produce very high concentrations of ethanol.

The final ethanol concentrations with 25 and 50 $g \cdot l^{-1}$ of glucose were 7.6 and 16.73 v/v, respectively. A high glucose concentration of 150 $g \cdot l^{-1}$ was used in continuous fermentation with immobilised *S. cerevisiae;* the obtained data for sugar consumption and ethanol production with retention time are shown in figure 5.11. As the retention time gradually increased the glucose concentration dropped, while the ethanol concentration profile showed an increase. The maximum ethanol concentration of 47 $g \cdot l^{-1}$ was obtained with a retention time of 7 hours. The yield of ethanol production was approximately 38% compared with batch data, where only an 8% improvement was achieved.

Continuous ethanol production in an ICR was successfully done with high sugar concentrations. In batch fermentation, when the concentration of glucose was 50 $g \cdot l^{-1}$, substantial substrate inhibition occurred. The advantage of the ICR was that the inhibition of substrate and product were not apparent even with a 150 $g \cdot l^{-1}$ glucose solution in the fresh feed (data not shown). The ICR system exhibited a higher yield of ethanol production (38%) than the batch system. The ICR column gave a high performance to processing feed with concentrated sugar. Most of ICR experimental runs resulted in a glucose consumption of 82-85%.

The results indicate that the immobilisation of *S. cerevisiae* possesses the capacity not only to utilise high concentrations of sugar but also to yield higher ethanol productivities during the course of continuous fermentation.

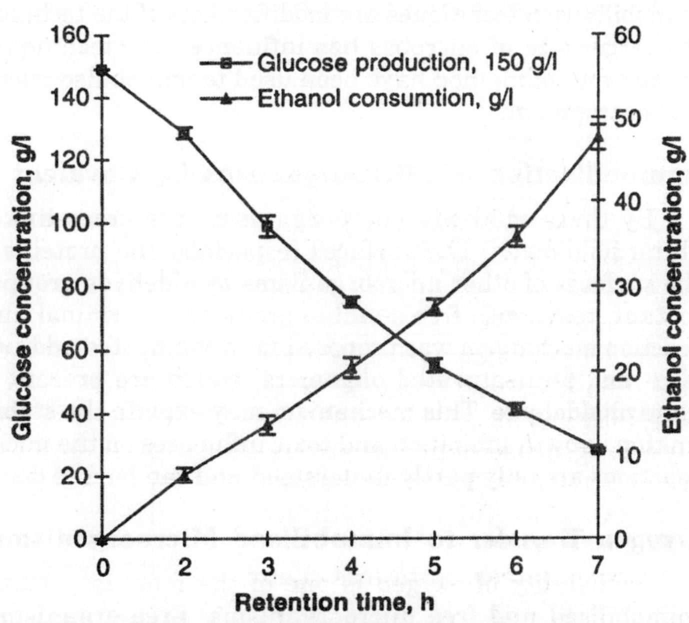

Fig. 5.11. Glucose concentration and ethanol production versus retention time in immobilised cell reactor with initial substrate concentration of 150 $g \cdot l^{-1}$ glucose.

Ethanol production in the ICR column increased by fivefold, as the glucose concentration was doubled from 25 to 50 g·l^{-1}. It is clear: the new findings in the present investigation would be the application of concentrated feed with a higher rate of ethanol production, as the cell loaded into the gel matrices of sodium alginate shown in the SEM micrographs had eliminated the free cells from the ethanol product stream.

Nomenclature

S Substrate concentration, g·l^{-1}

C Microorganisms cell concentration, g·l^{-1}

K_S Monod constant, g·l^{-1}

μ Specific growth rate, g·l^{-1} · h^{-1}

μ_m Maximum specific growth rate, h^{-1}

T Time, h

A_t Activity of bacterium at time *t*, U/g cell

A_o Initial activity of bacterium, U/g cell

k_d Dissociation constant, h^{-1}.

FUNDAMENTALS OF IMMOBILISATION TECHNOLOGY

It is well known that pure enzymes change their behaviour and stability when they are immobilised. In the past two decades the immobilisation of microorganisms, cells and parts of cells has gradually been introduced into microbiology and biotechnology. The cell immobilisation techniques are modifications of the techniques developed for enzymes. However, the larger size of microbes has influenced the techniques. As for immobilised enzymes, two broad types of method have been used to immobilise microorganisms: attachment to a support and entrapment.

Immobilisation of Microorganisms by Covalent Bonds

By these methods microorganisms are cross linked by chemical substances, *e.g.* by glutardialdehyde. The surfaces (especially the proteins) of microorganisms are linked with the surfaces of other microorganisms by aldehyde groups of glutardialdehyde. Yeast cells, for instant, react with free ε-amino group or N-terminal amino groups to form imines. Another reaction mechanism was proposed for a conjugated addition of amino groups to double linkages of α- and β-unsaturated oligomers, which are present in commercial aqueous solutions of glutardialdehyde. This mechanism may explain the stability of the linkages. By this chemical linking, growth inhibition and toxic influences on the microorganisms are very intensive. These reactions are only partly understood and can lead to decay or death of the microorganisms.

Oxygen Transfer to Immobilised Microorganisms

Availability of oxygen is one of the most important parameters that are different for immobilised and free microorganisms. Free organisms can get oxygen directly from the surrounding air or, in most technical processes, from the liquid, especially water, which contains dissolved oxygen. The molar transfer of oxygen with respect to time, dO_2/dt, can be described by the following equation:

$$\frac{dO_2}{dt} = K_L a(C_g - C_L) \quad ...(5.7)$$

$K_L a$ is the volumetric oxygen mass transfer coefficient, owing to the oxygen transfer from the gas phase or air, c_g, the surface of the cells, c_x or to the transfer of oxygen dissolved in water to the surface of the cells. In principle, the same formula can be applied to immobilised microoraganisms, but here the conditions are quite different.

The adsorbed cells form microbial films or varying thickness during a short incubation time. Oxygen has to be transferred into these microfilms, and because of this, zones of different oxygen concentration in the films exist, in which the growth of the microorganisms varies in direct relation to the oxygen concentration. By using the unsteady-state model, the maximum oxygen penetration depth for highly packed immobilised cells has been reported to be in the range of 50-2001 μm.

Substrate Transfer to Immobilised Microorganisms

In an immobilised cell bioreactor, significant substrate concentration gradients may exist. Cells are located close to the nutrient supply, which are likely to maintain higher quality and activity than cells located relatively further away, leading to differentiation in the quality or activity of the immobilised cell population. This differentiation is more pronounced if there are starvation regions (practically zero substrate concentration) inside the reactor. Typical approaches for measuring diffusivities in immobilized cell systems include bead methods, diffusion chambers and holographic laser interferometry. These methods can be applied to various support materials, but they are time consuming, making it onerous to measure effective diffusivity (D_{eff}) over a wide range of cell fractions.

Owing to the mathematical models involved, the deconvolution of diffusivities can be very sensitive to errors in concentration measurements. There are mathematical correlations developed to predict D_{eff} as a function of diffusivities in broth and inside the cells (D_0, D_c). The relation can also be used to extrapolate D_{eff} measurements from one cell fraction to any other cell fraction. Various values of D_0/D_c have been compiled in Table 5.4 for a variety of immobilised cell systems. These diffusivity ratios for specific systems (*i.e.* diffusing species, cell type and gel material) facilitate the prediction of D_{eff} values for a wide range of operating conditions.

Table 5.4. Methods of immobilisation

Attachment	
Without support	Aggregation of floc for formation of cross-linking
With support	Covalent binding
	Adsorption to ion-exchangers or inorganic
	Biofilm formation
Entrapment	Organic polymer
	Inorganic polymer
	Semi-permeable membrane

For a few systems, the diffusivity ratios are very large (∞). The infinite diffusivity ratio corresponds to a diffusing solute that does not enter the cells (*i.e.* $D_c = 0$); such cases were observed with large molecules of disaccharide (galactose), polysaccharides and cells of Z. *mobilis*.

Growth and Colony Formation of Immobilised Microorganisms

It was pointed out by several scientists that immobilised microorganisms differ in their growth rates and show altered morphological forms of colonies. Adsorbed cells form micro- and macro-films in which the microorganisms in the outer region have a different morphology than in the inner region. These microfilms often show different colony forms in relation to their density. Thick films have a slimy character, thin film often show the presence of individual microorganisms. These characteristics can be observed with bacteria, yeast and with moulds. They are caused by differing oxygen concentrations and limited concentrations of nutrients.

By referring to our previous work on ICR, a mathematical model for ICR performance may be obtained by applying a mass balance over a differential of the column:

$$\varepsilon\frac{\partial C_A}{\partial t} + u\frac{\partial C_A}{\partial z} = r_A \qquad ...(5.8)$$

where ε is the void volume of the packed column (ml), C_A the substrate concentration ($g{\cdot}l^{-1}$), u the bulk fluid velocity ($cm{\cdot}hr^{-1}$), z the axial reactor length (cm), and r_A the rate of substrate transfer to microbial film ($g{\cdot}l^{-1}\cdot h^{-1}$).

Assuming plug-flow and steady-state behaviour (5.8), reduces to $u\frac{\partial C_A}{\partial z} = r_A$ Plug-flow behaviour has been shown in this type of reactor by the use of tracer studies.

Table 5.5. Microbial Cells Covalently Linked to Various Supports.

Species	*Support*	*Product*
Actobacter	Metal hydroxides	Acetic acid
Aspergillus niger	Glycidyl methacrylate	Gluconic acid
Micrococcus luteus	CM-cellulose	Urocanic acid
Saccharomyces cerevisiae	Aminopropyl silica	Ethanol
Saccharomyces cerevisiae	Hydrocaylkyl methacrylate	Killer toxin
Saccharomyces cerevisiae	Cellulose	Ethanol
Zygosacharomyces lactis	Hydroxyalkyl methacrylate	β-galactosidase

The reaction rate for simple fermentation systems is normally given by the Monod equation. This model indicates that the specific conversion rate is constant when applied to an immobilised cell system. If a first-order rate equation for sugar consumption is used, yields:

$$u\frac{\partial C_A}{\partial z} = kC_A \qquad ...(5.9)$$

Table 5.6. Experimental Studies of Diffusion in Immobilised Cell Systems and their Associated Values of D_0/D_c.

Cell type	*Immobilisation*	*Solute*	D_0/D_c
Saccharomyces cerevisiae	Ca-alginate	glucose	0.1
Baker's yeast	Ca-alginate	glucose	10.6
Ehrlich ascites tumor	agar, collagen	glucose	2.4
Zymomonas mobilis	K-carrageenan, Ca-alginate	glucose	∞
Pseudomonas aeruginosa	Ca-alginate	glucose	2.3
Saccharomycess cerevisiae	Ca-alginate	glucose	2.8
Plant	Ca-alginate	sucrose	∞
Baker's yeast	Ca-alginate	galactose	15.8
Zymomonas mobilis	Ca-alginate	galactose	∞
Baker's yeast	Ca-alginate	lactose	∞
Ehrlich ascites tumor	agar, collagen	lactic acid	6.7
Clostridium butyricum	polyarylamide, agar collagen	hydrogen	3.2
Escherichia coli	natural aggregates	nitrous oxide	3.9
Saccharomyces cerevisiae	fermentation media	oxygen	2.3
Saccharomyces cerevisiae	Ca-alginate, Ba-alginate	oxygen	0.1
Escherichia coli	fermentation media	cxygen	2.2
Penicillium chrysogenum	fermentation media	oxygen	5.6
Bacillus amilaliquefaciens	Ca-alginate, PVA-SbQ gel	oxygen	1.8
Saccharomyces cerevisiae	Ca-alginate	ethanol	0.1
Baker's yeast	Ca-alginate	ethanol	∞

Equation (5.9) is linear first-order differential equation for concentration and reactor length. After separation of variables, the equation can be integrated as

$$\int_{C_{A_0}}^{C_A} \frac{dC_A}{C_A} = \int_0^z \frac{kdz}{u} \qquad \text{...(5.10)}$$

Integration of the above differential equation yields:

$$\ln(C_A/C_{A_0}) = \frac{kz}{u} \qquad \text{...(5.11)}$$

Thus a linear relation between ln (C_A/C_{A_0}) and the reactor length should exist if the model accurately describes the immobilised cell reactor. The experimental data fitting the model was discussed earlier.

Immobilised Systems for Ethanol Production

The most significant advantage of immobilised cell systems is the ability to operate with high productivity at dilution rates exceeding the maximum specific growth rate (μ_{max}) of the microbe. Several theories have been proposed to explain the enhanced fermentation capacity of microorganisms as a result of immobilisation. A reduction in the ethanol concentration in the immediate microenvironment of the organism owing to the formation of a protective layer or specific adsorption of ethanol by the support may act to minimise end product inhibition.

Table 5.7. Ethanol Productivity from Immobilised Systems.

System	*Feed sugar conc* ($g \cdot l^{-1}$)	*% of Feed sugar utilization*	*Dilution rate* (h^{-1})	*Max. ethanol productivity* ($g \cdot l^{-1} \cdot h^{-1}$)
S. Cerevisiae	Glucose 100			
Carrageenan		86	1.0	43
S. Cerevisiae	Glucose 127			
Ca-alginate		63	4.6	53.8
S. Cerevisiae	Molasses 175			
Ca-alginate		83	0.3	21.3
S. Cerevisiae	Molasses 197			
Carrier A		74	0.35	25
Z. mobilis	Glucose 150			
Ca-alginate		75	0.85	44
Z. mobilis	Glucose 100			
Ca-alginate		87	2.4	102
Z. mobilis	Glucose 150			
Carrageenan		85	0.8	53
Z. mobilis	Glucose 100			
Flocculation				120
Z. mobilis	Glucose 50			
Borosilicate				85
Z. mobilis	Whey-lactose 50			
Carrageenan-locust bean gum		98		178

Alternatively, substrate inhibition may be diminished in the case of a gel matrix, if the rate of fermentation meets or exceeds the rate of glucose diffusion to the cell. A third possibility is that alteration of the cell membrane during immobilisation provides improved transfer of substrate into and product out of the microbe. The effect of temperature on the rate of ethanol production is markedly different for free and immobilised systems. Thus while a constant increase in rate is observed with free *S. cerevisiae* as temperature is increased from 25 to 42 °C, a maximum occurs at 30 °C with cells immobilised in sodium alginate. The lower temperature optimum for immobilised systems may result from diffusional limitations of ethanol within the support matrix. At higher temperatures, ethanol production exceeds its rate of diffusion so that accumulation occurs within the beads).

The achievement of inhibitory levels then causes the declines observed in the ethanol production rate. Significant differences are also apparent for the effect of pH on the fermentation rate. The narrow pH optimum characteristic of a free cell system is replaced by an extremely broad range upon immobilisation. This effect stems from the gradient pH that exists within the bead.

CHAPTER

6 Antibiotics

New antibiotics reported this year have included several groups of closely related pseudo disaccharides; fortimicin C, a carbamoyl derivative of fortimicin A (1), fortimicin D, the 6′-demethylated analogue of fortimicin A, and fortimicin KE, the 6′-demethylated derivative of fortimicin B (2), have been isolated from the culture broth of Mcromcwospora o/ivoasferospora, which also yields fortimicin A and B and sporaricin A (3, $R^2 = NH_2CH_2CO$-) and sporaricin B (3, R^2 = H) have been obtained from the culture broth of *Saccharopolyspora hirsuta;* these are all 1,4-diamino-cyclitol α-glycosides of 6-*epi*-purpurosamine B.

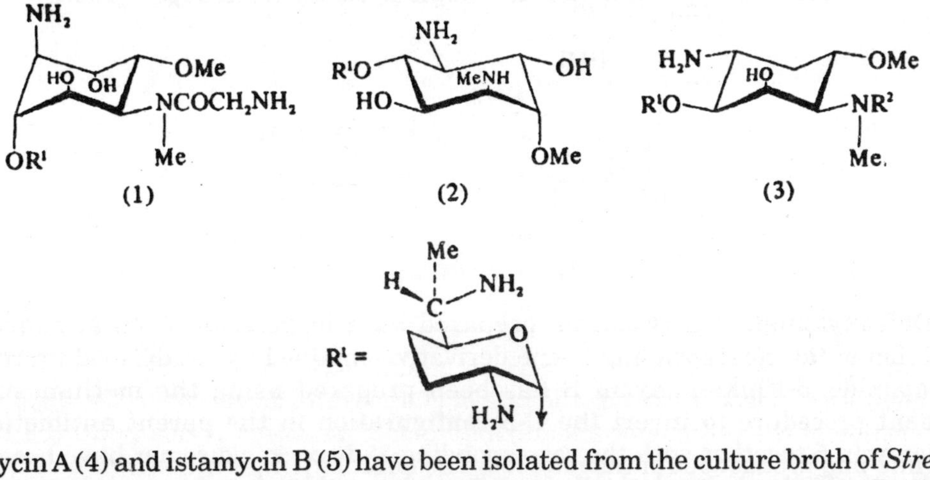

(1) (2) (3)

Istamycin A (4) and istamycin B (5) have been isolated from the culture broth of *Streptomyces tenjimariensis,* and sannamycin A (identical with istamycin A) and sannamycin B (6) from the culture broth of *Streptomyces sannanensis*; these are likewise 1,4-diamino-cyclitol α-glycosides of 6-*N*-methyl-purpurosamine C. Derivatives of the amino-sugar and cyclitol constituents of fortimicin B have been prepared by benzyloxycarbonylation and alcoholysis of the antibiotic.

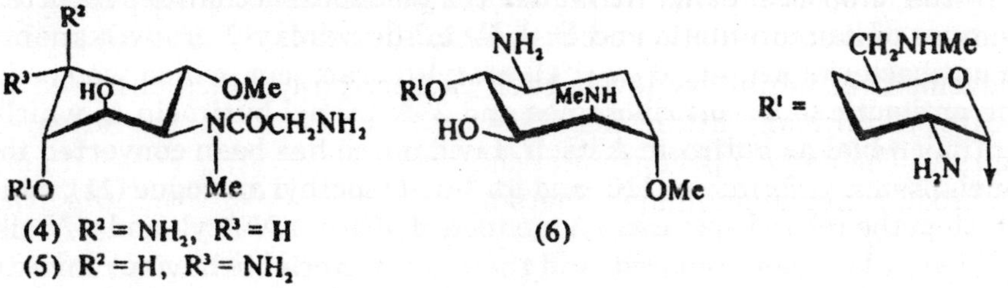

(4) $R^2 = NH_2$, R^3 = H
(5) R^2 = H, $R^3 = NH_2$

(6)

The structure (7) has been proposed for the heteropentasaccharide, virido-pentaose B, isolated from the antibiotic sporoviridin, which also contains two other pentasaccharides. A tetrasaccharide and a trisaccharide (8) composed of quinovose and viosamine were obtained by partial methanolysis of the pentaose (7) by removal of the 2-deoxy sugar units, the latter being characterized as the 3-amino-sugar, D-acosamine.

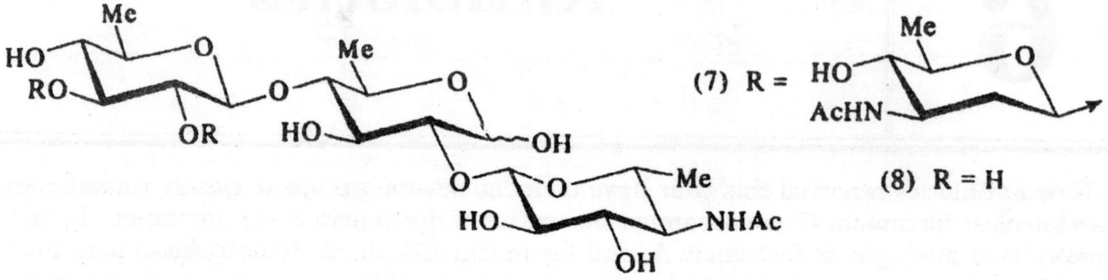

Recent advances in the synthesis of aminoglycoside antibiotics have been reviewed. Derivatized dihydrostreptomycin has been hydrolysed to give a chiral streptidine derivative (9), which could be condensed with a dihydrostrepto-biosaminyl chloride derivative to give α- and β-glycosides, the α-form yielding dihydrostreptomycin again on deprotection.

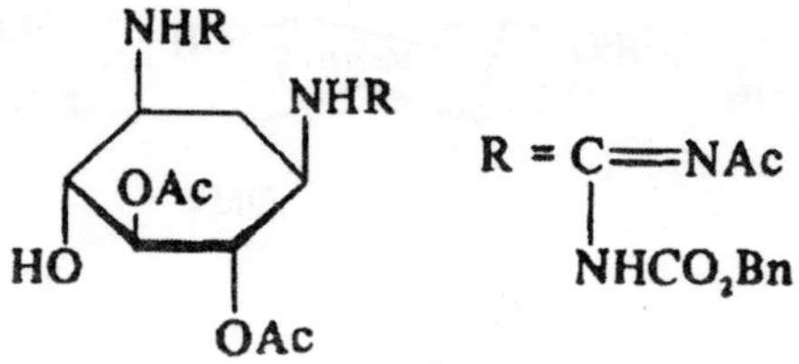

(9)

3′,4′-Dideoxykanamycin B can be prepared on a large scale from kanamycin B by hydrogenation of the corresponding 3′-ene derivative obtained by established procedures *via* the 3′,4′-epoxide. 5-Epikanamycin B has been prepared using the methanesulphonate displacement procedure to invert the C-5 configuration in the parent antibiotic, and 3′-deoxyseldomycin 5 together with the corresponding 2′-deoxy regioisomer have been obtained by reduction of a 2′,3′-epimino derivative prepared from seldomycin 5.

4′-Deoxy-6′-*N*-methyl-amikacin has been prepared from 1,3-ureido-kanamycin A by Raney nickel treatment of its 4′-thioacetate followed by 6′-*N*-methylation; the activity of this derivative suggests that the 4′-hydroxy- and 6′-amino-groups in amikacins play little part in their defence against inactivating enzymes. Other derivatives of pseudotrisaccharides reported include *NN*-dialkyl derivatives of kanamycins A and B, 1-*N*-(1,3-dihydroxy-2-propyl)kanamycin B, which showed good antibacterial activity 6′-*N*-[^{14}C] acetyl-tobramycin, which was used to study the binding of the antibiotic to *E. coli* ribosomes and 2′-*N*-propyl-butirosin A, which proved to be as good an antimicrobial as butirosin A itself. Lividamine has been converted to 2-deoxy-4-*N*-glycyl-6-*O*-α-nebrosaminyl-fortamine (10) and its 3-de-*O*-methyl analogue (11), which are weaker anti-bacterials than the related sporaricin A mentioned above. 4-*N*-Acyl- and 4-*N*-alkyl derivatives of fortimicin B have also been prepared, and their antibacterial behaviour investigated.

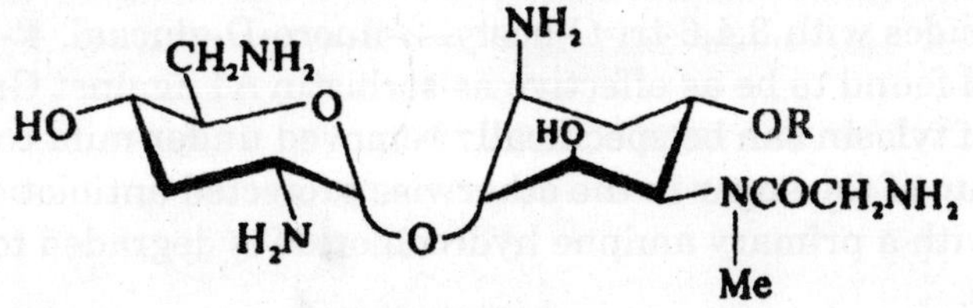

(10) R = Me
(11) R = H

Neamine has been converted to its eight possible 3′- and 4′-amino-deoxy and deoxy-hydroxy analogues by conversion to 3′,4′-anhydro-D-allo and D-galacto intermediates, which were then reduced to 3′- and 4′-deoxy compounds and appropriately transformed (Scheme 1). Some 4′-branched-chain sugar analogues were also prepared. A thio analogue of neamine (12) has been prepared by coupling the nitrosyl chloride adduct of the glycal (13) to the thio-streptamine (14) followed by reduction and deprotection.

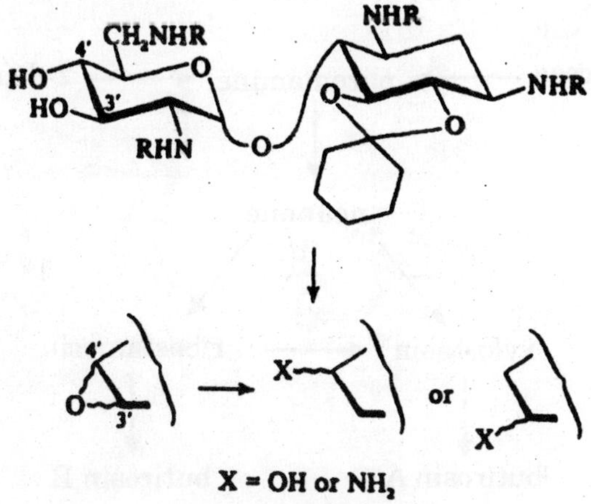

X = OH or NH_2

Scheme 1

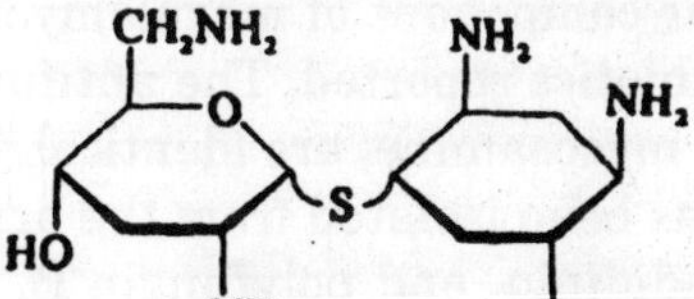

(12) 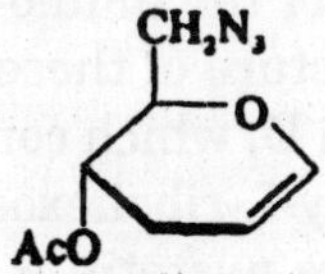 (13) 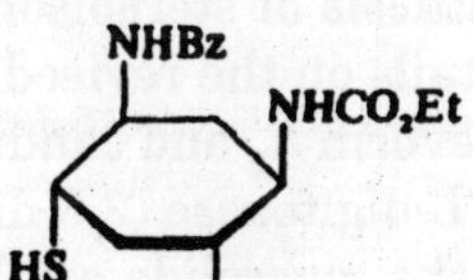 (14)

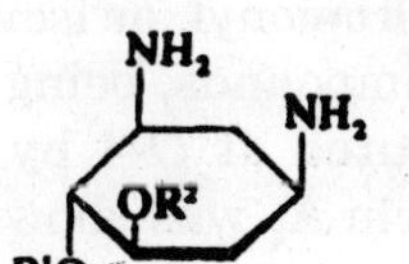

(15) $R^1 = R$, $R^2 = H$; $R^1 = H$, $R^2 = R$ R =

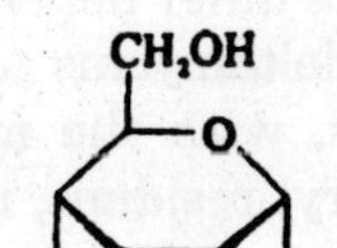

2′,3′-Dideoxy-2′-fluoro-pseudodisaccharides (15) have been synthesized using the glycal addition route to glycosides with 3,4,6-tri-*O*-acetyi-2-fluoro-D-glucaal. 4′-N-propyl sorbistin A_1 has been prepared, and found to be as effective as sorbistin A1 against Gram positive bacteria. The mycinose moiety of tylosin can be specifically removed under mild conditions by oxidation of the free hydroxy-group of the sugar in the otherwise protected antibiotic to the corresponding glycos-4-ulose, which with a primary amipne hydrochloride is degraded to a pyrrole derivative, liberating the aglycone.

Selective *N*-acylation of kanamycin A has been studied, and conditions found for the preferential formation of the mono-6″-*N*-acyl or the 3-*N*-acyl derivative. Base treatment of 3′-chloro-3′-deoxy-pseudotrisaccharides derived from butirosin have given new cyclic derivatives containing a 3′,5″-anhydro bridge. An improved procedure for the preparation of triacetyloleandomycin has been reported.

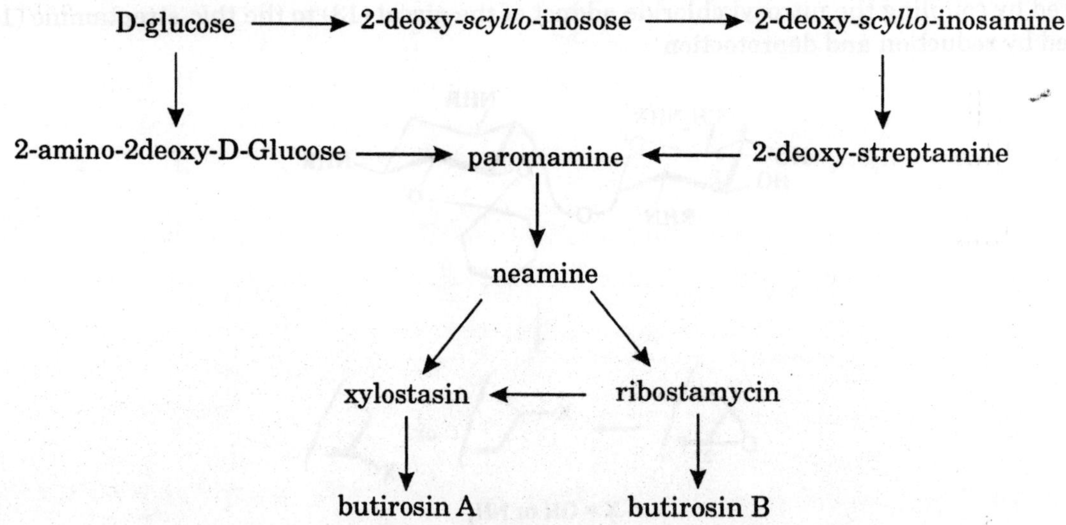

MACROLIDE ANTIBIOTICS

The synthesis of stereoisomers of the amino-sugar component of megalomycins, together with full details on the revised structure of these antibiotics reported. The antifungal polyene antibiotics levorin A_2 and candicidin D, which contain mycosamine, are identical. The unusual deoxysugar L-digitoxose (2,6-dideoxy-L-ribohexose) has been isolated from the acid hydrolysis products of the macrolide antibiotics nystatin A_3, candidinin, and polyfungin B. Maridomyin III, in which both ester groups are propionyl derivatives, is the principal component of the maridomycin complex elaborated by a valine resistant mutant of *Streptomyces hydroscopicus* (the six components differ depending on whether acetyl, propionyl, or isovaleryl ester groups are present). The deltamycins A_t-A_4 are closely related compounds, being 9-oxo analogues of the maridomycins, with the mycarose residue substituted at *O*-4 by acetyl, propionyl, butyryl, or isovaleryl residues, respectively, and deltamycin A_4 was shown to be identical to carbomycin A.

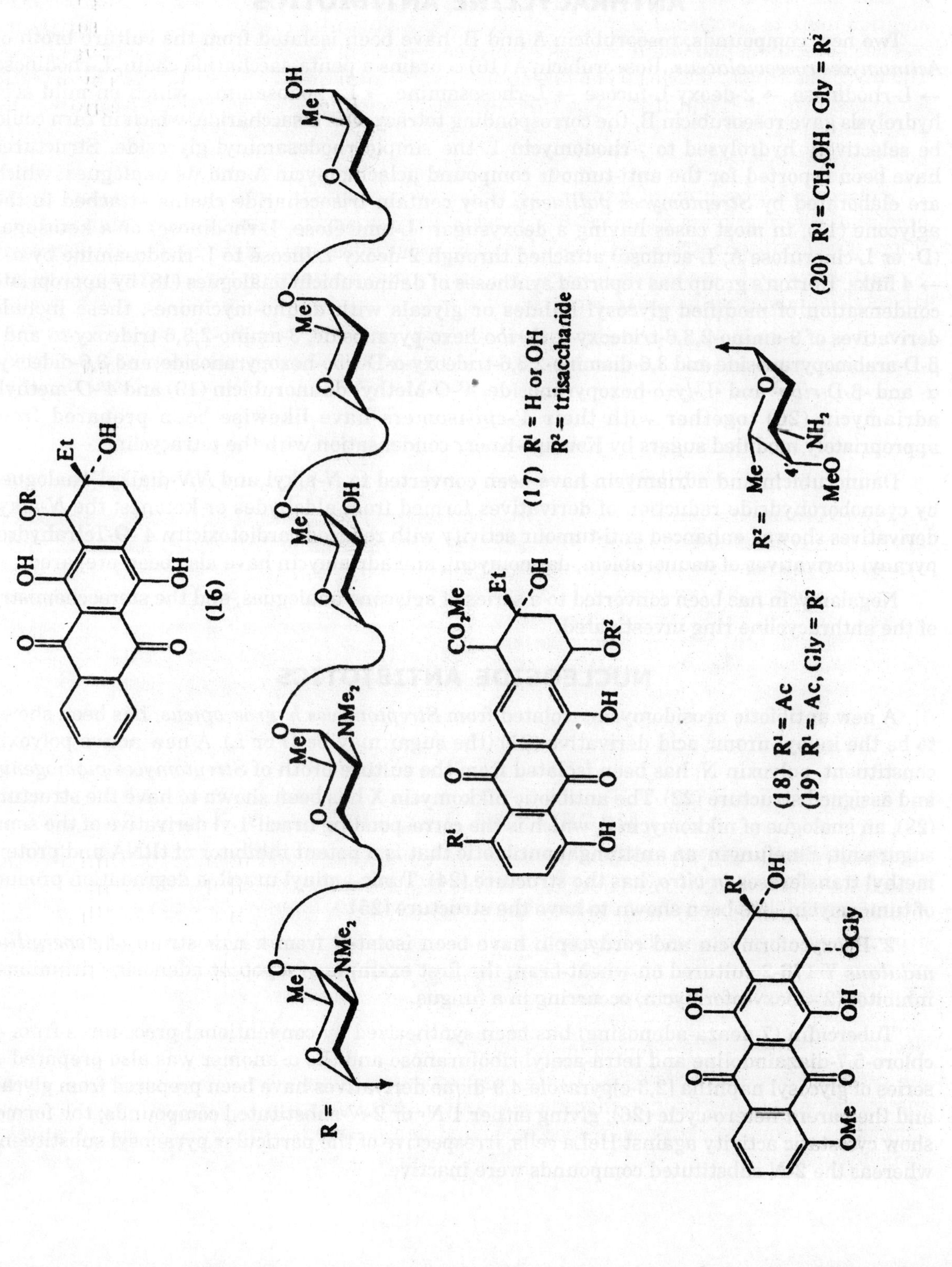

(16)

(17) R^1 = H or OH
R^2 = trisaccharide

(18) R^1 = Ac
(19) R^1 = Ac, Gly = R^2

(20) R^1 = CH_2OH, Gly = R^2

R^2 = Me, MeO, NH_2

R =

ANTHRACYCLINE ANTIBIOTICS

Two new compounds, roseorubicin A and B, have been isolated from the culture broth of *Actinomyces roseoviolaceus.* Roseorubicin A (16) contains a penta-saccharide chain, L-rhodinose → L-rhodinose → 2-deoxy-L-fucose → L-rhososamine → L-rhodosamine, which on mild acid hydrolysis gave roseorubicin B, the corresponding tetracycline disaccharide, which in turn could be selectively hydrolysed to γ-rhodomycin 1, the simple rhodosaminyl glycoside. Structures have been reported for the anti-tumour compound aclacinomycin A and its analogues, which are elaborated by *Streptomyces galilaeni;* they contain trisaccharide chains attached to the aglycone (17), in most cases having a deoxysugar (L-amicetose, L-rhodinose) or a ketosugar (D- or L-cinerulose A, L-aculose) attached through 2-deoxy-L-fucose to L-rhodosamine by α-1 → 4 links. Horton's group has reported syntheses of daunorubicin analogues (18) by appropriate condensation of modified glycosyl halides or glycals with dauno-mycinone-, these include derivatives of 3-amino-2,3,6-trideoxy-α-D-*ribo*-hexo-pyranoside, 3-amino-2,3,6-trideoxy-α- and -β-D-arabinopyranoside and 3,6-diamino-2,3,6-trideoxy-α-D-*ribo*-hexopyranoside, and 3,6-dideoxy-α- and -β-D-*ribo*- and -L-*lyxo*-hexopyranoside. 4′-*O*-Methyl-daunorubicin (19) and 4′-*O*-methyl-adriamycin (20) together with their 4′-*epi*-isomers have likewise been prepared from appropriately modified sugars by Koenigs-Knorr condensation with the tetracycline.

Daunorubicin and adriamycin have been converted to *N*-alkyl and *NN*-dialkyl analogues by cyanoborohydride reduction of derivatives formed from aldehydes or ketones; the *N*-alkyl derivatives showed enhanced anti-tumour activity with reduced cardiotoxicity. 4′-*O*-Tetrahydro-pyranyl derivatives of daunorubicin, daunomycin, and adriamycin have also been prepared.

Nogalamycin has been converted to a series of aglycone analogues, and the stereochemistry of the anthracycline ring investigated.

NUCLEOSIDE ANTIBIOTICS

A new antibiotic neosidomycin, isolated from *Streptomyces hygroscopicus,* has been shown to be the indoyl-uronic acid derivative (21) (the sugar may be D or L). A new active polyoxin constituent, polyoxin N, has been isolated from the culture broth of *Streptomyces piomogenus* and assigned structure (22). The antibiotic nikkomycin X has been shown to have the structure (23), an analogue of nikkomycin Z, which is the corresponding uracil-1-yl derivative of the same sugar unit. Sinefungin, an antifungal antibiotic that is a potent inhibitor of tRNA and protein methyl transferases *in vitro,* has the structure (24). Tunic-aminyl uracil, a degradation product of tunicamycin, has been shown to have the structure (25).

2′-Deoxycoformycin and cordycepin have been isolated from a new strain of *Aspergillus nidulans* Y 176-2 cultured on wheat-bran, the first example of a potent adenosine deaminase inhibitor (2′-deoxycoformycin) occurring in a fungus.

Tubercidin (7-deaza-adenosine) has been synthesized by conventional procedures from 4-chloro-5,7-diazaindoline and tetra-acetyl ribofuranose and the α-anomer was also prepared. A series of glycosyl naphtha [2,3-c]pyrazole-4,9-dione derivatives have been prepared from glycals and the parent hetero-cycle (26), giving either 1-*N*- or 2-*N*-substituted compounds; the former show cytostatic activity against HeLa cells, irrespective of the particular pyranosyl substituent, whereas the 2-*N*-substituted compounds were inactive.

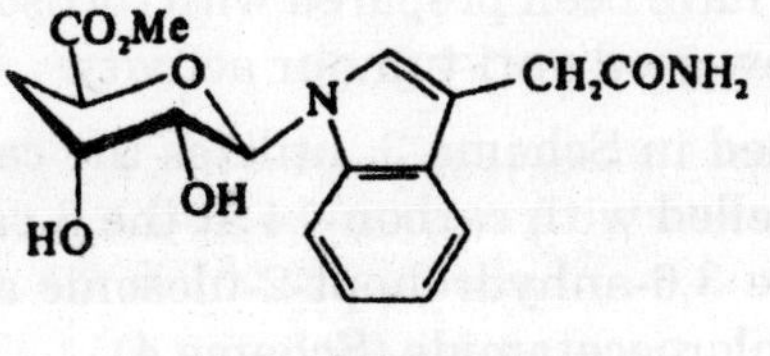

(21)

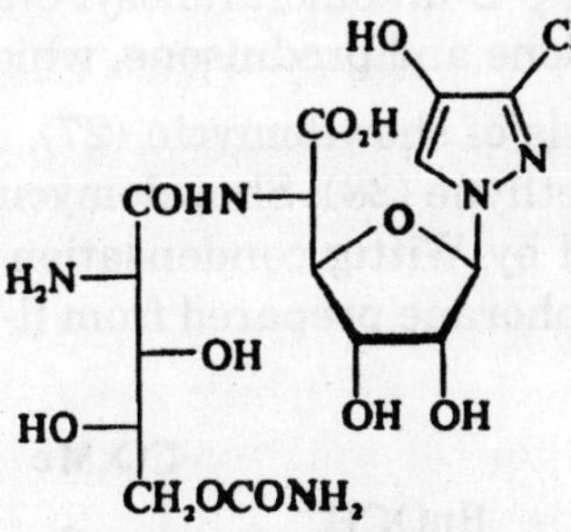

(22)

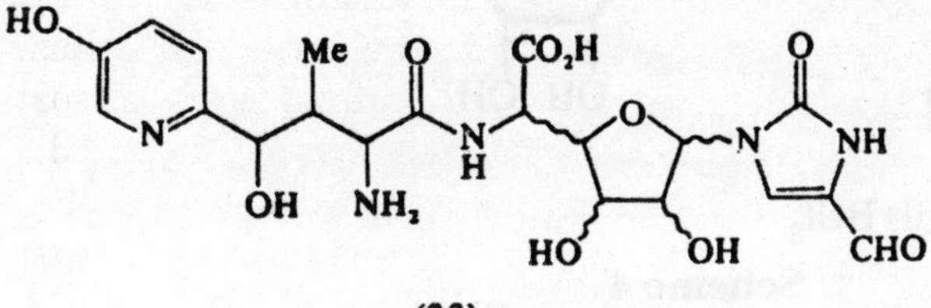

(23)

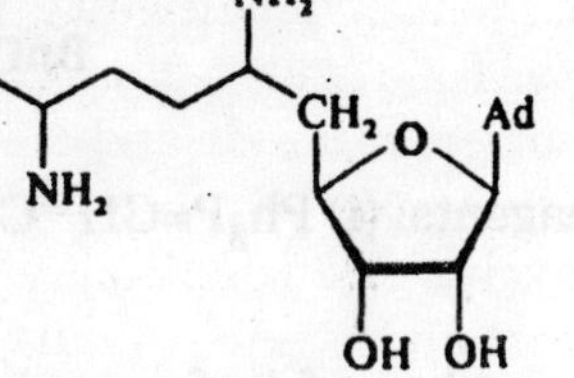

(24)

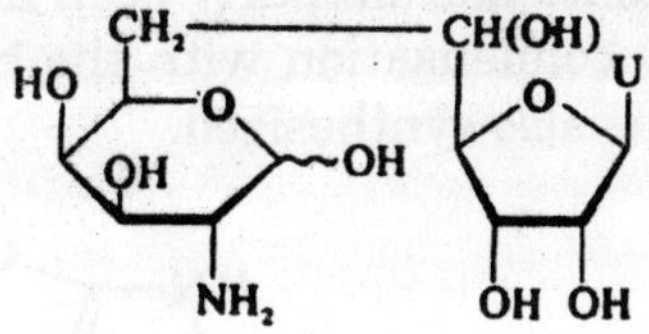

(25)

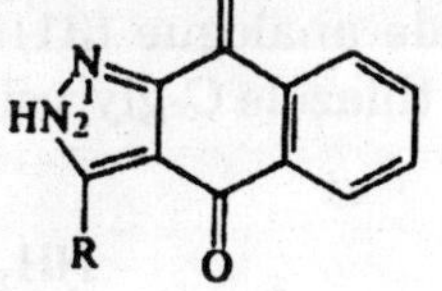

(26) R = H or CO_2Et

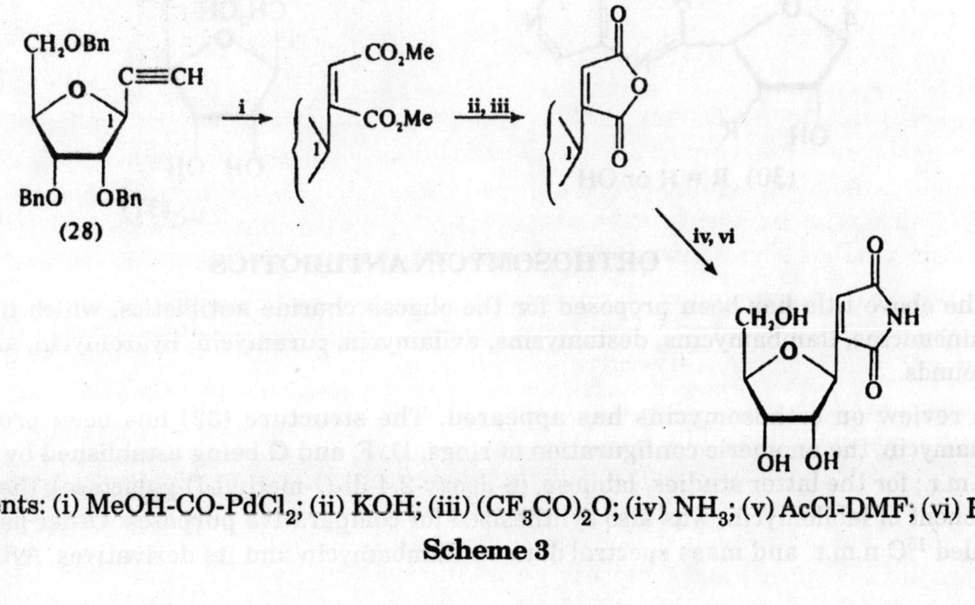

(28)

Reagents: (i) MeOH-CO-$PdCl_2$; (ii) KOH; (iii) $(CF_3CO)_2O$; (iv) NH_3; (v) AcCl-DMF; (vi) H_2-Pd

Scheme 3

Conjugates of 1-β-D-arabinofuranosyl-cytosine have been prepared with cortisol and cortisone and with prednisolone and prednisone, which show good anti-tumour activity.

A new synthesis of showdomycin (27), outlined in Scheme 3, utilizes the carboxylation of the ribofuranosyl ethyne (28). Showdomycin, labelled with carbon-14 at the β-carbonyl group, has been prepared by Wittig condensation of the 3,6-anhydrohept-2-ulosonic acid derivative (29) with the phosphorane prepared from [l-^{14}C]chloroacetamide (Scheme 4).

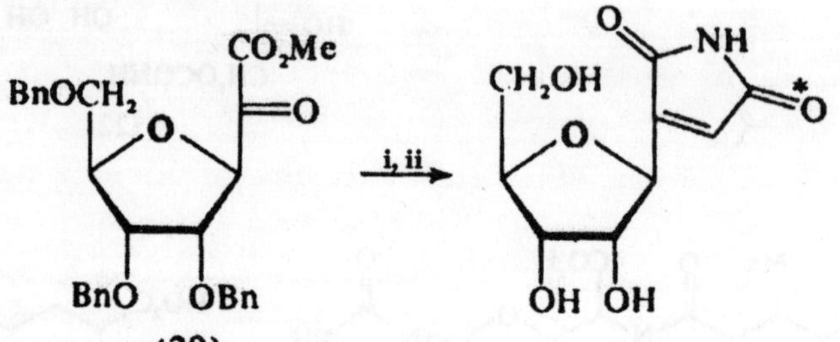

CO_2Me BnOCH$_2$ =O O BnO OBn (29) —i, ii→ CH$_2$OH O NH O* OH OH

Reagents: (i) $Ph_3P{=}CH^{14}CONH_2$; (ii) BCl_3

Scheme 4

A number of isoformycins (30) have been prepared from appropriate glyco-furanosyl thioformidate derivatives by condensation with 3-cyano-4-amino-pyrazole, and their conformations were investigated. glycosyl thioimidate has similarly been used to prepare the triazole C-glycoside analogue (31) of ribavirin by condensation with the hydrazone of ethyl oxamate, and the thiazole *C*-glycoside analogue was also synthesized.

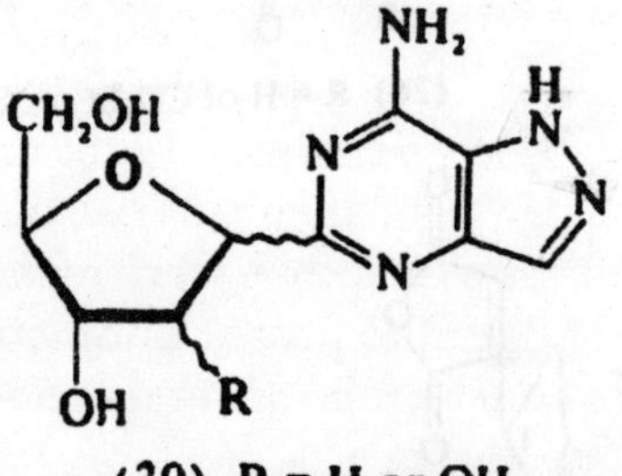

CH$_2$OH O NH$_2$ N H N N N OH R

(30) R = H or OH

CONH$_2$ HN N N CH$_2$OH O OH OH

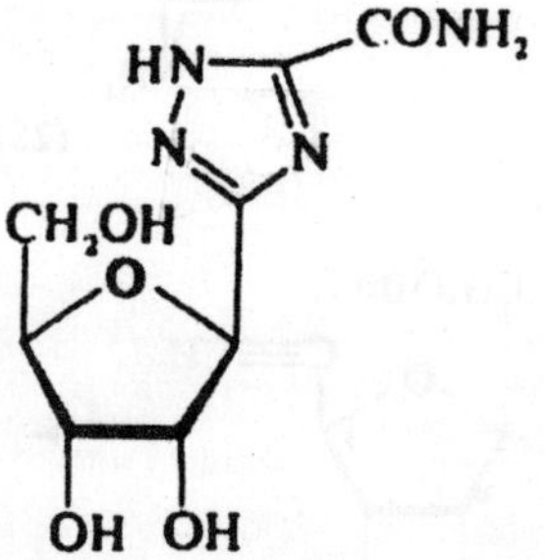

(31)

ORTHOSOMYCIN ANTIBIOTICS

The above title has been proposed for the oligosaccharide antibiotics, which include the everninomicins, flambamycins, destomysins, avilamycin, curamycin, hygromycin, and related compounds.

A review on orthosomycins has appeared. The structure (32) has been proposed for flambamycin, the anomeric configuration of rings, D, F, and G being established by high field ^{13}C n.m.r.; for the latter studies, labilose, (6-deoxy-2,4-di-*O*-methyl-D-galactose), the glycoside component of labilomycin, was also synthesized for comparative purposes. Other papers have recorded ^{13}C n.m.r. and mass spectral data on flambamycin and its derivatives. Avilamycin A

(33) has the same structure as flambamycin except that the sugar evermicose (2,6-dideoxy-3-*C*-methyl-D-*arabino*-hexose) (unit D) in the latter is replaced by evalose (6-deoxy-3-C-methyl-D-mannose). Avilamycin C differs from avilamycin A in having a 1-hydroxy ethyl side-chain instead of acetyl in unit H. A structure has been proposed for curamycin A (34) based on chemical and spectroscopic studies and comparison with flambamycin and avilamycin.

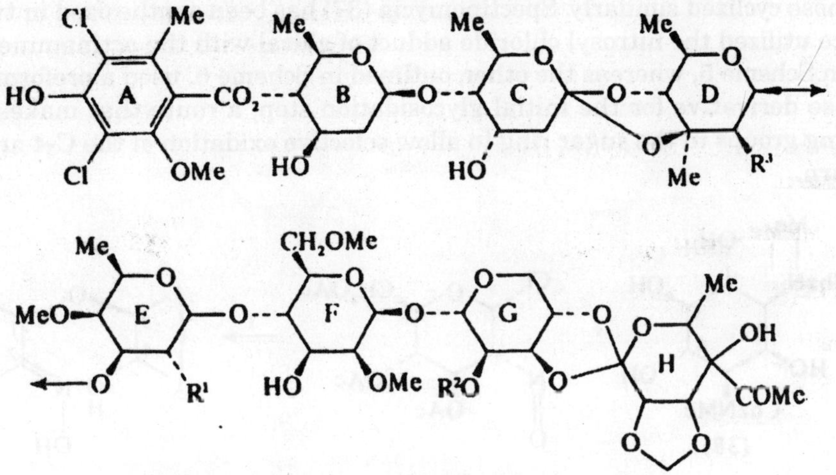

(32) $R^1 = OH, R^2 = Me_2CHCO^-, R^3 = OH$
(33) $R^1 = H, R^2 = Me_2CHCO\text{-}, R_3 = OH$
(34) $R^1 = OH, R^2 = Ac, R^3 = H$

MISCELLANEOUS ANTIBIOTICS

Methods for synthesizing anisomycin and pentenomycin and the chemistry and biological transformations of bleomycin and phleomycin have been reviewed. D-Ribo-furanosyl and -pyranosyl derivatives of *N*-methyl-*N*-nitroso-urea have been prepared as analogues of streptozotocin [2-deoxy-2-(*N*-methyl-*N*-nitrosoureido)-D-glucose]; they were reported to be more active against L1210 leukaemia in mice and less toxic than the parent antibiotic. Likewise *N*-(2-chloroethyl)-*N*-nitrosoureido derivatives of cyclopentane tetrols (35) and cyclo-hexane tetrols (36) have been synthesized for comparison with streptozotocin.

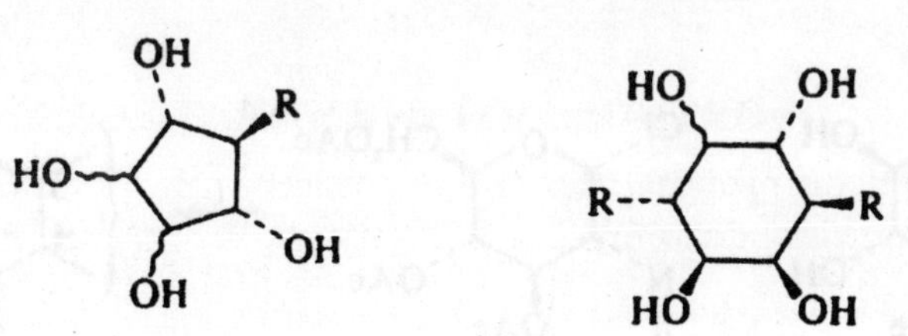

(35) (36)

$R = -NHCON(NO)CH_2CH_2Cl$

Biosynthetic studies have established that D-glycosamine is incorporated into streptozotocin, and L-citrulline and methionine furnish the amide side-chain and N-methyl group, respectively, but the source of the nitroso-group was not established. 2-Deoxy-2-ureido-D-glucose, which shows weak antimicrobial activity against some Gram negative bacteria and fungi, has been isolated from *Streptomyces halstedii;* in aqueous solution it slowly cyclized to give the corresponding 1,2-diazolidone derivative. Isomeric ureido derivatives of D-galactose and D-mannose cyclized similarly. Spectinomycin (37) has been synthesized in two laboratories; one sequence utilized the nitrosyl chloride adduct of glucal with the actinamine derivative (38) illustrated in Scheme 5, whereas the other, outlined in Scheme 6, used a preformed 4,6-dideoxy-D-*ribo*-hexose derivative for the initial glycosidation step; a route that makes elegant use of cyclic blocking groups in the sugar ring to allow selective oxidation of the C-4 and C-3 hydroxy-groups in turn.

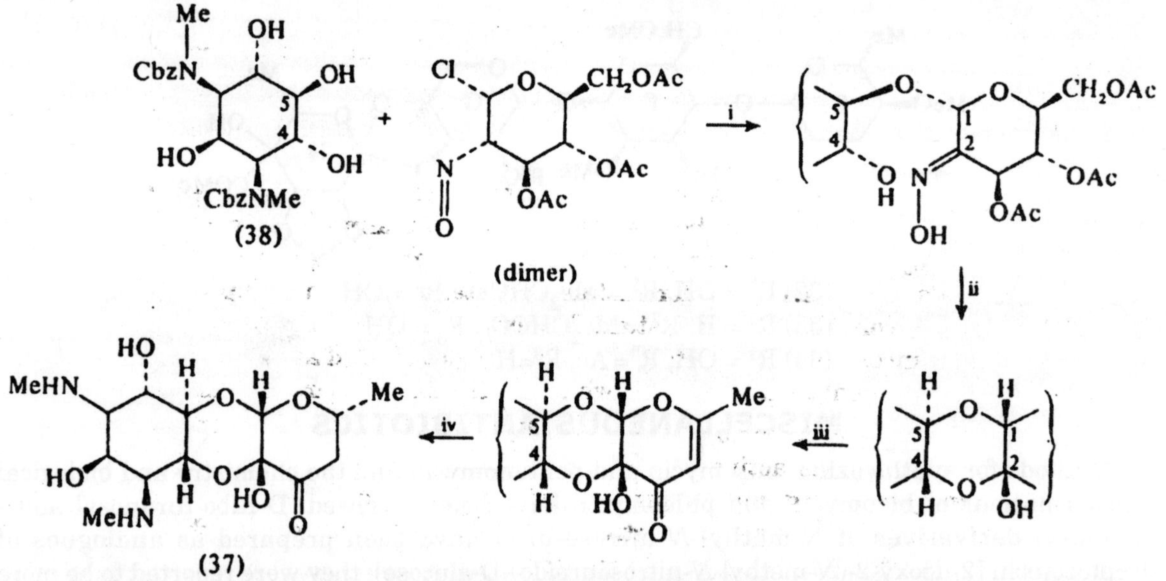

Reagents: (i) DMF; (ii) MeCHO-MeCN-HCl; (iii) $KHCO_3$-MeCN; (iv) H_2-Pd

Scheme 5

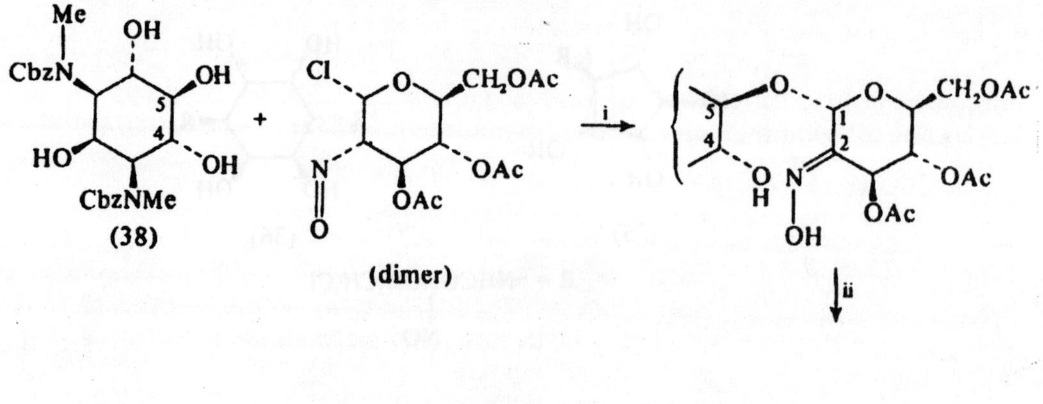

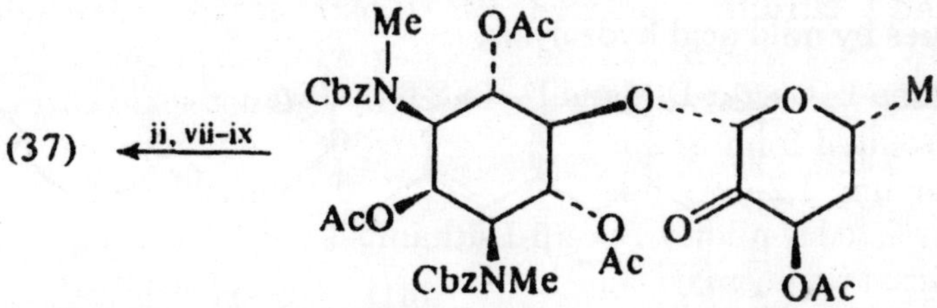

(37) ← ii, vii–ix

Reagents: (i) CF_3SO_3Ag; (ii) NaOMe; (iii) $MeC(OMe)_3$–H^+; (iv) Ac_2O-py-DMAP; (v) HOAc-H_2O; (vi) CrO_3-py-HCl; (vii) Bu_2SnO; (viii) Br_2-Bu_3SnOMe; (ix) H_2-Pd

Scheme 6

7-Deoxyspectinomycin (39) and 7-deoxy-8-*epi*-4-(R)-dihydro-spectinomycin (40) have been prepared by chromic oxide oxidation of a dihydro-spectinomycin derivative at C-9, with concomitant elimination of a C-7 acetoxy-group, to give an unsaturated ketone intermediate, which could in turn be reduced by sodium borohydride to the products as indicated in the part-sequence shown in Scheme 7.

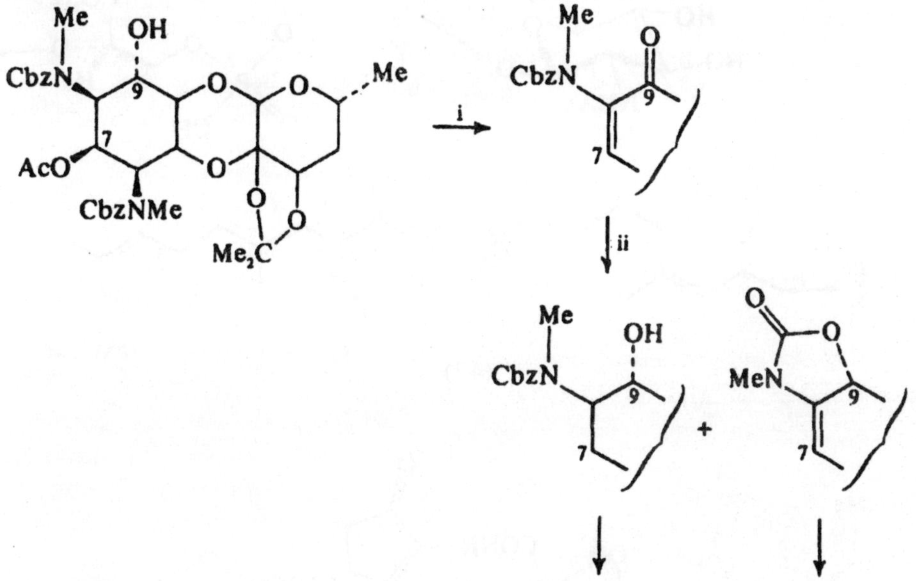

(39) (40)

Reagents: (i) CrO_3-py; (ii) $NaBH_4$

Scheme 7

A complete structure has been proposed for the antibiotic ristocetin A, which contains the sugar units β-D-mannosyl, α-L-ristosaminyl, and the tetrasaccharide (41) attached as aryl or benzyl glycosides to a complex macrocyclic core of phenolic amino-acids. Much of the evidence came from a 270 MHz ^{1}H n.m.r. study on the ristosaminyl-core obtained after removal of the other sugar residues by mild acid hydrolysis.

$$\beta\text{-D-Ara}p\text{-1} \longrightarrow 2\text{-}\alpha\text{-D-Man}p\text{-1} \longrightarrow 2\text{-}\beta\text{-D-Glc}p \overset{6}{\underset{\beta\text{-L-Rhamp-1}}{\uparrow}}$$

(41)

Two further degradation products have been obtained from moenomycin A, suggesting that when the part structures (42) and (43), are linked together they form the complete antibiotic. Another report describes the glucose-glucosamine disaccharide present in fragment (42), which was obtained from moenomycin A by acid hydrolysis. Earlier work on the structure and properties of moenomycin A has been reviewed.

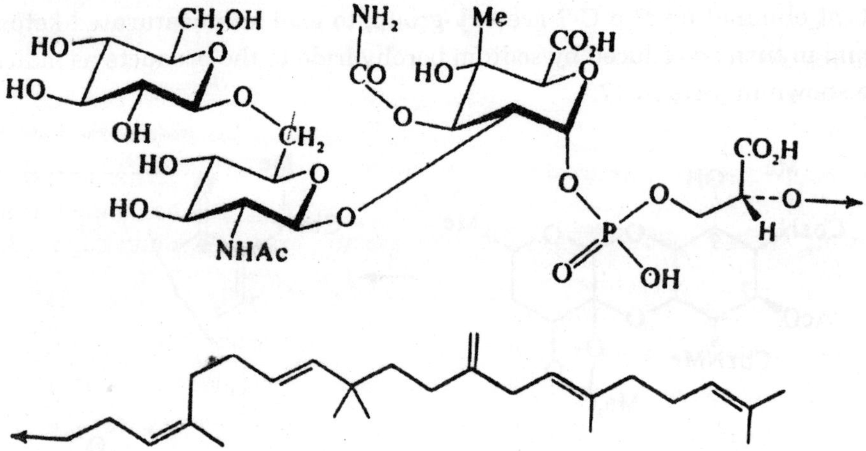

(42)

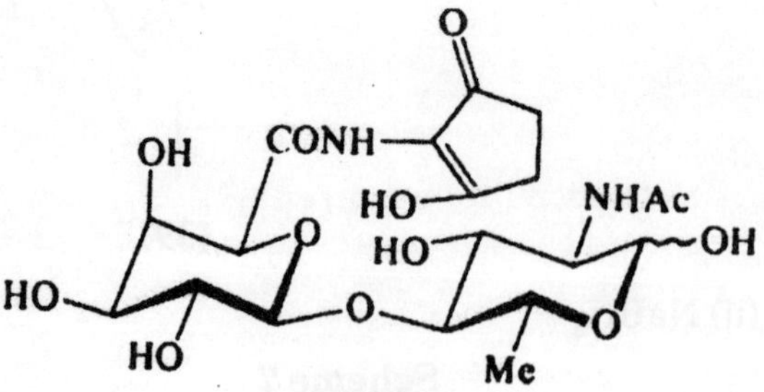

(43)

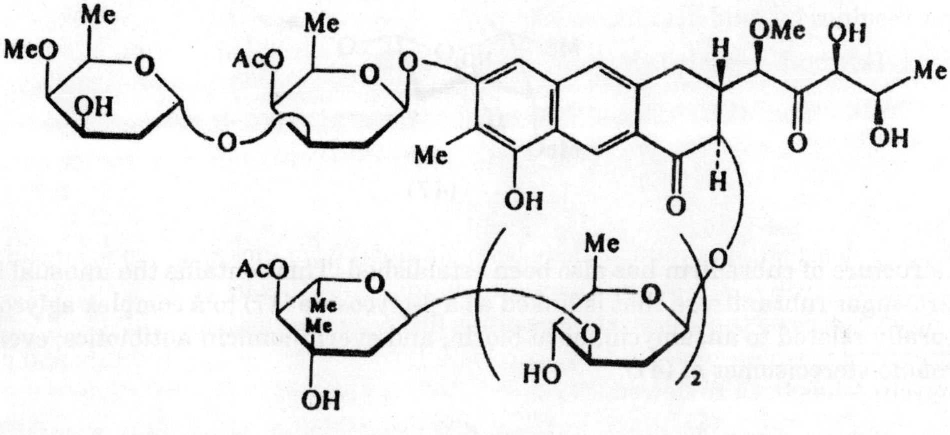

(44)

A combination of ^1H n.m.r. and ^{13}C n.m.r. techniques has been used to con-firm the gross structure of chromomycin A_3, but with the positions of the acetate and glycosidic groups on the disaccharide non-terminal sugar unit inter-changed to give the revised structure (44).

X-Ray crystal analysis has established the structure of the polyether anti-biotic carriomycin, which contains 2,3,6-trideoxy-4-*O*-methyl-D-*erthro*-hexose, and has shown that hedamycin is the interesting di-*C*-glycoside (45) involving amino and branched-chain sugars, kidamycin being the closely related analogue (46).

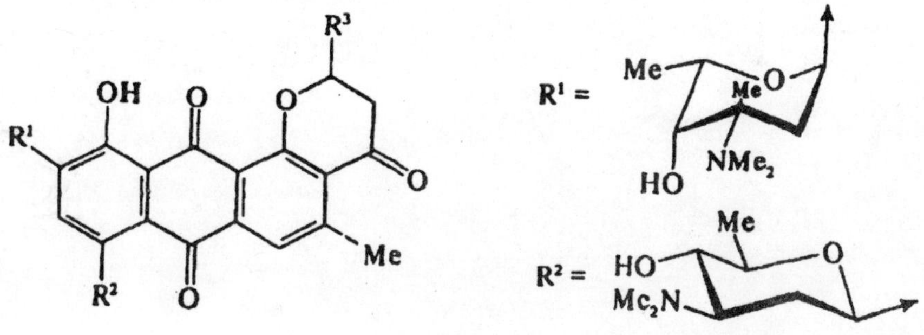

R^1 =

R^2 =

(45) R^3 =

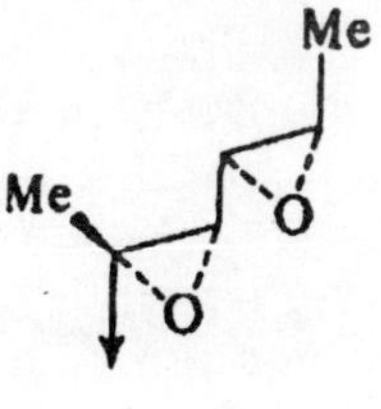

(46) R^3 =

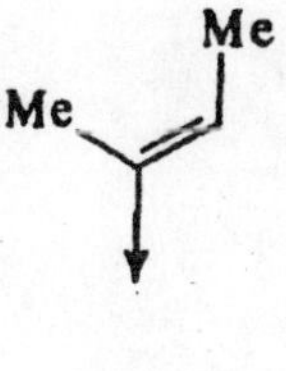

(47)

The structure of rubradirin has also been established. This contains the unusual branched-chain nitro-sugar rubranitrose that is linked as a β-glycoside (47) to a complex aglycone, which is structurally related to ansamycin, novo-biocin, and everninomicin antibiotics; evernitrose is the *L-arabino* stereoisomer of (47).

CHAPTER

7 Synthesis of Polysaccharides

TEICHOIC ACIDS

The ^{13}C n.m.r. spectra of a number of ribitol teichoic acids substituted with glycosyl residues, and of their dephosphorylated repeating units, have been recorded. Identical spectra are obtained for teichoic acids substituted with 2-*O*-and 4-*O*-glycosyl residues, so it is not possible to differentiate between these positional isomers using this technique. Investigation of the cell-wall composition and associated properties of strains of *Staphylococcus aureus,* which are resistant to methicillin, has not revealed evidence for any unusual wall polymers.

The ribitol teichoic acid isolated from *Staphylococcus hyicus* contains 2-acetamido-2-deoxy-D-glucosyl residues. Interaction of the teichoic acid with concanavalin A, and its susceptibility to α-, but not to β-D-2-acetamido-2-deoxyglucosidase, showed that the aminosugar is in the α-configuration. The cell walls of *Streptococcus pneumoniae* contain a teichoic acid (C polysaccharide) composed of 2-acetamido-2-deoxy-D-galactose, 2-acetamido-4-amino-2,4,6-trideoxy-D-hexose, ribitol phosphate, choline phosphate, and D-glucose. From the results of periodate oxidation and methylation analysis, together with an examination of the ^{13}C n.m.r. spectra, the structure (1) has been proposed. It is speculated that the C-3 hydroxy-group of the chain terminal diaminotrideoxy-D-hexose unit is glycosylated with isomaltose. Pneumococcal cell-wall teichoic acid, rather than peptidoglycan, is considered to be responsible for the activation of the alternative complement pathway.

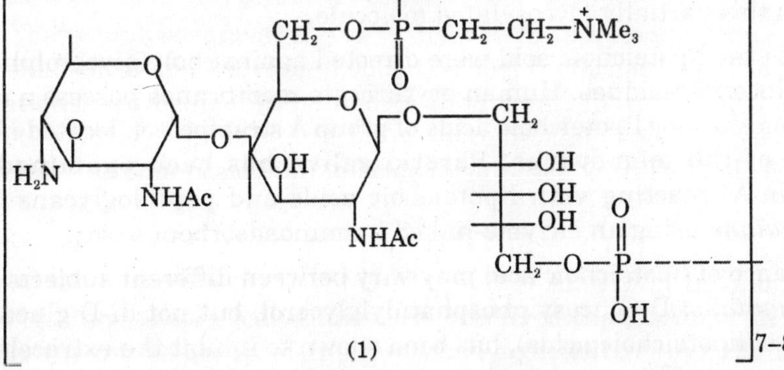

(1)

This contrasts with group A streptococci in which the greatest activity was found associated with a heat labile protein of the cell membrane. The addition of a pulse of phosphate to a phosphate-limited chemostat culture of *Bacillus subtilis* has been found to lead to the synthesis of teichoic acid and the consequent development of bacterial ability to bind phage SP50 within about one generation line after the addition of the pulse. The time taken for newly synthesized wall material to become maximally exposed at the surface of the cells was directly related to their growth rate and corresponded to about three-quarters of a generation time.

The newly synthesized material does not become exposed exclusively at any single highly localized zone, but appears near the middle and along the cylindrical length of the bacteria at newly formed polar caps. The regulation of teichoic acid synthesis during phosphate limitation in *B. subtilis* has been shown to be due to the inactivation of the enzyme catalysing the formation of P^1-2-acetamido-2-deoxy-D-glucopyranosyl P^2-polyisoprenyl pyrophosphate. Conditions have been described for the continuous culture of a derivative of *Staphylococcus aureus* H, with L-cysteine as the sole amino-acid and under various nutrient limitations.

The proportion of ribitol teichoic acid present in the wall and the extent to which it is substituted with 2-acetamido-2-deoxy-D-glucose varies on growth of the organism under different conditions, as does the extent of cross-linking of the peptidoglycan. Neither the derivative nor the original strain H produced teichuronic acid when grown under phosphate limitation. Phosphate-repressible phosphodiesterases, isolated from *Bacillus subtilis,* have been reported to be responsible for the depolymerization of teichoic acids. Lipoteichoic acid from *Streptococcus rhutans* has been purified by a combination of gel filtration, hydrophobic chromatography, and adsorption on to phospholipid vesicles. Plasma membranes from *Staphylococcus aureus,* purified on columns of immobilized human immunoglobulin G, have been characterized as lipoteichoic acids, composed of D-glucosylglycerol teichoic acid containing ester-linked L- 7 alanine with pentadecanoic acid as the major fatty acid.

Growth conditions for the production of extracellular lipoteichoic acid by *Streptococcus mutans* have been investigated. The greatest levels were produced by slow-growing organisms. A lipoteichoic acid fraction and a presumed deacylated form were separated by column chromatography. The effect of growth rate on lipid and lipoteichoic acid composition in *Streptococcus faecium* has been reported. Lipoteichoic acids from *Bacillus subtilis* 168 and its *gta* mutants have been purified and shown to contain glycerol and phosphorus in equimolar amounts, together with D-glucose and fatty acids. The D-glucosyl residues have the α-configuration on this partially glycosylated molecule.

Antibodies to the lipoteichoic acid were directed against poly(glycerolphosphate) and not against the D-glucosyl residues. Human erythrocyte membranes possess a single population of specific binding sites for lipoteichoic acids of group A streptococci, located almost exclusively at the surface of the membranes. Parotid saliva has been examined for antibodies (immunoglobulin A) reacting with lipoteichoic acids and peptidoglycans of whole cells of *Streptococcus mutans* using an enzyme-linked immunoadsorbent assay.

The significance of lipoteichoic acid may vary between different subjects and for different serotypes of *S. mutans.* D-Glucosylphosphatidylglycerol, but not di-D-glucosyldiacylglycerol (the head group of lipoteichoic acids), has been shown to inhibit the extracellular autolysin of *Staphy locoecus aureus.* This suggests a requirement for the phosphate esters of the lipoteichoic acid molecule, together with the fatty acid ester group, for inhibition. Scanning electron microscopy has revealed concentric circular structures on newly exposed surfaces of the walls and isolated cross-walls of *Staphylococcus epidermidis*. Extraction of the walls with trichloracetic acid removes most of 'the phosphorus and about half of the 2-acetamido-2-deoxy-D-glucosyl residues, and destroys this type of structure. Hence, teichoic acid and possibly peptidoglycan (since these polymers are covalently linked) may be arranged circularly on the surface of the wall.

$$
\begin{array}{l}
\rightarrow 4)\text{-}\beta\text{-D-Glc}p\text{-}(1 \rightarrow 3)\text{-}\alpha\text{-L-Rha}p\text{-}(1 \rightarrow 4)\text{-}\alpha\text{-L-Rha}p\text{-}(1 \rightarrow \\
\qquad\qquad\qquad\qquad\qquad\qquad\qquad\qquad\qquad\quad 3 \\
\qquad\qquad\qquad\qquad\qquad\qquad\qquad\qquad\qquad\quad \uparrow \\
\qquad\qquad\qquad\qquad\qquad\qquad\qquad\qquad\qquad\quad 1 \\
\qquad\qquad\qquad\qquad\qquad\qquad\qquad\qquad\quad \beta\text{-D-Glc}p\text{A}
\end{array}
$$

(2)

The primary structure (2) of the teichuronic acid of *Bacillus megaterium* M46 has been elucidated, using a combination of partial acidic hydrolysis, enzymic hydrolysis, periodate oxidation, and Smith degradation studies. The teichuronic acids of *B. megaterium* are unusual in that they resemble capsular polysaccharides in their size (mol. wt. 5×10^5) and composition, yet are structurally and functionally integral components of thc cell wall. The proton-decoupled ^{15}N. n. m. r. spectra of ^{15}N-labelled intact cells, isolated cell walls, and cell-wall digests have been described. Each of five Gram-positive bacteria studied displayed a unique set of cell wall ^{15}N resonances, which reflected variations in the primary structure of peptidoglycans and in the amounts of teichoic acid and teichuronic acid in the cell wall.

Specific assignments of cell wall ^{15}N resonances assigned to teichoic acid D-alanyl residues, teichuronic acid and acetamido groups, and to peptidoglycan acetamido, peptide amido and free amino-groups, have been made on the basis of a combination of specific isotope labelling and dilution experiments, and from the results of cell wall fractionation. In *Micrococcus various* all three glycerol phosphate residues in the linkage unit are derived from CDP-glycerol and are therefore not transferred from a lipoteichoic acid carried.

PEPTIDOGLYCANS

A range of helical conformations for the glycan chains of bacterial peptidoglycans has been proposed on the basis of X-ray diffraction studies of cell walls and peptidoglycans from several bacterial species. These studies support the view that the peptide conformation has proportions of the residues in the helical and hydrogen-bonded β-sheet conformations. The balance between the two conformations is presumably related to the mean separation of the glycan chains, which varies with the water content by a factor of approximately two. The pentapeptide chains of peptidoglycans have a sequence of amino-acids terminated with D-Ala-D-Ala at positions 4 and 5.

Possible conformations of D-Ala-D-Ala and its analogues L-Ala-D-Ala, D-Ala-L-Ala, D-butyryl-L-Ala, D-Ala-D-butyric acid, D-Val-D-Ala, and D-Ala-D-Val have been analysed by theoretical methods. From theoretical studies it is predicted that L-Ala or D-Val at the 4 or 5 position of the pentapeptide group of the peptidoglycan will reduce the cross-linking in peptidoglycan biosynthesis, whereas the effect of D-butyric acid will be marked at the 4 position and moderate at the 5 position. This is in agreement with experimental results. Possible conformations of the disaccharide-peptide sub-units of the peptido-glycans of *Staphylococcus aureus* and *Micrococcus luteus* have been studied using an energy-minimization procedure.

Contrary to earlier reports the favoured conformation of the disaccharide 2-acetamido-2-deoxy-β-D-glucosyl-(1 → 4)-*N*-acetylmuramic acid is different from that of cellulose or chitin. Three types of conformation, two compact and one extended, are postulated with all three being stabilized by intramolecular hydrogen bonds.

Two different models were proposed for the three-dimensional arrangement of peptidoglycan in the cell wall- X–Ray crystallographic studies of the binding of the trisaccharide β-MurNAc-(1 → 4)-β-D-Glc*p*NAc-(1 → 4)-MurNAc to sub-sites B, C, and D of lysozyme indicate no distortion of sugar residues at sub-site D. The specificity of antisera to oligoglycylalbumin and to L-alanyl-tetraglycyl-albumin has been used to demonstrate serological differentiation of some bacterial strains on the basis of the primary structure of their cell-wall peptidoglvcan. Numerous synthetic glycopeptide analogues of peptidoglycans have been analysed by m.s. The immunoadjuvant activity of a number of synthetic glycopeptide analogues of peptidoglycans appears to be dependent on the type of substitution of the γ-carboxy-group of the D-isoglutamine residue. The peptidoglycan of *Acetobacterium woodii* contains D-ornithyl residues, which function as interpeptide bridges between the γ-carboxy-group of D-isoglutamic acid and the carboxy-group of the terminal D-alanyl residue of an adjacent peptide sub-unit.

The usual D-alanyl residue in position 1 of the peptide sub-unit is replaced by an L-seryl residue. The primary site of action of the antibiotic amphomycin has been demonstrated to involve inhibition of the phospho-*N*-acetylmuramylpentapeptide translocase, an enzyme involved in the first step of a lipid cycle of peptidoglycan synthesis. Although *Bacillus cereus* 569 is known to be resistant to lysis by lysozyme, because of the presence in the peptidoglycan of residues of deacetylated 2-amino-2-deoxy-D-glucose, the action of lysozyme on isolated cell walls has been reported to release some free reducing groups, indicating a limited breakage of the polysaccharide chains of the peptidoglycan.

The action of the enzyme modifies the peptidoglycan and makes it more susceptible to autolysin (s) and also enhances the rate of septum separation. A carboxypeptidase-transpeptidase, which incorporates free diaminopimelic acid into previously formed cell walls, has been isolated from *Bacillus megaterium.* Membranes from *B. megaterium* have been solubilized and then reconstituted in a form capable of synthesizing peptidoglycan. Using this reconstituted system, an assay has been developed for a factor necessary for peptidoglycan synthesis. The factor may be a polymerase, which assembles the disaccharide-pentapeptide sub-units of the peptidoglycan into polymers. A peptidoglycan complex, possessing a linear, non-crosslinked structure composed of disaccharide-pentapeptide units, has been isolated from a penicillin-treated mutant of *Brevibacterium divaricatum* requiring biotin for growth.

The peptidoglycan fragments appear to be synthesized after incubation of the cells with penicillin. A peptidoglycan, of different type from that of other species of *Cellulo-monas,* has been isolated from *C. cartalyticum. L-Lysine* replaces ornithine and the interpeptide bridge consists of D-aspartyl-D-serine. The same peptido-glycan type has been identified in *Arthrobacter luteus, Brevibacterium liticum,* and *Corynebacterium manihot.* The peptidoglycan isolated from the photosynthetic organelles of *Cyanaphora paradoxa* contains 2-acetamido-2-deoxy-D-glucose, *N*-acetylmuramic acid, L-alanine, L-glutamic acid, and diaminopimelic acid (1 : 1 : 1.6 : 1 : 1) and may be present as a lipoprotein-peptidoglycan complex.

Close similarities to the structures of other Gram-negative bacteria and cyanobacteria were observed preferential orientation of the glycan chains perpendicular to the axis of the cell has been observed in an electron microscopic study of the sacculus of *Escherichia coli.* Selective partial hydrolysis of inter-peptide bridges was achieved with an *endo*-peptidase. The regulation of peptidoglycan synthesis in *rel* A^+ and *rel* A^- strains of *Escherichia coli* during diauxic growth on D-glucose and lactose has been reported. In both strains, the biosynthesis of peptidoglycan, lipid intermediates, and nucleotide precursors abruptly halted at the onset of diauxic lag from D-glucose to lactose, with the concomitant accumulation of guanosine 5'-diphosphate and guanosine 3'-diphosphate.

This is consistent with the proposal that the accumulation is involved in the inhibition of the incorporation of disaccharide-pentapeptide into peptidoglycan and in regulating nucleotide precursor synthesis. Lanthionine, a monosulphated analogue of diaminopimelic acid, has been isolated from the peptidoglycan of *Fusobacterium nucleatum,* which contained all the other normal constituents, with the exception of diaminopimelic acid. Glycopeptides have been isolated from the cell walls of *Lacto bacillus plantarum.* Arthritogenic and immunoadjuvant activities are associated with peptidoglycan fractions having an average glycan chain length of more than five Uisaccharide units.

The peptidoglycan of *L. vaccinostercus* contains *meso*diaminopimelic acid, an unusual component in heterofermentative lactobacilli. The turnover of cell walls during myxospore formation in *Myxococcus xanthus* has been studied by measuring the incorporation and release of *meso*-diamino[^{14}C]pimelic acid.

Following addition of glycerol to induce myxospore formation, there is a greatly increased rate of incorporation, resulting in substantial replacement of existing wall with newly synthesized wall. None of the representatives of four genera of methanogenic bacteria grown in pure culture was found to contain typical peptidoglycan as a cell-wall polymer. However, each of the four genera was found to contain different cell-wall polymers, indicating a very early divergence, not only of the methanogens from all other prokaryotes, but also of the various genera of the methanogens from one another. The appearance of soluble peptidoglycan fragments and concurrent turnover of the glycan and peptide region of peptidoglycan from growing Gonococci indicate that both glycan-splitting and peptidoglycan amidase activities are involved.

If released during gonococcal infections, these or similar soluble peptidoglycan fragments might influence the consequences of host-gonococcus interactions. Muramic acid has been detected in the cell walls of the anomalous alga *Prochloron* species, which although prokaryotic, shares several factors with the Chlorophyta. Lipoprotein, covalently linked to peptidoglycan, has been found absent from exponential phase cultures, but present in stationary phase cultures of *Proteus mirabilis.* The overall peptidoglycan structure did not change during transition from logarithmic to stationary growth. The covalently linked lipoprotein is not as essential for cell-wall stability in this organism as it apparently is in *Escherichia coli.*

It is possible that there is a higher degree of hydrophobicity in the rigid layer, owing to the presence of O-acetyl groups on some of the muramic acid residues, and that this suffices for the association of the outer-membrane components to the peptidoglycan later in exponentially growing cells. Compearison of the structural features in normalpeptidoglycan and the peptidoglycan of the unstable sphaeroplast L-form of *Proteus mirabilis* confirms earlier reports of a continuation of the peptide cross-linking reaction during L-form growth and peptidoglycan synthesis in the presence of benzylpenicillin.

However, the peptidoglycans formed, in the presence or absence of the antibiotic, were found to differ in the degree of *O*-acetylation of their muramic acid residues. The mechanism of peptidoglycan synthesis in *Pseudomonas aeruginosa* appears to be similar to that of *Escherichia coli,* although evidence has suggested that in the former the DD-carboxypeptidase and transpeptidase systems may have somewhat different specificity restrictions from those in *E. coli*. Some similarities in the effects of Bactam antibiotics on the biosynthesis of peptidoglycan in the two bacteria were also observed. A combined ^{15}N- and ^{13}C-n.m.r. spectroscopic study of the chain segmental motion of the peptidoglycan-pentaglycine chain of [^{15}N]glycine- and [2-^{13}C]-glycine-labelled *Staphylococcus aureus* cells and isolated cell walls has been reported.

Main-chain motions of the pentaglycine bridge have been interpreted in terms of flexible pentaglycine chains attached to immobile glycan strands, whose packing arrangements can vary as a function of temperature and growth conditions. Although cell surface components of *Staphylocoecus aureus,* including teichoic acid, peptidoglycan, and protein A, have all been shown to be capable of activating the classical complement pathway, only peptidoglycan consumed complement *via* the alternative pathway. Using a quantitative immunofluorescence assay, the peptidoglycan was shown to bind to the C-3 component of complement in both pathways and in the absence of immunoglobulins G and A antibodies. Peptidoglycan synthesis in *Staphylococcus aureus* is inhibited by 3-amino-3-deoxy-D-glucose. The primary mode of action appears to be the prevention of formation of 2-amino-2-deoxy-D-glucose 6-phosphate from D-fructose 6-phosphate. The physical state of the membrane lipid in *Staphylococcus aureus,* as measured by fluorescence and e.s.r. spectroscopy, has been shown to affect the activity of the phospho-*N*-acetylmuramylpentapeptide transferase (E.G. 2.7.8.13). A correction to conclusions from earlier studies on the biosynthesis of the peptidoglycan of *Staphylococcus aureus* has been occurrence of 2-amino-2-deoxy-D-glucosyl residues in the cell-wall peptidoglycan of group A *Streptococcus pyogenes* type 4 has been confirmed? The ultrastructural localization of peptidoglycan in streptococcal cell walls has been explored using an indirect immunoferritin technique.

The ferritin particles were bound predominantly to filamentous structures, which were produced both from surfaces of isolated peptidoglycan fragments and from isolated walls. The results contradict proposed models of the streptococcal wall in which the peptidoglycan is thought to form the innermost layer of the wall. The alternative view is of a mosaic structure in which peptidoglycan forms a network of peptidoglycan-polysaccharide complexes.

LIPOPOLYSACCHARIDES

A radioimmunoassay for bacterial lipopolysaccharide has been developed. The lipopolysaccharide could be derivatized and radiolabelled with. ^{125}I without apparent loss in its biophysical, immunological, or biological activities. From studies of the interactions between lipopolysaccharide and phosphatidyl ethanolamine in molecular monolayers at airwater interfaces, it was concluded that each lipopolysaccharide molecule is surrounded by approximately sixteen phosphatidyl ethanolamine molecules.

The hydrocarbon chains of lipopolysaccharides can undergo a reversible thermal order-disorder transition, as shown by thermal phase-transition studies. The ordered conformation of lipopolysaccharides interpreted from wide angle *X*-ray studies is considered less developed than that in normal phospholipid bilayers. Pulse-chase studies with D-galactose on suspensions of *Alteromonas haloplanktis* have been used to follow the progress of newly synthesized lipopolysaccharide through the three outermost layers of the cell wall. The lipopolysaccharide moves through the periplasmic space, through the outer membrane, and spreads over the outer surface of the organism; its presence in each of the three layers was confirmed by polyacrylamide gel electrophoresis. The lipopolysaccharide of *Fusobacterium nucleatum* has been coupled to horse radish peroxidase and used for locating the lipopolysaccharide binding site on erythrocytes.

Fluorescence analysis of dansylated derivatives of various lipopolysaccharides has been used to define the nature of metal binding sites of lipopolysaccharides. A high affinity binding site for Ca^{2+} and Mg^{2+}, which appears to be formed by the 3-deoxy-D-*manno*-octulosonate trisaccharide unit, is involved in the assembly and maintenance of the normal structural organization of the outer membrane.

Lipopolysaccharide-protein interactions, as well as protein-protein interactions, have been shown to be significant in maintaining ghost structures in strains of *Escherichia coli.* The recognition of unopsonized bacteria by mouse peritoneal macrophages is thought to involve capsular and lipo-polysaccharides.

Bacterial lipopolysaccharides have been degraded by enzymic preparations from *Helix pomatia,* resulting in an extensive loss of anticomplementary activity. The lipid A moiety of the lipopolysaccharide appeared to be the main site of attack with little evidence for the degradation of the polysaccharide component. Two fatty acyl amidases, amidases I and II, purified from *Dictyostelium discoideum,* act upon the disaccharide (3, $R = R^1 = H$, $n = 1$), which has been isolated from the lipopolysaccharide of a heptose-less mutant of *Escherichia coli* K-12.

Amidase I removes the 3-hydroxymyristoyl unit, which is esterified to the amino-group of the 2-amino-2-deoxy-D-glucosyl residue adjacent to the C-1 phosphate of the disaccharide. The product is then acted upon by amidase II, which removes the remaining 3-hydroxymyristoyl unit from the amino-group on the distal 2-amino-2-deoxy-D-glucosyl residaue. Amino-sugars derived from bacterial lipopolysaccharides have been identified by g.l.c.–electron impact-m.s. and g.l.c.-chemical ionization-m.s. The absence of amino-sugars from the lipopolysaccharide (mol. wt. 8.5×10^4) isolated from *Acholeplasma oculi* has been reported.

Neutral sugars (L-fucose, D-galactose, D-glucose, 1 : 12 : 3) were detected but the lipopolysaccharide elicited no antibody response in rabbits. Treatment of the lipopolysaccharide of a mutant of *Escherichia coli,* containing no heptose, with alkali released two phosphorylated components, LPS I and II, both consisting of a substituted 2-amino-2-deoxy-D-glucosyl disaccharide, but differing in the nature of the phosphate groups present at the glycosyl position. Their structures (3) were determined. LPS I contained two phospho monoester groups, one

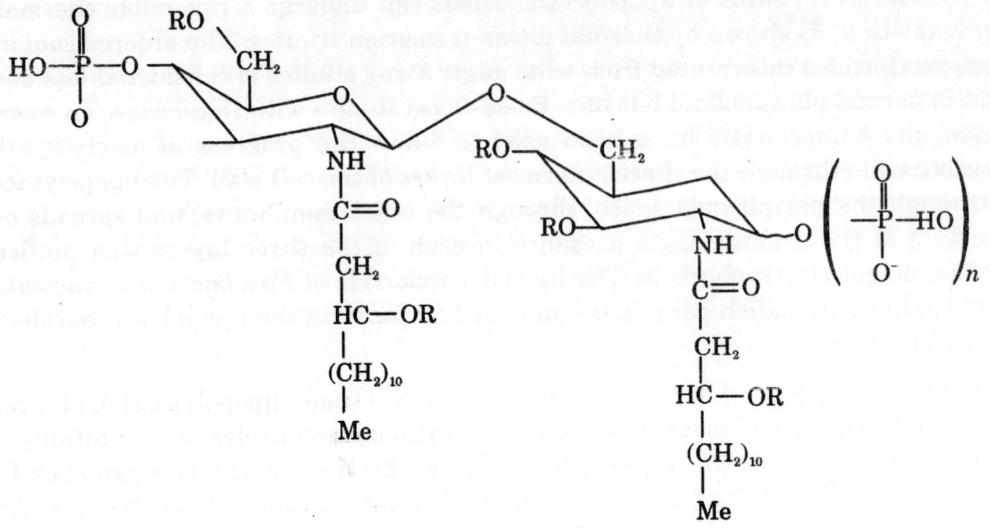

n = 1 or 2

R = fatty acyl residues or H (acyl groups are myristic, lauric, and β-hydroxymyristic acids)

R′ = 3 moles of 3-deoxy-D-Hmanno-2-octulosonic acid

(3)

at the 4′-hydroxy-position and one at the glycosyl position. The remaining hydroxy-groups were substituted with lauryl, myristoyl, and 3-hydroxymyristoyl groups although the exact assignment of the different fatty acyl groups to specific hydroxy-groups in the amino-sugar rinks was not determined.

Further information on the nature of the phosphate linkages has been obtained from P n.m.r. spectroscopic studies on the original lipopolysaccharide and on two products of alkaline treatment. The structure of the K-13 antigenic polysaccharide of *Escherichia coli* consists of a repeating sequence of 3-linked D-ribofuranosyl residues and 7-linked 3-deoxy-D-manno-octulosonic acid (KDO) residues, approximately 50% of which are O-acetylated at C-4 or C-5. The serological specificity of this polysaccharide is expressed through KDO and its O-acetylated derivative. The polysaccharide chain of the cell wall lipopolysaccharide of *Escherichia coli* O58 contains D-mannose, 2-acetamido-2-deoxy-D-mannose, 3-*O*-[(1′ *R*)-carboxyethyl)]-L-rhamnose (rhamnolytic acid), and *O*-acetyl groups in the molar ratios of 2 : 1 : 1 : 1.

Structural investigations show that this polysaccharide, which contains a substituted trisaccharide repeating unit, has an identical structure (4) to that of *Shigella dysenteriae* type 5.

$$\rightarrow 3)\text{-}\beta\text{-D-Glc}p\text{NAc-}(1 \rightarrow 4)\text{-}\alpha\text{-D-Man}p\text{-}(1 \rightarrow 4)\text{-}(\alpha\text{-D-Man}p\text{-2/3-OAc})\text{-}(1 \rightarrow$$

$$\begin{array}{c} 3 \\ \uparrow \\ 1 \\ \text{L-RhaLA} \end{array}$$

L-RhaLA = rhamnolytic acid

(4)

The O8- and O9-specific lipopolysaccharides of *Escherichia coli* lose their serological activity upon mild acid treatment to liberate their polysaccharide (mannan) moieties.

Restoration of the activity was achieved by substitution of one or two stearoyl groups per polysaccharide chain. The D-mannans obtained by *in vitro* biosynthesis were serologically active only when bound to the membrane-associated hydrophobic carrier. After removal of lipid A from the lipopolysaccharide of *Fusobacterium nucleatum,* the resultant high molecular weight fraction, which was devoid of phosphate and 3-deoxy-D-*manno*-octulosonic acid, was composed of *L-glycero-D-manno*-heptosyl-, 2-amino-2-deoxy-D-glucosyl-, and D-glucosyl-residues. Serological evidence has shown that this fraction contains the O-antigenic side-chains of the lipopolysaccharide. Alterations in the cell envelope of *Pseudomonas aeruginosa* resistant to polymyxin are associated with the alteration of the outer membrane through a diminution of lipopolysaccharide and outer membrane proteins.

Analysis of the lipopolysaccharides of *P. aeruginosa* PAO and three mutants selected for resistance to a virulent lipopolysaccharide-specific bacteriophage shows alterations in both O-specific side-chains and core regions. In a typical lipopoly-saccharide, isolated from the walls of *P. cepacia,* phosphate, L-rhamnose, D-glucose, heptose, and amino-sugars were found but there was no deoxy-octulosonic acid. The lipid A component of the lipopolysaccharide of *P. diminuta* has been reported to contain 2,3-diamino-2,3-dideoxy-D-glucose instead of 2-amino-2-deoxy-D-glucose. Some *Rhodopseudomonas* species also contain this unusual amino-sugar. The lipopolysaccharides of three strains each of *Rhizobium leguminosarum, R. phaseoli,* and *R. trifolii* have been purified and partially characterized.

The composition and immunodominant structures of the lipopolysaccharides were found to vary as much among strains of a single *Rhizobium* species as among the different species of *Rhizobium.* No obvious correlation between the nodulation group to which a *Rhizobium* species belongs and the chemical composition or immunochemistry of the lipopolysaccharides was observed. A component of the lipopolysaccharide of *R. trifolii,* previously designated as 'heptose I' is not *D-glycero-D-manno*-heptose but a 3-*O*-methylheptose. Heterogeneity of the lipopolysaccharide moiety of the intact rhizobium of free-living and bacterial forms of *R. leguminosarum* has been reported. A low molecular weight polypeptide exhibiting high affinity for lipopolysaccharides has been isolated from the cell surface of a heptose-less mutant of *Salmonella minnesota.*

Antiserum adsorption studies reveal that this protein also represents a common antigen of the Enterobacteriaceae. Antibodies against the synthetic disaccharide 3-*O*-α-abequosyl-L-rhamnose, representative of *Salmonella* O-antigen 8, have been used in the diagnosis of *Salmonella* bacteria by indirect immunofluorescence and co-agglutination using sensitized protein A-containing staphylococci. The octasaccharide (5), the synthesized disaccharide methyl

$$\begin{array}{l}
\alpha\text{-D-Gal}p\text{-}(1\rightarrow 2)\text{-}\alpha\text{-D-Man}p\text{-}(1\rightarrow 4)\text{-}\alpha\text{-L-Rha}p\text{-}(1\rightarrow 3)\text{-}\alpha\text{-D-Gal}p\text{-}(1\rightarrow 2)\text{-}\alpha\text{-D-Man}p\text{-}(1\rightarrow 4)\text{-D-Rha}\\
\qquad\qquad\qquad\quad 3 \qquad\qquad\qquad\qquad\qquad\qquad\qquad\qquad\qquad\qquad 3\\
\qquad\qquad\qquad\quad \uparrow \qquad\qquad\qquad\qquad\qquad\qquad\qquad\qquad\qquad\qquad \uparrow\\
\qquad\qquad\qquad\quad 1 \qquad\qquad\qquad\qquad\qquad\qquad\qquad\qquad\qquad\qquad 1\\
\qquad\qquad\quad \alpha\text{-Tyv}p \qquad\qquad\qquad\qquad\qquad\qquad\qquad\qquad\quad \alpha\text{-Tyv}p
\end{array}$$

Tyv*p* = tyvelopyranosyl residue.

(5)

3-*O*-α-tyvelopyranosyl-β-D-mannopyranoside, and methyl α-tyveloside inhibit the precipitation of *S. typhi* T2 alkali-treated lipopolysaccharide by O-factor 9 antibodies.

From the relative effectiveness of the oligosaccharides in inhibiting the precipitation reaction it appears that the 3-*O*-α-tyvelopyranosyl-D-mannopyranosyl structure is immunodominant in the *Salmonella* O9 antigen. 3-*O*-α-Tyvelopyranosyl-D-mannose coupled to bovine serum albumin elicited O-antibodies of higher specificity than those obtained by adsorption of antibacterial immune serum. The polysaccharide moiety of the lipopolysaccharide of *Schizothrix calcicola* contains 2-amino-2-deoxy-D-glucose, D-galactose, D-glucose, D-mannose, D-xylose, and L-rhamnose. In contrast to many enterobacterial lipopolysaccharides, the major fatty acid component has been identified as 3-hydroxy-palmitic acid and not 3-hydroxymyristic acid. A lipopolysaccharide has been isolated from *Shigella flexneri,* type 2, and some of its structural aspects have been elucidated. The specific polysaccharide isolated from the lipopolysaccharide of *Shigella newcastle (S. flexneri,* type 6) contains L-rhamnose, 2-acetamido-2-deoxy-D-galactose, D-galacturonic acid, and *O*-acetyl groups (in the molar ratio 2 : 1 : 1 : 1).

On the basis of ^{1}H and ^{13}C n.m.r. spectroscopy, methylation analysis, partial acid hydrolysis, and chromium oxide oxidation, the structure (6) has been assigned. Both structural and immunochemical evidence indicate that the lipopolysaccharide from *Shigella newcastle* belongs to the 'non-classical' type of somatic antigens with acidic O-specific polysaccharide chains. The hexose region of the core polysaccharides of lipopolysaccharides isolated from *Shigella flexneri* and some *Escherichia coli* species have a common structure (7).

$$[\rightarrow 4]\text{-}\alpha\text{-Gal}p\text{A-}(1 \rightarrow 3)\text{-}\alpha\text{-Gal}p\text{NAc-}(1 \rightarrow 2)\text{-}\alpha\text{-L-Rha}p\text{ 3Ac-}(1 \rightarrow 2)\text{-}\alpha\text{-L-Rha}p\text{-}(1 \rightarrow]_n$$

(6)

$$\begin{array}{c} \alpha\text{-D-Glc}p\text{-}(1 \rightarrow 2)\text{-}\alpha\text{-D-Glc}p\text{-}(1 \rightarrow 2)\text{-}\alpha\text{-D-Gal}p\text{-}(1 \rightarrow 3)\text{-}\alpha\text{-D-Glc}p\text{-}(1 \rightarrow \\ \qquad\qquad\qquad\qquad\qquad\qquad\qquad\qquad 3 \\ \qquad\qquad\qquad\qquad\qquad\qquad\qquad\qquad \uparrow \\ \qquad\qquad\qquad\qquad\qquad\qquad\qquad\qquad 1 \\ \qquad\qquad\qquad\qquad\qquad\qquad\qquad\qquad \alpha\text{-D-Glc}p\text{NAc} \end{array}$$

(7)

A lipopolysaccharide isolated from *Vibrio cholerae* 4715 (NAG) has been reported to be devoid of 2-amino-2,6-dideoxy-D-glucose (quinovosamine), a characteristic sugar of cholera vibrios, but to contain significant amounts of glycine in the lipid A moiety.

The occurrence of odd-numbered fatty acids in the lipopolysaccharides of a number of *Vibrio cholerae* strains has been reported. A high molecular weight heptose-free fraction representing the O-specific side-chain polysaccharide, and a low molecular-weight fraction containing heptose and representing the core polysaccharide region have been isolated from the lipopolysaccharides from *Vibrio cholerae* and *V. el-tor* (Indaba). The core polysaccharide is composed of residues of D-glucose, D-mannose, heptose *(D-glycero-L-manno*-heptose and *D-glycero-L-gluco*-heptose), 2-amino-2-deoxy-D-glucose, and D-glucuronic acid (in molar ratio 9 : 4 : 5 : 1 : 2 : 5).

A branched structure for the polysaccharide has been proposed, with (1 → 2)-D-manno-pyranosyl-, (1 → 4)-D-glucopyranosyluronic acid-, heptopyranosyl-. and 2-amino-2-deoxy-D-glucosyl-residues in the interior part of the molecule with D-glucopyranosyl- and heptopyranosyl-residues as the non-reducing end groups. Mild acid hydrolysis of the lipopolysaccharide released an oligosaccharide, representing the O-antigen of the lipopolysaccharide and composed of (1 → 2)-D-perosamine (4-amino-4,6-dideoxy-D-mannose) residues. The amino-groups are acylated with 3-hydroxypropionyl residues. Some of the toxic and immunological properties of the lipopolysaccharide and lipid A moiety of *V. el-tor* have been described. By use of different extraction procedures, two types of lipopolysaccharides have been isolated from a slightly virulent strain of *Yersinia pestis.*

Some structural information was obtained from methylation and periodate oxidation studies. The immunodominant moiety of lipid A has been shown to consist of a residue of 2-amino-2-deoxy-D-glucose, *N*-acylated with 3-hydroxytetradecanoic acid. A lipopolysaccharide isolated from *Y. enterolitica* exhibits serological cross-reactions with lipopolysaccharides from *Brucella abortus* and *Vibrio cholerae.* Common antigenic relationships between the lipopolysaccharides of *Escherichia coli, Kloechera africana,* and *Salmonella aberdeen* have been demonstrated. A survey of the composition of lipopolysaccharides and exopolysaccharides produced by *Cystobacter, Archangium, Sorangium,* and *Stigmatella* species has been reported. 3-*O*-Methyl-D-xylose was detected in several lipopolysaccharides. UDP-D-glucose, in addition to being required for the synthesis of the core oligosaccharide region of the lipopolysaccharide of *Escherichia coli* and for the synthesis of UDP-D-galactose, is now recognized to be required as a precursor for the D-glucosyl residues of the membrane-derived oligosaccharides. Mutant strains of *E. coli* specifically blocked in the synthesis of UDP-D-glucose were unable to synthesize membrane-derived oligosaccharides. A technique has been described for the selection of hybrid plasmids from *E. coli* and *Salmonella typhimurium,* containing genes for the glycosyl transferases involved in lipo-polysaccharide biosynthesis. Plasmids carrying the *rfa* G^+ gene, responsible for the synthesis of UDP-D-glucose: (heptosyl) lipopolysaccharide D-glucosyl-transferase, and the *rfa* M^+ gene, responsible for the synthesis of UDP-D-glucose: (D-glucosyl) lipopolysaccharide 1,3-D-glucosyltransferase have been identified. In addition, the identity of a plasmid carrying a gene capable of restoring the activity of UDP-D-galactose :(D-glucosyl)lipopolysaccharide α-1,3-D-galactosyl transferase in an *rfa* H mutant of *Salmonella typhimurium* has been reported.

The D-galactosyl transferase was also identified in *E. coli* K12 cells despite the apparent absence of the product of the enzyme-catalysed reaction in the cellular lipopolysaccharide of that organism. A glucolipid is considered to be an intermediate in the sequence of reactions involved in the biosynthesis of the O-9-D-mannan of *Escherichia coli.* The nature of the lipid, the characterization of the glucosylation reaction, and the sequence of the D-mannosyl transfer reaction have still to be clarified. The capacity of membranes derived from *E. coli* O8 and O9, when grown in the absence of D-glucose, to synthesize D-mannan, is stimulated by lipid-deriyed material from cells of either strain that had been grown on D-glucose.

It is possible that the first stage of the pathways of D-mannan synthesis may be the same, namely the synthesis of a hydrophobic D-glucosyl-D-mannose acceptor group. The incorporation of 2-acetamido-2-deoxy-D-glucose into the core region of the lipopolysaccharide of *E. coli* involves a 2-acetamido-2-deoxy-D-glucosyl-transferase that is localized in the inner membrane of the organism. The endogenous acceptor for the enzyme, which is also localized at the inner membrane, appears to be the basic polysaccharide structure of the organism. Subsequent to the incorporation of the amino-sugar, the polymer is assembled into a protein-lipopolysaccharide complex. Lactose, covalently attached to bovine serum albumin, has been used to elicit antilactose antibodies that are active against *Neisseria gonorrhoeae* lipopoly-saccharide.

Gonococcal cell-wall polysaccharide is immunogenic in BALB/c mice, but the kinetics of the response differ from those of *Salmonella typhosa* antigen and pneumococcal type III polysaccharide. Resistance against ascites tumour development and interferon-inducing activity has been demonstrated in the lipopolysaccharide derived from the protein-lipopolysaccharide complex of *Pseudomonas aeruginosa.* Both the lipid. A portion and the covalently linked polysaccharide-protein complex appear to be necessary for inhibition of tumour development, whereas the lipid A with amide-linked fatty acid is suficient to induce the *in vitro* formation of interferon. The lipopolysaccharide of *P. aeruginosa* has a more pronounced effect as immunogen, mitogen, and interferon inducer than the lipopoly-saccharide of *Brucella melitensis,* although they have similar toxicity levels. Since lipid A is responsible for most of these activities, the variations may be partially explained by structural differences in the lipopolysachartdis. Further evidence has supported the theory that the interaction between lectins of host plants and *Rhizobium* is the key to the mechanism of host specificity in *Rhizobium*-legume symbiosis and that the lectin receptor of Rhizobial cells exists in the cell surface lipopolysaccharide.

No evidence was obtained to show interaction of extracellular polysaccharides with the lectins. It is possible that a specific recognition system may involve the specific binding of, for instance, pea *(Pisum sativum)* root hairs to an antigen that is situated on the lipopolysaccharide of infective strains of *R. leguminosarum.* Some aspects of the mechanism of O-specific polysaccharide biosynthesis in *Salmonella newport* and *S. kentucky* have been reported. Cell envelope and soluble glycosyltransferase preparations from both organisms catalysed the formation of polyprenyl pyrophosphate oligosaccharides with carbohydrate units corresponding to the structures of the main-chain polysaccharides (Scheme 1). The D-glucosylation reaction occurs before the polymerization of the repeating units to the polysaccharide. The bindmg sites for lipopolysaccharides on mouse lymphocytes have been identified as membrane proteins.

Active fractions isolated by affinity chromatography from B cell membranes were identified as immunoglobulins, possibly IgM and IgD, whilst the histocompatibility-2 complex proteins (H-2D, H-2K, and la antigens) were found to be binding sites for lipopolysaccharides on both B and T cells. Some aspects of the chemistry and biology of bacterial lipopolysaccharides have been reviewed.

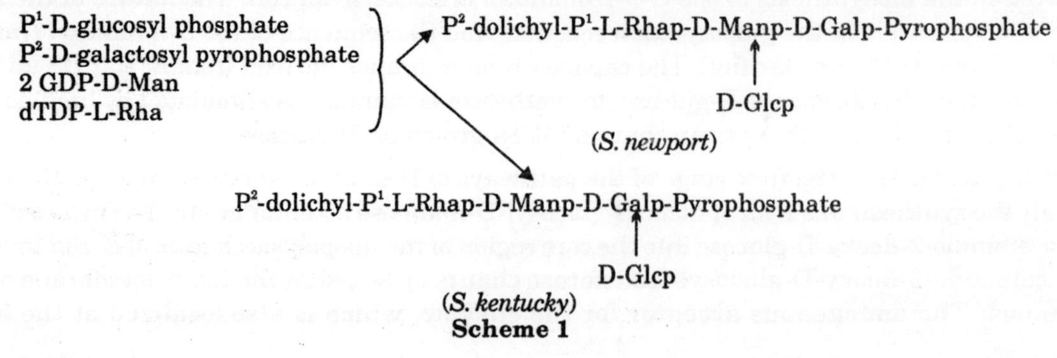

Scheme 1

CAPSULAR POLYSACCHARIDES

Colominic acid, a homopolymer of (2 → 8)-α-*N*-acetylneuraminic acid residues, is the *Escherichia coli* K1 capsular polysaccharide. Random *O*-acetylation at C-7 and C-9 of the polysaccharide, as determined by ^{13}C n.m.r. spectroscopy, has been shown to be responsible for the alternating antigenic change (form variation) associated with the organism. A serological method has been established for the identification and isolation of the form variants. *X*-Ray diffraction patterns for the capsular polysaccharides of *Escherichia coli* K29 and its M13 and M41 mutants have indicated that the molecular conformation and packing are the same for all the polysaccharides, suggesting that their primary structures are essentially unchanged.

Similar phage specificity and antibody reactions, together with the similarity of the diffraction patterns, indicated an overall chemical identity of the polysaccharides elaborated by the mutant strains with that of the parent strain. A model with anti-parallel two-fold helices at the corner and centre of the unit cell was proposed. The interaction of phage K29 with its host cell, *E. coli* K29, has been studied by use of combined virological and immunological methods.

The virion penetrates the capsule by a mechanism which involves receptor destruction and results in the release of DNA over a membrane-adhesion site. Fully methylated and fully acetylated disaccharides isolated from the capsular polysaccharide of *Escherichia coli* K100 and *Haemophilus influenzae* type b have been identified by m.s. as 2-*O*- and 1-0-D-ribofuranosyl ribitol, respectively. The capsular antigen of *H. influenzae* type c has been purified by ion-exchange chromatography. Structural investigations have identified the disaccharide (8) as the repeating unit of the polysaccharide. The highly branched capsular polysaccharide from *Cryptococcus neoformans* serotype c consists of a backbone of (1 → 3)-α-D-mannopyranosyl residues with single chain non-reducing (1 → 2)- and (1 → 4)-β-D-xylopyranosyl- and (1 → 2)-β–D-glucopyranosyl uronic acid residues.

```
                                                      O
                                                      ‖
→ 4)-β-D-GlcpNAc 3Ac-(1 → 3)-α-D-Galp-(1—O—P—O—
                                                      |
                                                      O⁻
```

(8)

```
   β-D-Xylp                                β-D-GlcpA
      1                                        1
      ↓                                        ↓
      2                                        2
→ 3)-α-D-Manp(1 → 3)-α-D-Manp-(1 → 3)-α-D-Manp-(1 →
      4                 2                      4
      ↑                 ↑                      ↑
      2                 1                      2
   β-D-Xylp          β-D-Xylp              β-D-Xylp
```

(9)

$$
\begin{array}{ccc}
\rightarrow 3)\text{-}\alpha\text{-D-Man}p(1 \rightarrow 3)\text{-}\alpha\text{-D-Man}p\text{-}(1 \rightarrow 3)\text{-}\alpha\text{-D-Man}p\text{-}(1 \rightarrow \\
\begin{array}{cc} 2 & 2 \\ \downarrow & \downarrow \\ 1 & 1 \\ \beta\text{-D-Xyl}p & \beta\text{-D-Glc}p\text{A} \end{array}
\end{array}
$$

(10)

Further examination of the polysaccharide (9) by alkaline degradation, followed by methylation, revealed that the D-glucuronic acid residues are located on D-mannosyl residues already substituted with a D-xylosyl residue. The capsular polysaccharide of *C. neoformans* serotype D is less highly branched, but with many structural similarities to (10). When high molecular weight pneumococcal capsular polysaccharide is injected intraperitoneally into rats, the antigen is excreted in the urine.

Examination of the molecular weight profile of the excreted product indicates that significant degradation has taken place. Pneumococcal type 19 polysaccharide is composed of residues of D-glucose, L-rhamnose, 2-acetamido-2-deoxy-D-mannose, and phosphate. Strong acidic hydrolysis has been found to be accompanied by migration of the phosphate groups. The structure of the capsular antigen of *Streptococcus pneumoniae* type 26 has been established (11). The only difference between the type 26 and the type 6 antigen is that the α-L-rhamnopyranosyl residue is linked to O-4 of the D-ribitol moiety in the former but to *O*–3 in

$$
\rightarrow 2)\text{-}\alpha\text{-D-Glc}p\text{-}(1 \rightarrow 3)\text{-}\alpha\text{-D-Gal}p\text{-}(1 \rightarrow 3)\text{-}\alpha\text{-L-Rha}p\text{-}(1 \rightarrow 4)\text{-D-ribitol-5—O—}\overset{\overset{\displaystyle O}{\|}}{\underset{\underset{\displaystyle O^-}{|}}{P}}\text{—O—}
$$

(11)

the latter. The configuration of the 4,6-0-pyruvic acid acetal in *S. pneumoniae* type 27 has been established from ^{1}H and ^{13}C n.m.r. spectral data of the natural product and model compounds.

Inhibition of the homologous precipitin reaction by synthetic diastereoisomeric pyruvic acid acetal showed that the configuration of the pyruvic acid acetal contributes to the serological specificity of the type 27 polysaccharide. As a result of the elucidation of the chemical structures of many *Pneumococci* and *Klebsiella* species, several previously uninterpretable serological cross-reactions have now been rationalized. The acidic capsular polysaccharides of *Klebsiella* species have been visualized by application of polycationic derivatives of ferritin which bind electrostatically to anionic sites on the surface of other glutaraldehy de-fixed or unfixed cells. Pyruvic acid has been identified in approximately half of seventy-seven *Klebsiella* species investigated.

On the basis of o.r.d. measurements and their interpretation from Moffitt-Yang plots, a helical superstructure is proposed for *Klebsiella* capsular polysaccharides of serotypes K1, K5, K6, K8, K11, K56, and K57. Measurements of the temperature dependence of the specific optical rotation reveal, in all cases, co-operative order-disorder transitions at temperatures between 25 and 40°C. Structures have been established for the capsular polysaccharides of *Klebsiella* K30 (12), K31 (13), K33 (14), K61 (15), K63 (16), and K64 (17). Limited *X*-ray studies on the *Klebsiella* K30 capsular polysaccharide suggest the existence of a trisaccharide-repeating structure with all the glycosidic link-ages disposed 1,4-diequatorial and producing a two-fold helix with a ribbon-like appearance. The antigenic determinant of hyporesponsiveness in BALB/c mice to *K. pneumoniae* type 47 has been reported to be L-rhamnose.

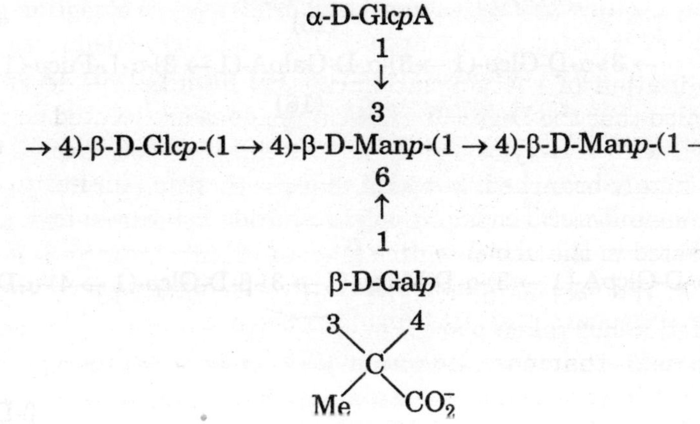

α-D-GlcpA
1
↓
3
→ 4)-β-D-Glcp-(1 → 4)-β-D-Manp-(1 → 4)-β-D-Manp-(1 →
6
↑
1
β-D-Galp
3 4
C
Me CO_2^-

(12)

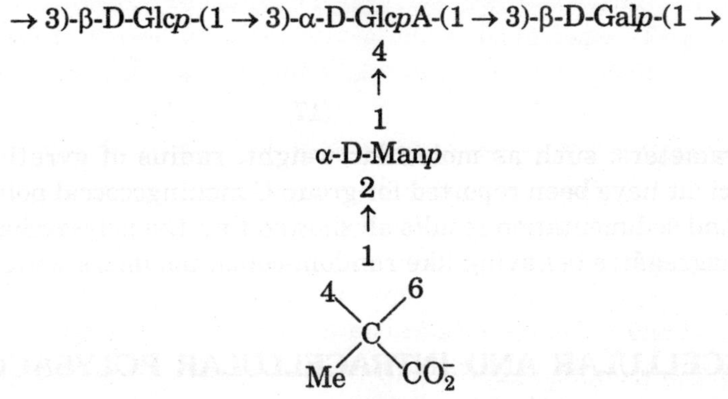

→ 3)-β-D-Glcp-(1 → 3)-α-D-GlcpA-(1 → 3)-β-D-Galp-(1 →
4
↑
1
α-D-Manp
2
↑
1
4 6
C
Me CO_2^-

(13)

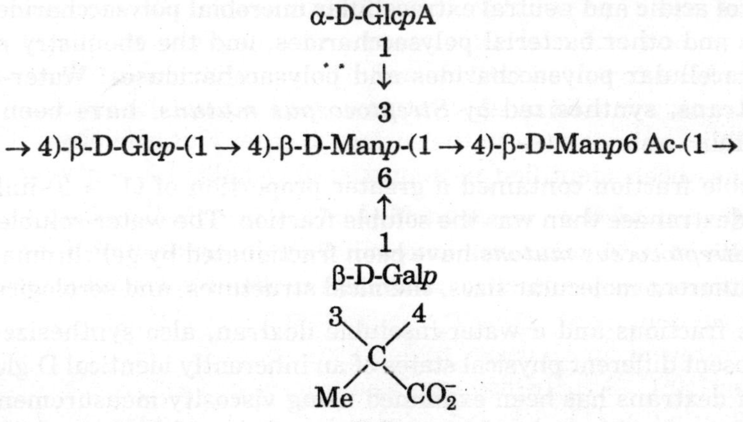

α-D-GlcpA
1
↓
3
→ 4)-β-D-Glcp-(1 → 4)-β-D-Manp-(1 → 4)-β-D-Manp6 Ac-(1 →
6
↑
1
β-D-Galp
3 4
C
Me CO_2^-

(14)

$$
\begin{array}{c}
\rightarrow 4)\text{-}\beta\text{-D-Glc}p\text{A-}(1 \rightarrow 2)\text{-}\alpha\text{-D-Man}p\text{-}(1 \rightarrow 3)\text{-}\beta\text{-D-Glc}p\text{-}(1 \rightarrow 6)\text{-}\alpha\text{-D-Glc}p\text{-}(1 \rightarrow \\
3 \\
\uparrow \\
1 \\
\alpha\text{-D-Gal}p
\end{array}
$$

(15)

$$\rightarrow 3)\text{-}\alpha\text{-D-Glc}p\text{-}(1 \rightarrow 3)\text{-}\alpha\text{-D-Gal}p\text{A-}(1 \rightarrow 3)\text{-}\alpha\text{-L-Fuc}p\text{-}(1 \rightarrow$$

(16)

$$
\begin{array}{r}
\alpha\text{-L-Rha}p \\
1 \\
\uparrow \\
3 \\
\rightarrow 4)\text{-}\alpha\text{-D-Glc}p\text{A-}(1 \rightarrow 3)\text{-}\alpha\text{-D-Man}p\text{-}(1 \rightarrow 3)\text{-}\beta\text{-D-Glc}p\text{-}(1 \rightarrow 4)\text{-}\alpha\text{-D-Man}p\text{-}(1 \rightarrow \\
6 \\
\downarrow \\
1 \\
\beta\text{-D-Gal}p \\
4 \quad 6 \\
C \\
Me \quad CO_2^-
\end{array}
$$

(17)

Physical parameters, such as molecular weight, radius of gyration, and translational diffusional coefficient have been reported for group C meningococcal polysaccharide. Viscosity, light scattering, and sedimentation results all showed that the polysaccharide forms aggregates in solution, the ággregates behaving like random-coiled polymers with no measurable shape factors.

EXTRACELLULAR AND INTRACELLULAR POLYSACCHARIDES

Several review articles on extracellular polysaccharides have dealt with such topics as the structures of acidic and neutral extracellular microbial polysaccharides, the production of xanthan gums and other bacterial polysaccharides, and the chemistry and biochemistry of microbial extracellular polysaccharides and polysaccharidases. Water-soluble and water-insoluble dextrans, synthesized *by Streptocorpus mutans,* have been fractionated from solutions of alkali.

The insoluble fraction contained a greater proportion of (1 → 3)-linkages and was less susceptible to dextranase than was the soluble fraction. The water-soluble dextrans produced by a strain of *Streptococcus mutans* have been fractionated by gel chromatography into three types having different molecular sizes, chemical structures, and serological properties.

One of the fractions and a water-insoluble dextran, also synthesized by the organism, appear to represent different physical states of an inherently identical D-glucan. The molecular compression of dextrans has been examined using viscosity measurements. Marked osmotic compression of the flexible molecular structures occurs within the semi-dilute concentration range. Decreases in molecular volumes may have important effects on transport processes such as the multi-component systems found in living organisms. Chemically defined, highly branched dextrans have been studied by a ^{13}C spin-lattice relaxation method.

Correlations were made to the positions of carbon atoms associated with branching residues, permitting the values of *O*-substitution to be established for D-glucopyranosyl residues. Fourier-transform-i.r. difference spectra of dextrans have been correlated to the type and degree of branching, which had been established previously by permethylation for these polysaccharides. A combination of ^{1}H n.m.r.- and Fourier-transform-i.r.-spectroscopy has been used to give complementary information on the structure of dextrans. The results of the high temperature enhancement of ^{13}C n.m.r. chemical shifts of dextrans have been correlated with those of methylation analysis. The diagnostic nature of the 70 – 75 p.p.m. spectral region with regard to the type of dextran branching, and an increase in resolution of the polysaccharide spectra at higher temperatures, are reported.

A combination of chemical-ionization m.s., ammonia m.s., and ^{13}C n.m.r. spectroscopy has been used for the unambiguous identification of peracetylated aldononitrile derivatives of methylated dextrans. The method has been used to distinguish α-D-glucosyl residues substituted at O-3 in the linear chain and branch-point positions, or only in the branch-point positions. Several ^{13}C n.m.r. signals that are diagnostic for 4,6-di-O-substituted α-D-glucopyranosyl residues have been identified. Analytical data for g.l.e.-m.s. methylation fragmentation and periodate oxidation for numerous dextrans have shown that results obtained using the two techniques are in general agreement for slightly branched dextrans, but are divergent for highly branched dextrans.

Hitherto unrecognized effects of (1 → 2)-, (1 → 3)-, (1 → 4)-linkages in the dextrans were observed:. A water-insoluble α-D-glucan has been isolated from growing cultures of *Streptococcus salivarius*. The results of methylation analysis and periodate oxidation studies indicated the presence of about 80% (1 → 3)-α-D-glucosidic linkages, together with short side-chains attached to the backbone by (1 → 4)-, (1 → 3)-, and (1 → 2)-α-D-glucosidic linkages. A linear (1 → 3)-α-D-glucan was recovered after periodate oxidation of the parent polysaccharide. Further studies have confirmed that immunoglobulin A (W3129) binds only to the non-reducing end of dextrans and not to any intercatenary segments of five D-glucopyranosyl residues. Extended, *X*-ray absorption fine structure analysis of ferritin and imferon (a synthetic complex of iron oxide coated with dextran) show that the near-neighbour environment around the iron atom in the two macromolecules is identical. The action patterns of an intracellular dextranase and three α-D-glucosidases (oligo-D-glucanases) from *Pseudomonas* UQM733 have been used to determine the structure of two branched isomalto-oligosaccharides obtained by enzymic degradation of dextran.

The adherence of heat-killed *Streptococcus mutans* cells to glass, induced by incubation with sucrose and cell-free dextransucrase (EC 2.4.1.5), is markedly inhibited by low concentrations of dextranases. The dextranases repress the production of water-insoluble α-D-glucan and reduce adhesiveness and cell-agglutinating activity of the D-glycans, by lowering the content of (1 → 6)-linkages and molecular weights, thus inhibiting the formation of D-glucan film and adherence of *S. mutans* cells on glass. Three extracellular proteins having affinities for binding to *Streptococcus mutans* serotype *c* cells have been isolated by affinity chromatography on immobilized α-D-glucans containing (1→ 3)- and (1→ 6)-linkages.

Dextransucrase activity was found to be associated with two of the proteins. Neither of the proteins had a unique binding specificity for (1 → 3)-α-D-glucosyl residues, and they were able to bind to (1 → 6)-α-D-glucans, other dextrans, and also to the (1 → 3)-α-D-glucosyl regions in pseudonigeran. The role of antibodies against cell-wall antigens of *Streptococcus mutans* in the protection against dental caries in humans has been reviewed. Soluble antigen preparations containing dextransucrase and from one serotype of *S. mutans* have been shown to elicit a protection immune response against infection with cariogenic *S. mutans* from many or possibly all serotypes.

Significant protection from experimental dental caries has been achieved by oral administration of soluble dextransucrase in animal feeding experiments. A survey of anti-dextransucrase activities in rabbit antisera against whole cells of *S. mutans* in terms of inhibition of cell-free and cell-associated dextran-sucrases has been reported. Inhibition of antiserum against serotype *c* was restricted to dextransucrases of serotypes *c, e,* and *f* but the dextransucrases of serotypes *a, d,* and *g* were not affected. The production, purification, and properties of dextransucrase from *Leuco nostoc mesenteroides* NRRL B-512F have been reported. The glycosyltransferase complexes [dextransucrase and levansucrase (EC 2.4.1.10)] from two strains of *Streptococcus mutans* have been resolved by polyacrylamide gel electrophoresis.

An improved method for the purification of dextransucrase involves the use of the non-ionic detergent Tween 80 to prevent aggregation of the enzyme, and affinity chromatographic absorption of the enzyme on insoluble dextran. Ammonium sulphate precipitation of dextran-sucrase appears to favour aggregation of the enzyme to a form which catalyses the synthesis of insoluble α-D-glucan containing mainly (1 → 3)-linkages. Gel filtration (hence dilution of the enzyme) promotes disaggregation to a low molecular weight form that catalyses only the synthesis of water-soluble a-D-glucans, which contain predominantly (1 → 3)-linkages. An intracellular polysaccharide, mol. wt. 3×10^5, isolated from *Bacteroides fragilis,* has been characterized as being of the glycogen-type, as it stains with iodine and is degraded to D-glucose by glucoamylase?

The intracellular glycogen-like polysaccharide present in the cells of two strains *of Streptococcus mutans* at various stages of growth has been measured by quantitative electron microscopy. The nature of the binding sites for the substrates, activators, and inhibitors located on each of the four sub-units of adenosine diphosphate-D-glucose pyro-phosphorylase (EC 2.7.7.27) from *Escherichia coli* have been investigated. The binding of adenosine triphosphate and D-fructose 1,6-diphosphate to the enzyme is synergistic, showing heterotrophic co-operativity, and the two phosphorylated compounds must also be present together to inhibit effectively the binding of adenosine monophosphate.

A mechanism has been proposed to explain some of the kinetic and binding properties in terms of asymmetry in the distribution of the conformational states of the four identical sub-units. Quantitative correlation of the changes in the rates of synthesis of glycogen and the utilization of D-glucose with simultaneous changes in the cellular levels of both D-glucose 6-phosphate and D-fructose 1,6-diphosphate have been reported for *E. coli.* An equation has been developed which describes the *in vivo* velocity of ADP-D-glucose pyrophosphorylase, and the rate of glycogen synthesis. Conditions have been described which cause significant changes in glycogen accumulation during periods of nitrogen starvation.

Differences have been observed in the values of velocity of ADP-D-glucose synthetase in the intact *E. coli* cell, when the carbon source is varied from D-glucose to D-fructose, glycerol, or succinate.

These differences are a result of the different cellular levels of adenosine 3',5'-phosphate in the cells using different carbon sources. A kinetic mechanism describing the action of the D-glucose 6-phosphate-dependent form of glycogen synthase (EC 2.4.1.11) of *Neurospora crassa* has been derived from initial velocity experiments and inhibition studies and has proved to be consistent with a rapid equilibrium random Bi-Bi mechanism. *X*-Ray diffraction studies on the (1 → 3)-β-D-glucan, curdlan, from *Alcaligenes faecalis* have shown the existence of microfibril formation, and a triple helical structure is proposed for the crystalline form of this polysaccharide. The maximum production of the D-fructan, levan, has been shown to occur during the mid-exponential to the late exponential phase of growth of *Actino-myces viscosus* and is followed by a rapid decline as a result of the production of levan hydrolase activity.

The levan produced by *Streptococcus mutans* OMZ 176 appears to have a structure similar to other known D-fructans and inulin. On the basis of chemical shift displacements of the ^{13}C n.m.r. spectra, resulting from *O*-substitution at specific carbon atoms, resonances have been assigned to the carbon atoms of the β-D-fructofuranosyl residues in levans. A variety of different levans show almost identical ^{13}C n.m.r. spectra in contrast to a wide divergence in the corresponding spectra of dextrans. The extracellular polysaccharide isolated from *Arthrobacter stabilis* NRRL B3225 contains residues of D-glucose, D-galactose, pyruvate, succinate, and acetate (in molar ratio 6 : 3 : 1 : 1 : 1.5).

The viscosity of aqueous salt-free solutions is relatively low, but atypical of anionic polysaccharides, increasing rapidly in the presence of salts, acids, and alkali. The improvements of the rheological properties of microbial exopolysaccharides through mutagenesis and mutant selection has been demonstrated for *Escherichia coli* and *Enterobacter (Klebsiella aerogenes).* The chemical composition of each group of polysaccharides produced by the mutants did not show any significant differences from those originating from the parent strain. Antigenic polysaccharides from *Escherichia coli* 29M and *E. coli* 72M have been isolated from chicken intestine and the structures of their repeating units, (18) and (19) respectively, have been established using the standard methodology. When grown on methanol an obligate methylotroph, *Methylocystis parvus,* originally isolated as a methane-utilizing bacterium, elaborates an extracellular viscous heteropolysaccharide composed of residues of D-glucose and L-rhamnose (6 : 1).

The extracellular polysaccharide of a methylotrophic culture contains residues of D-glucose, D-galactose, D-mannose, L-fucose, and L-rhamnose when grown on methane as the sole carbon source. Transfer of the culture from methane to methanol results in the loss of its ability to utilise methane as the sole carbon source and is accompanied by a loss of its ability to synthesize 6-deoxyhexoses in the extracellular polysaccharide. An acidic polysaccharide that exhibits unusual properties, such as formation of a brittle water-insoluble gel in the presence of Ca^{2+}, has been isolated from a *Pseudomonas* species when grown on methanol.

The polysaccharide is composed of a main chain of (1 → 3)-linked β-D-galactopyranosyl residues, with additional residues of D-glucose, D-mannose, D-glucuronic acid, and D-allose in undetermined positions. The slime polysaccharide synthesized by *Pseudomonas* PB1 contains

```
                        D-GlcpA
                           1
                           ↓
                           6
→4)-D-Manp-(1 → 4)-D-Manp-(1 → 4)-D-Manp-(1→
                           2
                           ↑
                           1
                        D-Galp
                           4
                           ↑
                           1
                        D-Manp
                           3
                           ↑
                           1
                        D-Glcp
                           4
                           ↑
                           1
                        D-Manp
                           3
                           ↑
                           1
         D-GlcpA-(1→ 3)-D-Galp
                           6
                           ↑
                           1
                        D-Galcp
                         3   4
                          \ /
                           C
                          / \
                        Me   CO2-
                         (18)
```

```
                        D-GlcpA
                           1
                           ↓
                           6
→4)-D-Manp-(1 → 4)-D-Manp-(1 → 4)-D-Manp-(1→
                           2
                           ↑
                           1
                        D-Manp
                           4
                           ↑
                           1
                        D-Galp
                           3
                           ↑
                           1
                        D-Glcp
                           4
                           ↑
                           1
                        D-Manp
                           3
                           ↑
                           1
         D-GlcpA-(1→ 3)-D-Galp
                           6
                           ↑
                           1
                        D-Galcp
                         4   6
                          \ /
                           C
                          / \
                        Me   CO2-
                         (19)
```

D-glucose, D-galactose, acetate, and pyruvate and therefore differs from many exopolysaccharides formed by other pseudomonads in not containing residues of hexuronic acid. Solutions of the polysaccharide were pseudo-plastic, in that the measured viscosity was shear rate dependent, and formed gels in the presence of Fe^{3+} and Al^{3+}. D-Fructose favours the production of exopolysaccharides by *Rhizobium meliloti*. In contrast, D-glucose is metabolized to 2-ketogluconic acid, an inhibitor of the exopolysaccharide biosynthesis.

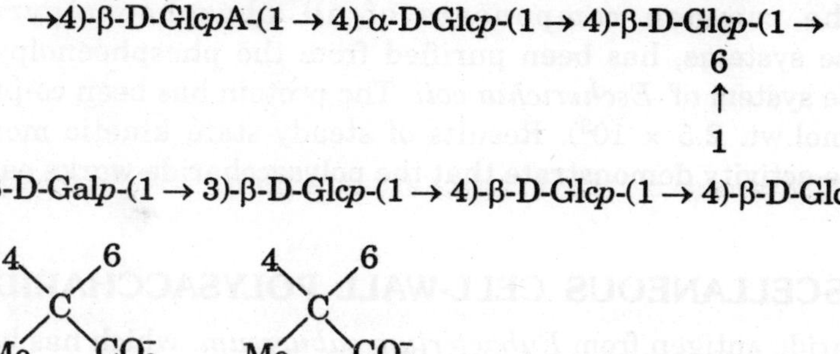

→4)-β-D-Glc*p*A-(1 → 4)-α-D-Glc*p*-(1 → 4)-β-D-Glc*p*-(1 →
6
↑
1
β-D-Gal*p*-(1 → 3)-β-D-Glc*p*-(1 → 4)-β-D-Glc*p*-(1 → 4)-β-D-Glc*p*

4, 6 C Me CO_2^- 4, 6 C Me CO_2^-

(20)

→4)-β-D-Glc*p*-(1 → 4)-β-D-Glc*p*-(1 → 3)-β-D-Glc*p*-(1 → 4)-β-D-Glc*p*-(1 →
6
↑
1
β-D-Glc*p*-(1 → 3)-β-D-Glc*p*-(1 → 4)-β-D-Glc*p*-(1 → 4)-β-D-Glc*p*

4, 6 C Me CO_2^-

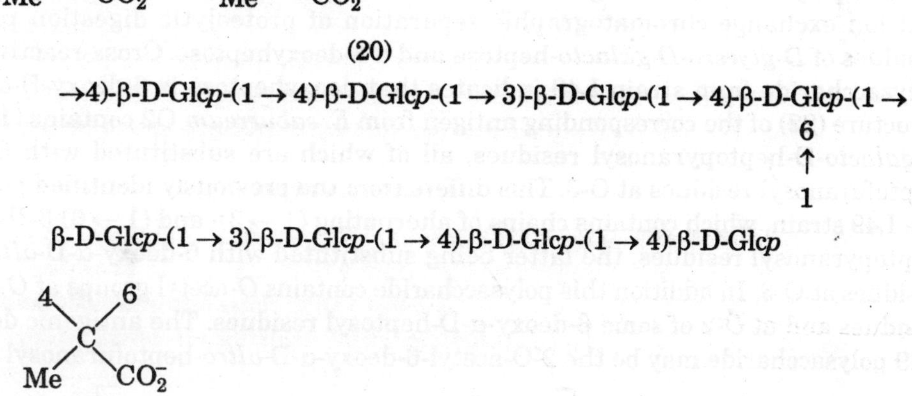

(21)

Growth yields, polysaccharide production, and energy conservation in chemostat cultures of *R. trifolii* have been reported. The extracellular polysaccharides of different strains of *R. meliloti* have been found to induce the production of polygalacturonase by roots of *Medicago sativa* seedlings. A comparative study of polysaccharide production and monosaccharide composition of partially purified fractions capable of inducing the enzyme demonstrated differences according to their origins. The structure (20) of the heptasaccharide repeating unit of an extracellular polysaccharide from *Rhizobium trifolii* has been established by a combination of methylation analysis, sequential oxidation, and elimination of oxidized residues. This is not a repeating unit in the true sense as approximately one third of the units lack the terminal β-D-galactopyranosyl residues. Comparative studies on the polysaccharides elaborated by *R. meliloti, Alcaligenes faecalis,* and *Agrobacterium radiobacter* indicate that they have a common structure (21) except for the position of acylation. A D-gluco-D-mannan (mol.wt. 1.4 × O^6) has been isolated from *Rhodococcus erythropolis*. The nutritional requirements for the production of xanthan by *Xanthomonas campestris* have indicated that D-glucose and sucrose are the best sources of carbon for the production of the polysaccharide.

Xanthan has been shown to possess a rigid rod-like conformation, 6000 Å in length and independent of ionic strength $>10^{-2}$ M . The conformation has been studied as a function of temperature, ionic strength, and polymer concentration. C.d. measurements indicate that the polysaccharide exists in a combination of random and ordered forms regardless of these conditions. The virulence of the plant pathogen *Erwinia amylovora* has been correlated with the ability of the organism to produce extracellular polysaccharide. A soluble carrier protein (HPr), one of the common components of all phosphoenolpyruvate-dependent phosphotransferase systems, has been purified from the phosphoenolpyruvate-dependent phosphotransferase system of *Escherichia coli*. The protein has been co-purified with a (1 → 6)-α→-D-glucan (mol.wt. 2.5×10^3). Results of steady-state kinetic measurements of the phosphotransferase activity demonstrate that the polysaccharide works as an activator of the enzymic system.

MISCELLANEOUS CELL-WALL POLYSACCHARIDES

The polysaccharide antigen from *Eubacterium saburreum,* which has been isolated by gel- and ion exchange-chromatographic separation of proteolytic digestion products, contains residues of *D-glycero-D-galacto*-heptose and a 6-deoxyheptose. Cross-reactivity with a related polysaccharide from strain L49 indicates that deoxyheptose is 6-deoxy-D-*altro*-heptose. The structure (22) of the corresponding antigen from *E. saburreum* O2 contains (1 → 6)-β-*D-glycero*-D-*galacto*-D-heptopyranosyl residues, all of which are substituted with 6-deoxy-α-D-*altro*-heptofuranosyl residues at O-3. This differs from the previously identified polysaccharide from the L49 strain, which contains chains of alternating (1 → 3)- and (1 → 6)-β-*D-glycero-D-galacto*-heptopyranosyl residues, the latter being substituted with 6-deoxy-α-D-*altro*-heptofuranosyl residues at O-3. In addition this polysaccharide contains *O*-acetyl groups at *O*-7 of some heptosyl residues and at *O*-2 of some 6-deoxy-α-D-heptosyl residues. The antigenic determinant on the L49 polysaccharide may be the 2-O-acetyl-6-deoxy-α-D-*altro*-heptofuranosyl residues.

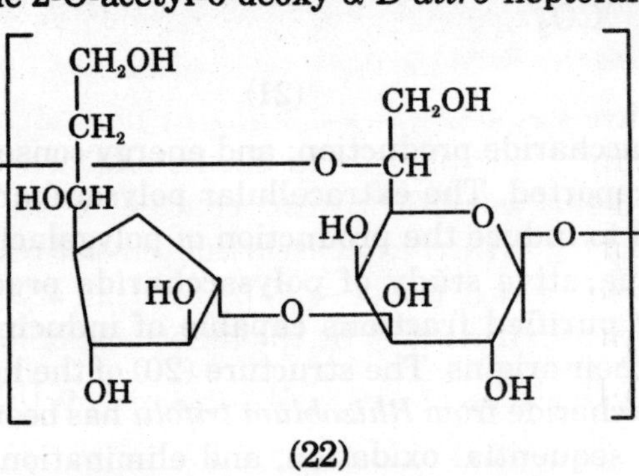

(22)

One common characteristic of the cell walls of halococci so far investigated is the high content of glycine. A 2,4-dinitrophenyl derivative of 2-deoxy-2-glycylamido-D-glucose has been isolated after hydrolysis of 2,4-dinitrophenylated cell walls.

It is postulated that glycine may play a role in connecting glycan strands through peptide linkages between the amino-group of 2-amino-2-deoxy-D-glucose and the carboxy-group of a uronic acid residue. When incubated under appropriate conditions with tritiated L-methionine, *Mycobacterium smegmatis* accumulates significant amounts of labelled oligosaccharides that are related to known 3-*O*-methyl-D-mannose-containing polysaccharides. A homologous series of oligosaccharides has been isolated, consisting predominantly of (1 → 4)-3-*O*-methyl-α-D-mannopyranosyl residues, terminated at the non-reducing end by D-mannopyranosyl residues or 3-*O*-methyl-α-D-mannopyranosyl residues and terminated at the reducing end of the -chain by a methyl aglycone.

These oligosaccharides are probably biosynthetic precursors of the larger 3-*O*-methyl-D-mannan homologues. ^{1}H N.m.r. spectroscopic studies suggest that the 3-*O*-methyl-D-mannan undergoes a major change in conformation on complexation with palmitic acid; the change is suggested to result from the tightening of a loosely coiled polymer. The fatty acid is orientated, at least part of the time, with its carboxy-group near the *O*-methyl groups of the polysaccharide. The phylogenetically closely related bacterium, *Streptomyces griseus,* also produces an *O*-methyl-D-mannose-containing polysaccharide, which in addition contains approximately four esterified acetyl groups.

These *O*-acetyl groups in the polysaccharide are located at position 6 of approximately half of the ten contiguous (1 → 4)-3-*O*-methyl-α-D-mannosyl residues. The affinity of this polysaccharide for palmitic acid suggests that it plays a role similar to the 3-*O*-methyl-D-mannose-containing polysaccharide of *M. smegmatis* in its regulation of the bacterium's fatty acid synthase. An acidic arabino-D-mannan isolated from *M. smegmatis* also contains two phosphate, six monoesterified succinate, and four ether-linked lactate groups per molecule.

Saponification of the polysaccharide releases a phosphorylated and a non-phosphorylated polysaccharide, the main features of which are the presence of chains of contiguous (1 → 5)-arabinofuranosyl residues attached to *O*-4 of L-arabinopyranosyl residues. A serologically active D-arabino-D-mannan, isolated from the cells of *M. tuberculosis,* is composed of (1 → 5)-α-D-arabino-furanosyl residues and (1 → 6)- and (1 → 2)-D-mannopyranosyl residues. Methylation analysis and enzymic degradation studies established the presence of a structure (23) containing short side-chains of (1 → 2)-α-D-mannosyl residues attached to a main backbone of (1 → 6)-α-D-mannopyranosyl residues.

In addition, a serologically inactive α-D-mannan whose structure (24) resembles that of the core L-arabino-D-mannan has been isolated. In addition to a high molecular weight lipopolysaccharide, the cell surface of *Rhizobium* contains a low molecular weight (1 → 2)-β-D-glucan. Curie point m.s. of a polysaccharide isolated from the cell walls of the cyanobacterium *Spirulina platensis* resembles that of the (1 → 2)-β-D-glucan of *Agrobacterium tumefaciens.* It is speculated that this polysaccharide may be the innermost of four layers of the cell wall.

→ [5-α-D-Ara*f*-1]$_m$ →– α-D-Ara*f*-(1 → 6)-α-D-Man*p*-(1 → 6)-α-D-Man*p*-(1 → 6)-α-D-Man*p*-(1 →

3	2	2
↑	↑	↑
1	1	1
α-D-Ara*f*	α-D-Man*p*	α-D-Man*p*
2	2	
↑	↑	
1	1	

[α-D-Araf]$_n$ α-D-Man$p$
5
↑
1
α-D-Araf $m + n = 8 - 9; n > 5.$

(23)

→ 6)-α-D-Man$p$-(1 → 6)-α-D-Man$p$-(1 → 6)-α-D-Man$p$-(1 → 6)-α-D-Man$p$-(1 → 6)-α-D-Man$p$-(1 →
2 2 2 2
↑ ↑ ↑ ↑
1 1 1 1
α-D-Man$p$ α-D-Man$p$ α-D-Man$p$ α-D-Man$p$
2 2
↑ ↑
1 1
α-D-Man$p$ α-D-Man$p$

(24)

Chemical and immunochemical studies have been used in the assignment of the structures (25)-(27) of the group-specific carbohydrates of Groups A, A-variant, and C streptococci, respectively. A close relationship among the L-rhamnan structures is revealed. Human antibodies to Group A and Group C carbohydrate have been measured in a radioimmunoassay system. The Group-specific antibodies arise after infection or persistent colonization of the pharynx with either Group A or Group C streptococci.

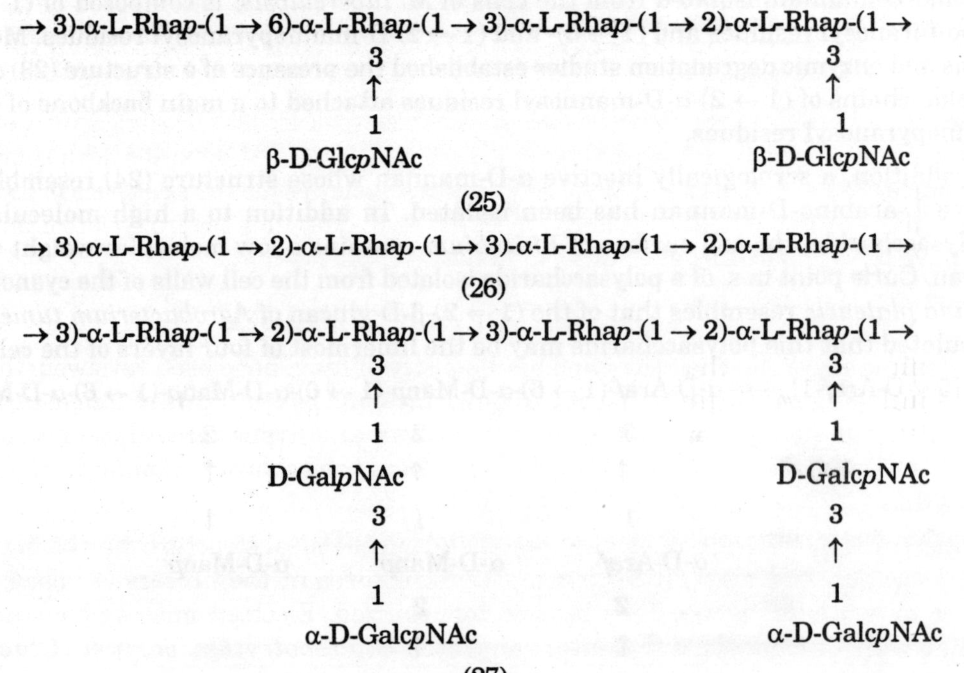

Several characteristic differences in the chemical composition of the cell walls among some serological types of *Streptococcus mutans* have been investigated through their susceptibility to *Flavobacterium* L-11 enzyme.

Generally, non-peptidoglycan components of the walls are responsible for serological differences in bacterial strains. The glycan isolated from *Streptococcus faecalis,* the structure of which (28) has been previously established, behaves abnormally on periodate oxidation.

Cyclic acetals are formed following periodate oxidation and this leads to the protection of hydroxy-groups during methylation. Inter-residue hemiacetals were identified between the aldehyde groups at C-2 or C-3 of oxidized residues and hydroxy-groups at C-3 or C-2 of an adjacent unoxidized residue.

$$
\begin{array}{c}
\rightarrow 4)\text{-D-Glc}p\text{-}(1 \rightarrow 4)\text{-D-Glc}p\text{-}(1 \rightarrow 4)\text{-D-Gal}p\text{-}(1 \rightarrow \\
6 \\
\uparrow \\
1 \\
\text{D-Glc}p \\
4 \\
\uparrow \\
1 \\
\beta\text{-D-Gal}p
\end{array}
$$

(28)

The purification of b-antigen of *Streptococcus sanguis* has been achieved by separation from teichoic acid by affinity chromatography on immobilized anti-poly(glycerol phosphate) columns. The antigen, which contains residues of D-glucose, L-rhamnose, and 2-acetamido-2-deoxy-D-glucose, appears to contain a single antigenic determinant related to isomaltose. A new Group-specific antigen, designated W-antigen, has been characterized for five strains *of Streptococcus sanguis.*

Two routes exist in *Streptococcus faecalis* for the synthesis of TUP-D-galactose, an inducible and a constitutive route. Incorporation studies with TDP-D-[^{14}C]hexoses have shown that the sugar nucleotide serves as a donor for D-galactosyl moieties for the synthesis of immunogenic glycans in the cell wall of *Streptococcus faecalis.* Antibodies to a heteroglycan of *Streptococcus bovis* have been purified by an affinity chromatographic method.

FUNGAL POLYSACCHARIDES

Specific areas of the chemistry and biochemistry of fungi have been reviewed: the autolysis of fungi including the degradation of carbohydrates, the biosynthesis of fungal cell walls, the biochemistry of dimorphism in *Mucor,* and cell wall and hyphal growth. Fractions from the cell wall and protoplasts of *Aspergillus fumigatus* have been shown to contain common antigenic components.

Differences in both immunological reactivity and chemical structure may be linked to their cellular location and may reflect incomplete structures in the process of biosynthesis and transfer to wall sites where they become immobilized. Electron microscopic studies of the conidial wall surface of the deutero-mycetes *Cladobotrytum varium, Cladosporium macrocarpum, Aspergillus niger,* and *Penicillium* species, and of the sporangiophore surface of the zygomycete *Mycotypha microspora* have been reportedf. The wall layers, isolated from *A. niger,* exhibited distinct differences in total carbohydrate, monosaccharide. amino-sugar, peptide, and lipid composition.

In *Phycomyces blakesleeanus,* during the formation of the germ-tube cell-wall, the composition shifts from that of a D-glucan, in the formant spore wall, to that of a chitin-like polymer in the mycelial wall. Heat-induced germination of the sporangiophore results in an increase in the content of chitin and a decrease in the D-glucan contents. In the presence of 2-deoxy-D-arabinohexose, *Rhodosponridiumtoruloides* synthesizes higher proportions of D-glucan and chitin, while the relative contents of D-mannan and D-galactan decrease drastically.

D-Glucans. — A study designed to ascertain whether nigeran might serve in determining phylogenetic relationships within the large and complex genera *Aspergillus* and *Penicillium* has shown that the nigeran content, used in conjunction with traditional morphological characteristics, can provide a useful and convenient tool for the determination of natural groupings of these two species. An alkali-soluble D-glucan which contains minor proportions of D-xylose, D-mannose, 2-amino-2-deoxy-D-glucose, and amino-acids has a large proportion of (1 → 4)-α-D-glucopyranosyl residues and a small proportion of (1 → 3)- and (1 → 6)-β-D-glucosyl residues.

On the basis of the identification of 2-*O*-α-*D*-glucopyranosylerythritol, it appears that segments of the (1 → 4)-α-D-glucopyranosyl residues are joined by single (1 → 3)-linkages. The relationship between the chemical structure and anti-tumour activity of D-glucans prepared from *Grifora umbellata* has been investigated. The anti-tumour D-glucans appear to contain the basic structural unit (29) although the length of the branches, branching frequency, molecular size, and molecular

R → 6)-β-D-Glc*p*
1
↓
6
→ 3)-β-D-Glc*p*-(1 → 3)-β-D-Glc*p*-(1 →
R = other D-glucosyl residues

(29)

conformation are also important. An extracellular (1 → 3)-β-D-glucan (mol. wt. 7×10^4) with single (1 → 6)-β-D-glucosyl side chains on every second D-glucosyl residue of the main chain has been isolated from *Monilinia fructigena.* Alkali-soluble and alkali-insoluble polysaccharides from the fruiting bodies of the basidiomycetes *Laetiporus sulphureus* and *Piptoporus betulinus* have been examined by *X*-ray diffraction, u.v. spectroscopy, and chemical methods.

The presence of (1 → 3)-α-D-glucan, (1 → 3)-β-D-glucan, and chitin was established. Variations in the levels of these polysaccharides were noted in different parts of the fungal structure. Potato, red kidney bean, and soybean plants respond to highly purified β-D-glucan from *Phytophthora megasperma* by accumulating phytoalexinst. The plants appear to respond to the fungus following recognition of the same mycelial wall component and it is possible that plants may respond to a variety of fungi by recognizing the same or similar mycelial wall components. Cell-wall formation in *Saccharomyces cerevisiae* appears to be the result of two main patterns of deposition of wall material, *viz.*, around the whole periphery of the non-budding cells and mainly at the tip of the daughter cell, or at the cross-wall that separates dividing cells.

Synthesis and secretion of β-D-glucan is unaffected by inhibition of RNA and protein synthesis. Growth of *S. cerevisiae* under conditions of phosphate limitations results in a decreased content of D-glucan and increased protein content as compared with control cells. Papulacandin B, an antibiotic produced by the deuteromycete *Papularia sphaerosperma,* inhibits the growth of yeasts, but has no effect on other fungi, bacteria, or protozoa. The mode of action involves the inhibition of synthesis of alkali-insoluble β-D-glucan of *Saccharomyces cerevisiae* and *Candida albicans* in sphaeroplasts and causes lysis of cells by osmotic rupture.

Echinocandin B and a structurally related antibiotic aculeacin A also inhibit β-D-glucan synthesis in yeast, possibly in a similar manner. Papulacandin B has been shown to inhibit the transfer of D-glucose from UDP-D-glucose, but not D-mannose from GDP-D-mannose, into dolichyl phosphate. Some properties of the β-D-glucan synthetase from *Saccharomyces cerevisiae* have been described.

UDP-D-Glucose is a competitive inhibitor and D-glucono-1,4-lactone is a non-competitive inhibitor of this enzyme. High yields of laminaribiose (3-*O*-β-D-glucopyranosyl-D-glucose) have been achieved by treatment of the (1 → 3)-β-D-glucan *of Saccharomyces cerevisiae* with an *endo*-(1 → 3)-β-D-glucanase from *Kluyveromyces phaseolosporus.* Glycogen metabolism in resting and growing cells of *Saccharomyces carlbergensis* has been shown to be regulated by a variety of endogenous and environmental factors. A ^{13}C.n.m.r. study of the gel networks of the branched (1 → 3)-β-D-glucans lentinan (from *Lentinus edodes)* and schizophyllan (from *Schizophyllum commune*) has been reported.

Conformational changes of the helical to random-coil forms were induced by treatment with aqueous sodium hydroxide. The effect of low pH on the growth of *Aureobasidium pullulans* has been shown to prevent transformation of the cells into the yeast phase, with the elaboration of small amounts of pullulan. The main degradation product (75-90%) resulting from the action of selected α-amylases on the α-D-glucan of *Elsinoe leucospila* has been identified as the trisaccharide *O*-α-D-glucopyranosyl-(1 → 3)-α-D-glucopyranosyl-(1 → 4)-D-glucose. This confirms previous reports that the polysaccharide consists chiefly of maltotriose units and a small proportion of maltotetraose units joined by (1 → 3)-α-D-glucosyl linkages. Cell extracts of *Oyptococcus laurentii,* a yeast which synthesizes amylose when grown at low pH, contain an α-D-glucan phosphorylase (EC 2.4.1.1) which has been characterized. Evidence for a possible role of this enzyme in the biosynthesis of amylose was reported. The ability of yeast cells to liberate sphaeroplasts following treatment with (1 → 3)-β-D-glucanases has been proposed as a taxonomic criterion for differentiating basidiomycetous from ascomycetous yeasts and for classifying the yeasts of the fungi imperfecti. Purified preparations of extracellular β-D-glucans from *Sphacelia sorghi* and from the cell walls of *Claviceps purpurea* have been shown to contain enzymically active protein.

The binding of the protein, which exhibited β-D-fructo furanosidase and acid phosphatase activities, to the D-glucan appeared to be non-covalent. When the enzymes were liberated from the D-glucan they showed enhanced activity, although the K_m values remained the same, suggesting that the matrix acts in the same manner as a non-competitive inhibitor. The solubility characteristics of β-D-glucans in the fungal cell wall have been shown to undergo changes after specific removal of chitin. Two types of β-D-glucans were isolated, one water-soluble and highly branched, and the other alkali-soluble and with single residue branches.

D-Mannans. — Six (1 → 6)-α-D-manno-oligosaccharides have been isolated after partial acid hydrolysis of proteolytic digests of an extracellular glycan of *Absidia cylindrospora.* A pentasaccharide and higher molecular weight oligosaccharides inhibited significantly the precipitation reaction of intracellular glycans from *A. cylindrospora, Mucor hiemalis,* and *Rhizopus nigricans* with anti -A. *cylindrospora* serum. On the basis of the ^{1}Hn.m.r. spectra of the cell-wall D-mannans, and the base composition and sequence similarities of the deoxyribonucleic acids, no distinction could be made between strains of *Candida utilis* and *Hansenula jadinii* although differences could be detected between *H. jadinii* and *H. petersonii.* The catabolic and anabolic transformation of D-glucose by a *Candida* species, into cell wall D-glucan and D-mannan has been reported. Approximately 20-40% of the hexose of these cell wall polysaccharides has been derived from the resynthesis of hexoses *via* the pentose phosphate pathway. Methylation analysis and periodate oxidation studies have been used to characterize the extracellular β-D-mannan produced by *Rhodotorula rubra* as being composed of alternating (1 → 3)- and (1 → 4)-β-D-mannopyranosyl residues.

Of eight oligosaccharides isolated from the cell walls of *R. glutinis* after lysis with an enzyme from *Penicillium lilacinum,* two have been characterized as 3-O-β-D-mannopyranosyl-D-mannose and 4-*O*-β-D-mannopyranosyl-D-mannose. The remaining oligosaccharides, which contained both D-glucose and D-mannose, were composed of (1 → 3)- and (1 → 4)-β-D-mannopyranosyl residues with (1 → 6)-β-D-glucopyranosyl side chains. The transformation of D-glucose by a *Candida* species has been followed catabolically and anabolically.

Maximal incorporation into D-glucan and D-mannan is only achieved at a critical concentration of D-glucóse. Dolichyl monophosphate-D-mannose is not an intermediate in the synthesis of the outer chains of the D-mannan-protein of *Saccharomyces cerevisiae* and dolichyl diphosghate oligosaccharides are synthesized by different D-mannosyl-transferases. An osmotically-sensitive mutant of *Saccharomyces cerevisiae* S288 contains more β-D-glucan, less α-D-mannan, and less alkali-soluble glycogen than the parent strain. The D-mannan has been characterized as containing more short side-chains and fewer long side-chains than does the parent D-mannan and also contains a different proportion of (1 → 2)- and (1 → 3)-α-D-mannosyl residues.

Chitin. — The ability of lytic β-D-glucanases of *Arthrobacter* GJM-1 to dissolve the cell wall of *Saccharomyces cerevisiae* with the exception of the chitin component has been reported. Electron microscopic examination of such wall residues isolated from various stages of the budding cycle shows a correspondence to the formation of an annular structure found as part of the bud scar after cell division. This annular chitin-rich structure could not be isolated at either cell-cycle stages preceding the bud emergence or at the earliest stages of bud development.

Observations confirmed that the annular structure (chitin ring) formed during bud growth represents a major part of the total chitin present in the bud scar after septum closure. Chitin synthetase (EC. 7.4.1.16) has been located, mostly in an inactive form, associated with membranous fractions of *Neurospora crassa,* and could be activated either endogenously or by addition of exogenous proteases. Plasma membranes from *Schizophyllum commune* protoplasts have been stabilized against fragmentation by coating the protoplasts with concanavalin A. Approximately half of the chitin synthase activity is associated with the plasma membranes in the form of a zymogen and could not be detached from the membranes. It is suggested that this inactive form represents the cyto-plasmic transport form of the enzyme.

The nature and location of proteases of *Aspergillus nidulans* and the interaction of one of these with the zymogen of chitin synthetase have been described. A potential industrial method for the production and isolation of chitosan from the hyphal walls of *Mucor rouxii* as an alternative to the product derived from shellfish chitin has been reported.

Miscellaneous Polysaccharides

An extracellular polysaccharide of *Aspergillus parasiticus* has been characterized as being a homopolymer of 2-amino-2-deoxy-D-galactosyl residues, of which approximately 50% are *N*-acetylated and the remainder unsubstituted. The occurrence of the free amino-groups may result from enzymic *N*-deacetylation, since an enzyme isolated from the organism is capable of *N*-deacetylation of oligosaccharides containing fourteen or more 2-acetamido-2-deoxy-D-galactosyl residues, at the same rate as the polysaccharide degraded. A D-galacto-D-mannan, isolated from the mycelia of *A. fumigatus,* reacts with antibodies specific for circulating antigen in invasive aspergillosis in umans and rabbits.

The polysaccharide is composed of a backbone of (1 → 6)-D-mannopyranosyl residues with trisaccharide side chains terminated with residues of D-galactofuranose. An enzyme-linked immunoadsorbent assay (ELISA) for immunoglobulin (IgG and IgE) antibodies to protein and antigens of *A. fumigatus* has been reported. D-Mannans, D-gluco-D-mannans and D-galacto-D-mannans have been identified in phytopathogenic strains of *Ceratocystis paradoxa* and *C. fimbriata* isolated from different hosts, including tropical fruits and sugar cane.

Isolates of *Ceratocystis* within the same species were found to synthesize different polysaccharides, depending on their original hosts. Different polysaccharides could also be synthesized by the same strain at different temperatures. No correlation was made between host specificity and synthesis of a specific polysaccharide. It is possible that adherence of the fungus to the plant cells involves either specific or non-specific affinity reactions, which in turn do not correlate with the capacity of the micro-organism to cause disease. In a study of the carbohydrate composition of *Geotrichum, Trichosporon,* and allied genera, the intact cells of *Geotrichum* were characterized as containing predominantly D-galacto-D-mannans. L-Iduronic acid and D-glucuronic acid have been identified by g.l.c.-m.s. of the *O*-TMS ethers of the corresponding hexuronolactones prepared from the glycuronan 'protuberic acid' from *Kobayasia nipponica.* A family of glycopeptides of disperse molecular weights, charge, and sugar composition have been isolated from the culture filtrates of *Fulvia fulvia.*

D-Galactose, D-glucose, and D-mannose were identified as the major sugar components, together with smaller amounts of 2-amino-2-deoxy-D-glucose, 2-amino-2-deoxy-D-galactose, D-glucuronic acid, and diester phosphate. The molecular organization of these polymers was investigated using mild alkali- and acid-hydrolysis, and acetolysis of the residual polysaccharide. The proposed structure (30) is similar to that suggested for the peptidophosphogalactomannan isolated from *Penicillium charlesii.* The pathogenic fungus of rice blast disease *Piricularia oryzae,* contains a cell-wall heteroglycan-protein complex containing residues of D-mannose, D-glucose, and D-galactose in the molar ratio 6 : 2 : 1.

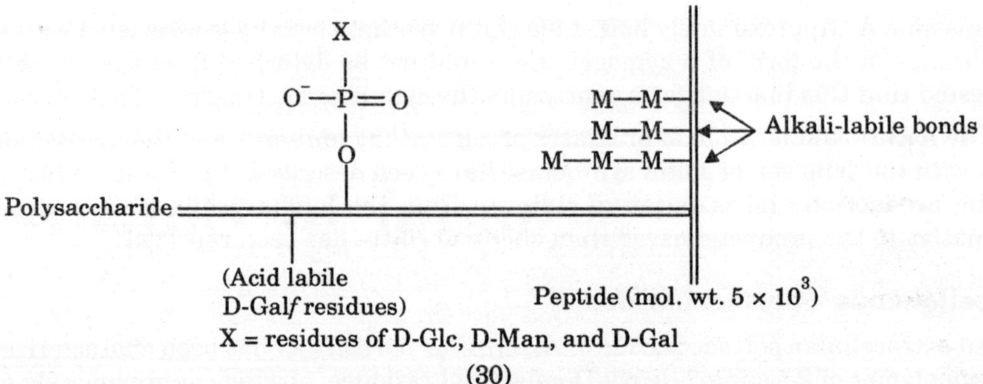

X = residues of D-Glc, D-Man, and D-Gal

(30)

Methylation analysis, partial acid hydrolysis, and digestion with an α-D-mannanase have shown the presence in the carbohydrate portion of a backbone of (1 → 6)-α-D-mannopyranosyl residues with side chains terminated by D-glucopyranosyl-, D-galactofuranosyl-, (1 → 2)- and (1 → 3)-α-D-mannopyranosyl residues. The heteroglycan is attached to the protein *via* O^3-D-mannopyranosyl-L-serine or O^3-D-mannopyranosyl-L-threonine linkages. Further studies have shown that the side chains comprise one to four D-mannosyl units, some of which are terminated by D-glucosyl-or D-galacto-furanosyl-residues. Histochemical staining at the cell surface of yeast-like forms, hyphae, and conidia of *Sporothrix schenckii* cell-types with iron(III) hydroxide and ferritin has been shown to correspond to the locations of concanavalin A-reactive peptido-L-rhamno-D-mannan complexes in the cells. Several acidic oligosaccharides, isolated from the acidic heteroglycan of the edible mushroom *Tremella fuciformis,* have been characterized by methylation analysis and ^{1}H n.m.r. spectroscopy and the structure (31) has been proposed for the parent polysaccharide.

A similar structure has been proposed for a cell-wall polysaccharide isolated from growing cultures of haploid cells of two strains of *T. fuciformis;* two extracellular polysaccharides showed only minor variations from this structure.

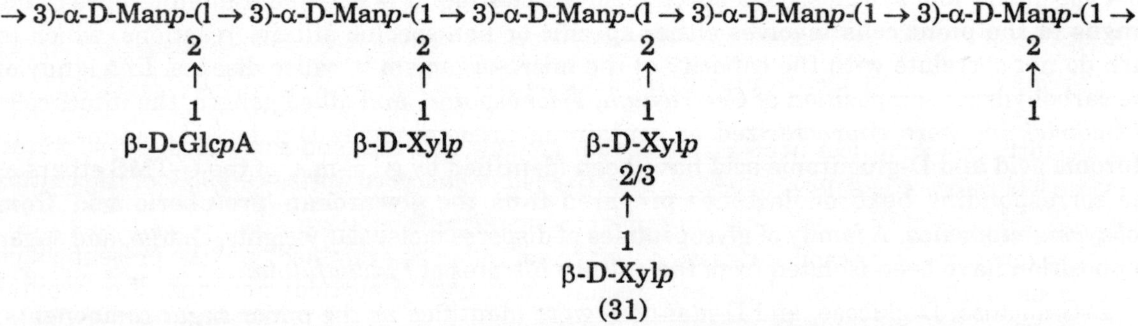

(31)

Water-soluble polysaccharides have been isolated from cultures of lichen mycobionts and phycobionts. The polysaccharides of the mycobionts show close similarities with those of the parent lichens, whereas the polysaccharides of the phycobionts possess different features. The mycobiont of *Parmelia caperta* contained an α-D-glucan having equal proportions of (1 → 3)- and (1 → 4)-linkages, and identical to a D-glucan of the parent lichen. The phycobiont of *Ramalina crassa* contains a major water-soluble D-galactan, whose properties are different from those of the intact parent lichen and its mycobiont.

CHAPTER

8 Ethanol in Culture

Studies relying on animal models demonstrate that the development and organization of the brain are profoundly effected by ethanol. In vivo models identify targets of ethanol action and provide context to the effects, however, they are limited in their utility as a substrate for exploring molecular, intracellular, and intercellular mechanisms. In vitro systems lend themselves to such analyses because the composition of the culture, the environment of the cells, and the manipulation of the environment can be carefully controlled. This is particularly important in examinations of growth factor actions, intracellular signal transduction, and genetic manipulations such as electroporation and blocking RNAs.

TWO CULTURE MODELS

Two types of cultures have been particularly useful in alcohol studies. These are 1) primary cultures in which samples are derived directly from animals and 2) cultures of cell lines (e.g., cancer cells and genetically engineered/selected -lines). The present chapter discusses the primary cultures. Various types of primary cultures have been used in alcohol studies: dissociated cell cultures and organotypic explants. Each has its uses and limitations.

Dissociated Cell Cultures

Primary cell cultures are composed of neurons, glia, or stem/progenitor cells. The constituents are relatively homogeneous, nontransformed cells. Common cell types used in these cultures are cortical neurons, cerebellar granule neurons, and astrocytes. These cultures are enriched by various procedures so that the cell of choice comprises 90-95% of the cultured cells.

The enrichment method varies with the type of cell to be examined, and the present chapter describes the method tor obtaining primary cultures of cortical neurons. For example, procedures used for generating cultures of astrocytes and cerebellar granule neurons can be found elsewhere. Dissociated cultures can be criticized for their nonphysiological arrangement, i.e., removing cells from their normal microenvironments; Neurons are intimately dependent on interactions with other cells, thus, it becomes particularly important to preserve in vivo cellular organization before extrapolating the effects of exogenous substances on neuronal development. This need drove investigators to develop systems in which quasi-normal relationships are maintained organotypic (slice) cultures.

Organotypic Slice Cultures

Many studies have focused on the developing cerebral cortex. Neurons in the cer-ebral cortex are derived from two proliferative zones, the ventricular zone (VZ) 10 Generation and Use of Primary Rat Cultures for Studies of the Effects of Ethanol and the *subventricular zone* (SZ).

Intrinsic activities of cells in the proliferative zones proceed normally in slice preparations. This includes interkinetic nuclear migration of VZ cells and the migration of postmitotic cells. Indeed, these metrics describing these activities are similar in slices and in vivo preparations. Neurogenesis in the neocortical VZ is influenced by both intrinsic intracellular factors and extrinsic intercellular factors. Using such explants has provided unique insight into the mechanism of ethahol toxicity. Chief criticisms of the cortical explant model are that not all relationships in the developing cortex are maintained.

The formation of the slice is likely to disrupt paths of migration or axonal trajectories. Long-distance projections such as the ascending monoaminergic and thalamic afferents are often severed in the production of the slice. Likewise, under normal circumstances, many cortical local circuit neurons are generated in another site, the ganglionic eminence (GE). In generating the slices, either the GE is eliminated from the preparation or the path-ways connecting the GE to the developing cortex may be eliminated. Thus, the contribution from the young local circuit neurons could be eliminated from the developing cortex of the slice.

A consequence of producing the slice is that cells on the surface of the slice are damaged. The response is the death of many of these cells and in the formation of a glial coat. These can lead to an altered microenvironment of cells near the surface of the slice and possibly a change in the microenvironment of other strata of the slice. Conceivably, the increased amount of glia are elaborating growth factors that are expressed in higher amount than would be expressed in vivo.

MATERIALS

Animals

Timed pregnant Sprague-Dawley rats: Depending upon the issue to be addressed, rats may be anywhere from 13 to 19 days pregnant. Commonly, fetal rats are harvested on gestational (G) day 16. The first day a sperm-positive plug is detected is designated as G1.

Primary Cultures of Cortical Neurons

Surgical Equipment

1. Operating scissors.
2. Microdissecting scissors.
3. Microdissecting forceps.
4. Microdissecting tweezers #5 (× 2).
5. Spatula.
6. 60-mm or 10-mm round tissue culture dishes.

7. Dissecting microscope.
8. 1.5-mL microcentrifuge tubes.
9. Glass aspirating pipets.
10. Alcohol burner.

Reagents

1. Hank's balanced salt solution (HBSS, Gibco).
2. 70% ethanol in water.
3. Poly-D-lysine molecular (PDL) weight 70 – 100kDa (Sigma, St. Louis, MO) is dissolved in sterile water at 10 mg/mL to make a 100 x stock solution that is aliquoted and frozen at –20°C. Working solutions of PDL are prepared by dilution in phosphate-buffered saline (PBS, Gibco), to a final concentration of 100 µg/mL.
4. Eagle's Minimal Essential Medium (MEM) containing glutamine (Gibco) and supplemented with 10% fetal bovine serum (FBS, Gibco), 33 mM dextrose, and penicillin/streptomycin (Gibco).

Culturing Methods

Preparation of the Culture Dishes

1. Coat the experimental tissue culture plates with PDL sufficient to cover the plating surface for a minimum of one hour at room temperature.
2. Chill two 100-mm tissue culture plates containing 10mL each of HBSS on ice.
3. Chill a 1.5-mL centrifuge tube containing 10mL of HBSS on ice.
4. Sterilize surgical instruments in 70% ethanol.

Dissection of the Rat Cortices

1. Anesthetize the timed pregnant G16 dams with a cocktail of ketamine (1.0 mg/kg) and xylazine (1.0 mg/kg). Open the abdominal cavity to expose the uterine horns and extract the pups, one at a time.
2. Hold the pup between the thumb and index finger and cut a shallow incision from the upper spine to the base of the brain, and then laterally around the head from the base of the brain to the eye. Use the forceps to peel the skin and skull from left to right up over the head and expose the brain. Scoop out the entire brain with a spatula and place in one of the culture dishes with chilled HBSS to rinse blood away. Repeat this pro-cedure for all pups. Transfer the brains to a fresh culture dish with chilled HBSS.
3. Under the dissecting scope, use the sharp #5 tweezers to remove the meninges and vasculature from the cortex as fully as possible. After removal of the meninges, use the tweezers to pinch off the cortical wall by removal of the olfactory bulb, the striatum, and thc hippocampus.
4. Place all cortices in a 1.5-mL centrifuge tube with BBSS. Use sterile glass pipets to gently triturate the cortices and dissociate the cells.

5. After the initial cell dissociation, allow heavier fragments (meninges, tissue clumps) to settle for a few minutes to the bottom of the conical tube. Carefully transfer the cell suspension to a 15 mL conical tube leaving the fragments in the bottom of the centrifuge tube. For optimal cell yield, add 1 mL of BBSS to the fragments and repeat the trituration, breaking up as many tissue chucks as possible, while leaving any meninges present following dissection to settle out of the cell suspension. Transfer the cell suspension to the 15 mL conical tube.
6. Centrifuge the cells at 600 x g for 3 min to pellet cells. Add 2.0 mL of 10% serum containing MEM and resuspend cells. Dilute the cell suspension to a concentra-tion suitable for counting.

Yield and Plating of the Cells

The total number of primary neuronal cells recovered after the dissociation of G16 cortices is highly dependent on a variety of factors. These factors include the number of pups in utero, the extent to which areas other than the cortical wall are included for culture, and the quality of dissociation. Two pregnant dams with 12 pups each generate approximately 100 million cells with considerable flux depend-ent on the variables stated. This cell yield would result in approximately four to six 100-mm tissue culture plates, if treatment with growth factors and ethanol with intent to generate a protein lysate, is planned. Plating 2×10^7 cells per 100mm tis-sue culture dish yields more than 1.0 mg of total protein and is approximately 70% confluent.

An Alternative Method for Obtaining Primary Cultures of Cortical Neurons

Reagents

1. HEBSS: 25mM HEPES, 13.8rMMaCl, 5.0nM KCl, 4.2mM $NaHCO_3$, 1.0mAf NaH_2PO_4, 0.01% phenol red.
2. $MgSO_4$ stock solution (3.82%): 1.14g $MgSO_4$ in 30mL of sterile water.
3. Solution A: 50 mL of HEBSS containing 2.5mg/ml dextrose, 0.30% bovine serum albumin fraction V (BSA), and 0.50 mL of $MgSO_4$ stock solution.
4. Solution B: 12.5mL of HEBSS containing 4.0% BSA and 0.10 mL of $MgSO_4$ stock solution.
5. Solution C: 10mL of Solution A containing 2.5mg trypsin (1000-1500U/mg; Sigma).
6. Solution D: 10mL of Solution A containing 0.10ml $MgSO_4$ stock solution, 7.5mg trypsin inhibitor (Sigma), 130kU/ml DNase (Sigma).
7. Solution E: Add 8.4 mL of Solution A to 1.6ml of Solution D.

Methods

1. Prepare all solutions described in Subheading 2.3 (A through E) before perform-ing dissection. Filter Solutions A to E through a syringe with a 0.22-io,m filter attached.
2. Dissect out the cerebral cortices of all pups with a spatula and place the corti-cal tissue in a 1.5-mL microcentrifuge tube containing 1.0 mL of solution A.3. In a sterile hood, carefully pour the contents of the tube onto the frosted side of a sterile frosted glass plate. Use a sterile razor blade to homogenize tissue.

3. Pipet the homogenized tissue into a 15-mL conical tube. Add Solution C to the conical tube and place in a hot water bath at 37°C for 15 min. Swirl occasionally to mix.
4. Add Solution E to the conical tube and swirl gently. Centrifuge at 2000rpm for 5 s to pellet cells.
6. Decant the supernatant and add Solution D. Triturate 10-15 times with a pipet for maximum cell yield. Let the solution sit for 5 min to allow debris and clumps to settle. Remove the supernatant and place in a fresh 15-mL conical tube.
7. Add Solution B to the bottom of the conical tube. There will be a visible separa-tion between Solution B and the cell-rich supernatant. Centrifuge at 2,000 rpm for 5 min to pellet the cells.
8. Decant the supernatant and add MEM medium containing 10% serum. Mix gen-tly and triturate to resuspend the cells for counting.
9. Plate cells

Organotypic Slice Cultures

Equipment

1. MacIlwain Tissue Chopper.
2. Razor blade.
3. Plastic disc for tissue chopper.
4. 70% ethanol.
5. Tube of commercial Krazy Glue.
6. Glass Pasteur pipet.
7. 4-inch microdissecting forceps with 0.8-mm tip width.
8. Fine-tip paintbrush with trimmed bristles.
9. Spatula.
10. Plastic transfer pipet.
11. Ice.
12. 60-mm tissue culture dishes (Falcon, Lincoln Park, NY).
13. Millicell filter inserts with 0.40 uAf pores.
14. Plastic container with lid.

Reagents

1. Neurobasal medium (Gibco, Carlsbad, CA) supplemented with B27 (Gibco), 200 µ*M* L-glutamine (Gibco), 100 µ*M* penicillin/streptomycin (Gibco), and 100 *mM* dextrose.
2. Krebs buffer (10x) containing'Nad (73.6g/L), KC1 (1.87g/L), NaH_2PO_4–H_2O (1.66g/L), MgCL, (2.44g/L), CaCl, (3.68 g/L). This can be stored at 4°C for 2-3 months.

3. Krebs buffer with additives containing 10 × Krebs buffer (final concentration): 126mM NaCl, 2.5 mM KCl, 1.2mM NaH_2PO_4, 1.2mM MgCL,, 2.5mA/ $CaCl_2$), 11 mM dextrose, 25 mM $NaHCO_3$, lOmM HEPES, double-distilled H_2O.
4. 4.0% (w/v) paraformaldehyde in PBS (pH 7.4) at room temperature, freshly prepared.

Methods

Dissection

Pregnant C57/B6 mice are anesthetized on G15 or G17 with a cocktail of keta-mine (1.0mg/kg) and xylazine (1.0mg/kg). The fetuses are collected and their brains are removed. The forebrains are dissected out with forceps and a spatula and placed into 60-mm tissue culture dishes containing Krebs buffer with additives on ice.

Preparation and Culture of Mouse Cortical Slices

A brain is removed from a tissue culture dish and prepared for cutting. Krazy glue is used to affix the rear of the brain to the plastic disc to prevent movement while slicing. A reservoir of Krebs buffer with additives is placed around the perimeter of the brain to allow slices to float away from the uncut brain during the slicing process. A trimmed paintbrush, which allows more control for gentle manipulation, is used to gently push down on and flatten out each hemisphere to ensure a uniform thickness of slices. The brain is cut into 300- to 400 μm coronal slices with the Macllwain Tissue Chopper. The slices are gently retrieved with a plastic transfer pipet containing Krebs with additives and transferred to the original Krebs buffer containing 60mm tissue culture dish. Approximately six coronal slices are obtained per fetal brain.

Slices from each brain are transferred to 3.0mL of neurobasal medium supple-mented with B27, L-glutamine, penicillin/streptomycin, and dextrose, as indicated previously, in a 60-mm tissue culture dish containing a 0.40 μ*M* Millicell filter insert. Slices are carefully arranged flat with forceps in the center of the insert. The residual Krebs buffer remaining in the insert, following the transfer of the slices, is slowly removed and replaced with neurobasal medium from the surrounding dish. The level of medium is then lowered to form a meniscus just above the slices. Slices are cultured in an incubator for 2h at 37°C with 95% O_2 / 5.0% CO_2 to allow for recovery before experimental treatments.

Preparation and Culturing of Rat Cortical Slices

There are slight variations in the preparation and culturing of rat cortical slices that are worthy of mention. Rat cortical slices do not tolerate Krebs buffer well. Therefore, to preserve the integrity of the cortical slices, artificial cerebrospinal fluid (ACSF) is substituted for Krebs. Brains are dissected out of fetuses and placed in ACSF containing 124mM NaCl, 5.0mM KCl, 1.23mM NaH_2PO_4~H_2O, 18.5mM$MgSO_4$, 2.0mM $CaCl_2$, 10mM dextrose, and 26 mM $NaCO_3$ in sterile water. Slices are cultured in neurobasal medium containing L-glutamine, penicillin/streptomycin and 20% FBS. Neurobasal medium supplemented with B27 or N2, 5.0mM L-glutamine, and peni-cillin/streptomycin without serum. After 2h, the serum-free medium is replaced with neurobasal medium supplemented with B27 or N2, 5.OmM L-glutamine and penicil-lin/streptomycin without serum. For shorter ethanol treatments (less than 90min), slices are cultured in a serum-free medium for the duration of the experiment.

Controlled Ethanol Exposure

A potential confounding factor for studies of ethanol exposure is the volatility of ethanol. This problem is overcome by incubating primary cultures within a sealed chamber containing ethanol. These chambers maintain a consistent vapor pressure established by ethanol evaporation, prevent loss of ethanol from the culture medium, and can stabilize the concentration of ethanol for more than 72 h.

Equipment

1. 1.6-liter Rubbermaid "Flex and Seal" food storage containers.
2. "Picnic table" plastic 24-slot microcentrifuge tube rack with the handles snapped off to make a flat surface for incubation of tissue culture dishes and plates.
3. Syringe (60 mL) with 6-inch tubing attached for CO_2 injection.
4. GM7 Micro-Stat Analyzer

Reagents

1. Medium with MEM supplemented with 33mM dextrose, penicillin/streptomycin, and serum as needed.
2. 95% ethanol.
3. Compressed carbon dioxide gas (CO_2).

Methods

1. Set up the chambers prior to initiating ethanol treatment. Each experimental concen-tration of ethanol requires a separate chamber. For example, to treat neuronal cultures with ethanol at 0,200, and 400mg/dl, a minimum of three chambers is required. Add 200 mL of water to each chamber and set a rack inside the chamber. The 1.60-liter chamber system described here holds up to two 100-mm culture dishes per rack.
2. The amount of ethanol added to the treatment medium (in ij,L/mL) is determined by the following equation: (molar concentration of ethanol treatment) x (46.07 g/ M)/ (ethanol concentration = 0.95) x (specific gravity of ethanol = 0.7750 g/mL). For exposure to 400mg/dL (86.6mA/) of ethanol, 5.43 uL/mL of 95% ethanol is added to both the culture medium and the water in the chamber (AUO NuUr4). Cell cultures not subject to ethanol exposure are incubated in identical chambers, but without ethanol addition.
3. Place the culture dishes securely on the rack in the ethanol concentration appro-priate chamber. Make sure the dishes are level and that medium adequately cov-ers the cells. Position the lid on the chamber and secure all but one corner. Fill a syringe with 60 mL of CO_2 by using a short tubing attachment to connect to the valve on the CO_2 tank. Pinch off the tubing with a hemostat and, removing the tubing from the valve, insert the tubing into the unsecured corner of the chamber. Release the constriction

on the tubing and plunge the CO_2 from the syringe into the closed chamber. Quickly close the open corner to tightly seal off the chamber and place the chamber in the incubator. These instructions assume the use of a 37°C incubator with 6.0% CO_2 and saturating humidity.

APPLICATION OF EXOGENOUS SUBSTANCES AND HARVESTING THE CELLS/TISSUE

Fresh Samples

Primary neuronal cell cultures or organotypic slice cultures are easily manipulated for a variety of biochemical assays. Growth factors, biological inhibitors, or other modulatory substances may be added to the medium alone, or in combination with ethanol, using the chambered treatment system. Conditioned medium from the cultures may be saved for detection of cell released cytokines by enzyme-linked immunosorbant assay (ELISA).

Brief exposure of cell cultures to 0.25% trypsin/EDTA (Gibco) generates a dissociated cell suspension suitable to process for cell sorting by flow cytometry. Protein expression profiles from slice cultures or dissociated primary cultures may be deter-mined by Western immunoblot or ELISA after lysis in an appropriate lysis buffer.

Fixed Samples

Cortical slices are best prepared for immunohistochemistry by fixation. Slices are fixed in freshly made 4% paraformaldehyde for 35 min at room temperature after experimental treatments. Fixative is gently removed from the filter and culture dish and slices are washed once with 0.10MPBS. Slices are transferred to a 12-well plate (each well labeled with the original treatment condition) containing 10% sucrose and incubated for 24 h. We have found it useful to use a spatula with the tip bent to a 90° angle to remove slices from the culture dish and transfer them to the 12-well plate. After 24 h in 10% sucrose, slices are incubated in 15% sucrose for at least 3 h before freezing in prerjaration for cutting. Qn preparation for cryosectioning, slicesare removed from the sucrose and trans-ferred to a precleaned Superfrosti'/Plus microscope slide.

Typically three slices or groups of slices are placed on each slide. Sucrose is aspirated off as much as possible. Slices are covered with just enough Optimal Cutting Temperature (OCT; Sakura Finetechnical, Tokyo, Japan) embedding medium to allow for complete coverage of the slices. The slide is pushed down firmly into a flat block of dry ice until completely frozen. Slides are wrapped in aluminum foil and placed in –20°C freezer until ready for cryosectioning in 12μm increments.

We have found that slice integrity and antigenicity diminish if left in the freezer for longer than a week. Before immunonistochemistry, sections are incubated in U.01UM citric acid in sterile ddH_2O (pH 6.0) for 20min followed by two rinses in PBS. Standard protocols are then used for the remaining immunolabeling process.

NOTES

1. Samples of neurons from pups and older animals are less likely to be viable, though allowances (mostly in the composition of the medium) can be made.
2. This takes practice since the cortex is thin and will tend to peel away with the meninges. One method that seems to work well is to start at the olfactory bulbs or the thicker striatal areas.
3. Overly vigorous trituration will result in low cell recovery. Heat the glass pipette tip over an alcohol burner to serially narrow the pipet tip for complete dissocia-tion of cells (cool the tip before triturating).
4. It is recommended that the concentration of ethanol in the medium be measured with an Analox GM7 Micro-Stat Analyzer (or another similar instrument) before and after treatment. Ethanol concentrations will vary with chamber selec-tion and culture conditions and may need to be optimized experimentally. The calculation given for ethanol addition is intended as a starting point to achieve the desired ethanol concentration. In practice, the mathematically determined ethanol concentration generally results in a lower measured ethanol concentra-tion for the chambers described. Increasing the ethanol added to the chamber water and the treatment medium by 10-20% attains the desired ethanol concen-tration. Samples collected for ethanol measurement may be frozen at -20°C for several weeks without detectable loss of ethanol.

CHAPTER

9 Oils and Fats

Phosphatides and mucilaginous substances are emulsifying agents and in excessive quantity may increase the loss of neutral oil when the oil is caustic refined. If the oil is going to be physical or steam refined, the content of phosphatides and metallic pro-oxidants must be lowered by a degumming process before being refined. Water degumming generally removes about 90 percent of the phospholipid material. There are, however, some non-hydratable phosphatides which do not react with water and are removed during caustic refining. Their presence tends to increase the neutral oil lost in the soapstock.

It is usually not economical to degum sunflower oil in a separate operation because the percentage of lecithin is so small. In addition to removing lecithin, degumming also removes some other foreign material. However, the quantity removed is usually so little that it can be easily removed during the alkali refining procedure. Additional equipment cost, labor expense and oil loss does not warrant degumming of sunflower oil if it is to be alkali refined.

NEUTRALIZATION OR REFINING

Neutralization removes nonglyceride impurities, which consist principally of free fatty acid (FFA), and significant quantities of mucilaginous substances, phosphatides, chlorophyll, and color bodies. The refining process removes the objectionable impurities in the oil with the least damage to either the glycerides, the tocopherols or other natural antioxidants.

Chemical (or caustic) refining converts the FFA to oil-insoluble soaps, removes all the substances, which become insoluble upon hydration, and removes some of the colour bodies present in the oil. Reactions are at a relatively low temperature, allowing for the refining equipment to be fabricated essentially from iron or mild steel.

The technology of caustic refining is concerned with: (*a*) the proper choice of alkali used, (*b*) the quantity of alkali used, and (*c*) the refining technique used to produce the desired purification (without excessive saponification of the neutral oil) and (*d*) the method used for efficient separation of the refined oil from the soapstock.

A disadvantage of the caustic refining method is the loss of a small portion of neutral oil, in addition to the FFA, as a result of saponification. This loss can be minimized, however, by utilizing good technology.

Sunflower oil can be alkali refined using either a batch method or a continuous method of refining. The batch method requires less capital investment, but; (*a*) the yield of refined oil is considerably less than via the continuous method, (*b*) the thruput rate is less and (*c*) there is a higher labor cost.

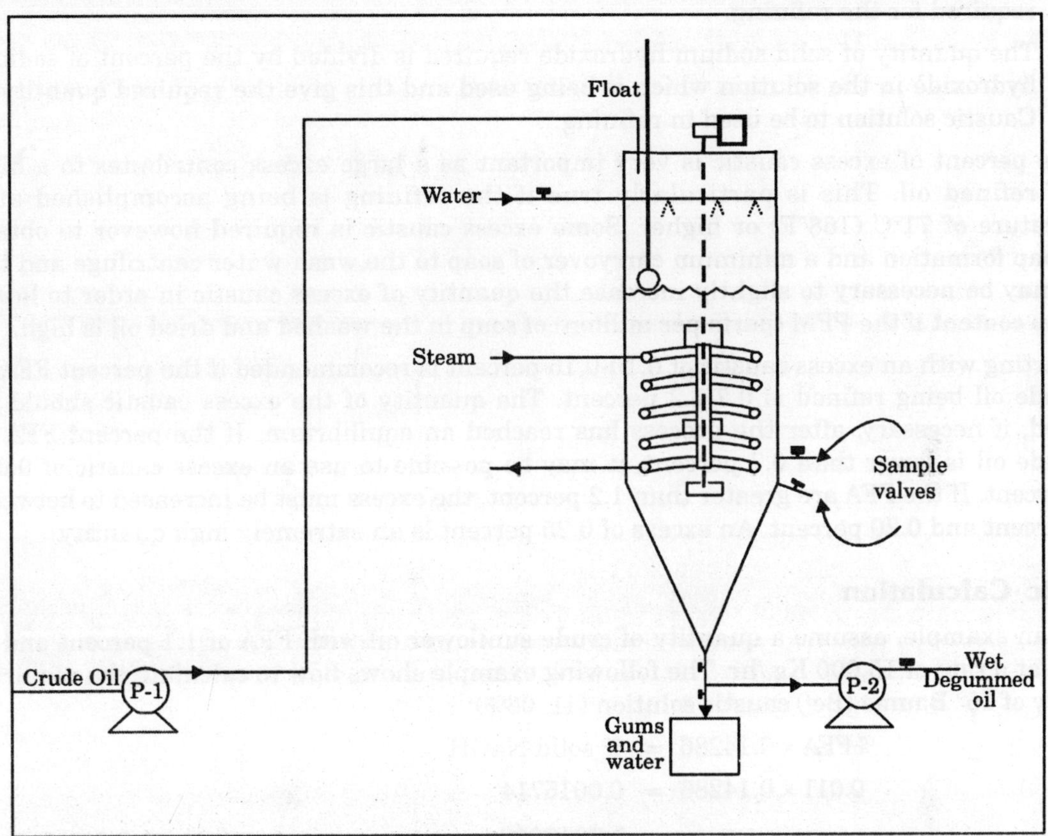

Fig. 9.1. Bath Degumming.

Calculation of Caustic

Caustic calculation is the method of determining the quantity and concentration of the caustic solution required for alkali refining as it is applicable to both the batch and continuous methods.

The quantity of caustic required for neutralization can be determined in the following manner:

(1) Multiply the percent FFA of the oil by 0.14286. This gives the percent of solid sodium hydroxide required for neutralization.

(2) To this percent, add a slight excess quantity, as the quantity calculated in Step 1 is the theoretical quantity. The excess usually ranges between 0.05 percent and 0.20 percent, depending upon the percent of FFA present in the oil and the quantity of non-oily constituents present in the crude oil.

(3) Add the percent calculated in (1) to the desired excess established in (2).

(4) The quantity of oil in pounds (or kilograms) is multiplied by the total percent arrived at in (3). This gives the number of pounds (or kilograms) of solid sodium hydroxide required for the refining.

(5) The quantity of solid sodium hydroxide required is divided by the percent of sodium hydroxide in the solution which is being used and this give the required quantity of Caustic solution to be used in refining.

The percent of excess caustic is very important as a large excess contributes to a high loss of refined oil. This is particularly true if the refining is being accomplished at a temperature of 71°C (168°F) or higher. Some excess caustic is required however to obtain good soap formation and a minimum carryover of soap to the wash water centrifuge and the oil. It may be necessary to slightly increase the quantity of excess caustic in order to lower the soap content if the PPM (parts per million) of soap in the washed and dried oil is high.

Starting with an excess caustic of 0.10-0.15 percent is recommended if the percent FFA of the crude oil being refined is 0.7-1.2 percent. The quantity of the excess caustic should be adjusted, if necessary, after the process has reached an equilibrium. If the percent FFA in the crude oil is lower than 0.7 percent, it may be possible to use an excess caustic of 0.05-0.10 percent. If the FFA are greater than 1.2 percent, the excess must be increased to between 0.15 percent and 0.20 percent. An excess of 0.25 percent is an extremely high quantity.

Caustic Calculation

As an example, assume a quantity of crude sunflower oil with FFA of 1.1 percent and is refined at a rate of 15,000 Kg./hr. The following example shows how to calculate the required quantity of 16° Baume (Be′) caustic solution (11 .06%):

$$\%\text{FFA} \times 0.14286 = \%\text{ solid NaOH}$$

$$0.011 \times 0.14286 = 0.0015714$$

$$\text{Plus excess of} = \frac{0.0010000}{0.0025714}(0.1\%)$$

$$15{,}000 \text{ Kgs/hr.} \times .0025714 = \text{solid NaOH}$$

$$\frac{38.571 \text{ Kgs. of solid NaOH}}{0.1106(16°\text{Be}')} = 384.74 \text{ Kgs. of } 16°\text{Be}' \text{ per hour}$$

It is suggested that a 16° Be′ caustic solution be used for refining sunflower oil. The concentration is not so high as to result in a significant amount of saponification of the neutral oil if the excess quantity was greater than the optimum. The Be′ of the caustic should be varied depending on various factors and conditions. If the percent FFA of the oil to be refined is greater than 1.5 and the crude oil is relatively free from non-oily constituents such as gums and lecithin, it would be advisable to use a caustic solution having a Be′ of 16.5-17.0. The reason for this is to avoid introducing an excessive amount of water which could inhibit the good separation.

If refining using the continuous method, adjust the quantity of caustic solution as the percent FFA changes or the quantity of non-oily constituents in the feed stock change. In a continuous refinery, a chart should be prepared which shows the quantity of caustic solution to be used as the percent FFA changes during the run.

If a phosphoric acid treat is used before caustic refining, additional caustic solution is required to neutralize the added phosphoric acid. The oil and acid should be mixed well, using good shearing agitation. The mixing should continue for a fairly long time (prior to the addition of the caustic solution) to prevent the formation of a sticky soap. Such soap is difficult to discharge and will foul the centrifuge discs.

Batch Refining

This method of refining is conducted in a mild steel, steep cone-bottom tank (about a 60° cone angle) equipped with an agitator (slow speed) having paddle-type blades which are pitched to lift the product. Also, it should have internal steam coils and a method of spraying the caustic solution and water on the surface of the oil. It is desirable to have a small draw-off valve for obtaining samples located about 24 cm (six inches) above the start of the straight side of the tank and another 24 cm. below the top of the cone bottom. A thermometer should be located about halfway up on the straight side of the tank.

Refining oil loss using the batch method is about six times the percentage of FFA in the crude oil.

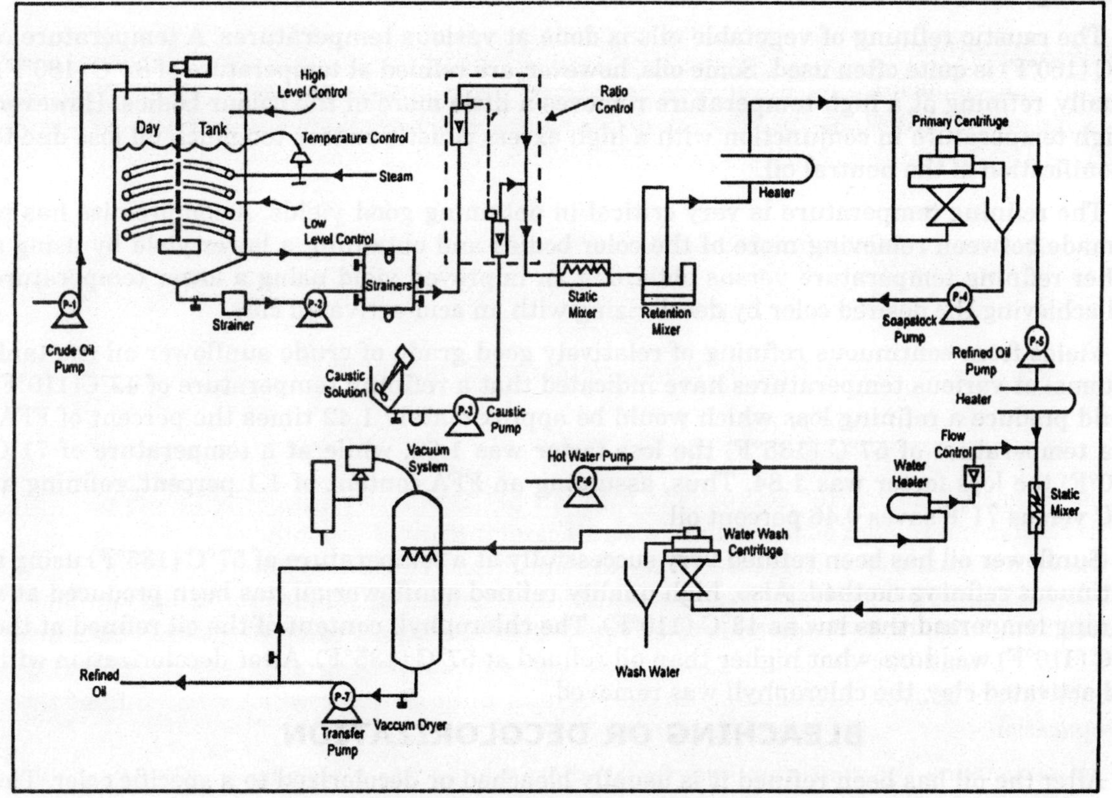

Fig. 9.2. Continuous Refining

Continuous Refining

A continuous method of refining requires a considerable investment in equipment. However, it is the most efficient and economical method.

The quantity of caustic soda solution to be used is calculated as previously shown for the batch method. The excess and °Be′ for the continuous refinery is very similar to what is used in the batch refining. The excess alkali should be determined occasionally during the refining to ensure the process is under control. It is desirable to determine the soap content after the water washing and vacuum drying. This is so small changes can be made in the refining conditions to minimize the soap level before further processing the oil.

Addition of a small amount of weak acid solution (*e.g.* citric acid) prior to vacuum drying may be desirable if the oil is going to be deodorized or further processed without decolorization with an acid activated clay. However, if the oil is to be bleached with bleaching clay, the acidity in the clay usually neutralizes the slight alkalinity of the oil.

The continuous method of refining permits varying the refining temperature considerably to obtain a good separation of the heavy phase using the centrifuge. A temperature as low as 43°C (110°F) has been used to produce a good quality refined oil. A neutralization temperature of 57"C (135°F) gives both very good refined oil and good yield.

Refining Temperature

The caustic refining of vegetable oils is done at various temperatures. A temperature of 71°C (160°F) is quite often used. Some oils, however, are refined at temperature of 82°C (180°F). Usually, refining at a high temperature removes a little more of the colour bodies. However, a high temperature in conjunction with a high excess caustic causes too much oil loss due to saponification of the neutral oil.

The refining temperature is very critical in obtaining good yields. A compromise has to be made between removing more of the color bodies and obtaining a lower yield by using a higher refining temperature versus obtaining an improved yield using a lower temperature and achieving the desired color by decolorizing with an acid-activated clay.

Yields from continuous refining of relatively good grade of crude sunflower oil (no tank bottoms) at various temperatures have indicated that a refining temperature of 43°C(110°F) would produce a refining loss which would be approximately 1.42 times the percent of FFA. At a temperature of 57°C (135°F) the loss factor was 1.62, while at a temperature of 71°C (160°F) the loss factor was 1.84. Thus, assuming an FFA content of 1.1 percent, refining at 43°C versus 71°F saves 0.46 percent oil.

Sunflower oil has been refined very successfully at a temperature of 57°C (135°F) using a continuous refining method. Also, high quality refined sunflower oil has been produced at a refining temperature as low as 43°C (110°F). The chlorophyll content of the oil refined at the 43°C (110°F) was somewhat higher than oil refined at 57°C (135°F). After decolorization with acid activated clay, the chlorophyll was removed.

BLEACHING OR DECOLORIZATION

After the oil has been refined it is usually bleached or decolorized to a specific color. The bleaching process should be done under vacuum and at an approximate temperature of 77°C

(170°F). The use of acid activated clay is recommended as the residual acid in the clay neutralizes the traces of soap that may be present. The chlorophyll removal is an important step as otherwise a green tint may appear in the oil after deodorization.

The oil should be dry before the addition of the clay if an acid activated clay is used. Any moisture present in the oil causes the acid to leach out of the clay, which causes an increase in the FFA content of the refined oil. If the filters are well blown, the oil loss in the bleaching process is about equal to the quantity of bleaching clay used in the process.

DEWAXING

The dewaxing process removes the waxy-type of the non-oily constituents that are not removed either by the alkali refining process or the degumming process. There are various methods, which can be used for dewaxing the oils such as a batch method or various types of continuous methods.

Batch Dewaxing Method

The batch method is relatively simple. It consists of an well-insulated tank of a given capacity, equipped with some means of agitation. A relatively slow speed mechanical agitator is preferred, however, it is possible to use a pump to circulate the oil from the bottom to the top. Under no circumstance should air be used for agitation of oil. Air bubbles will form in the oil which (a) inhibit the heat transfer causing an extended processing time and (b) increase the peroxide value of the oil.

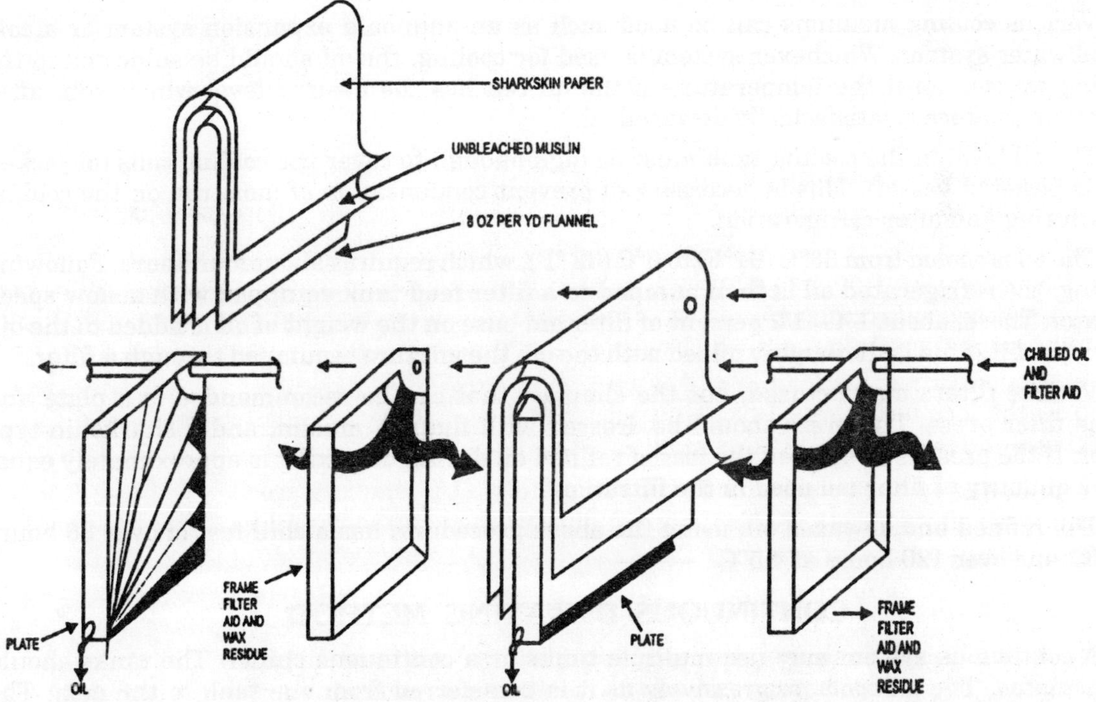

Fig. 9.3. Dressing the Filter Press-Dewaxing.

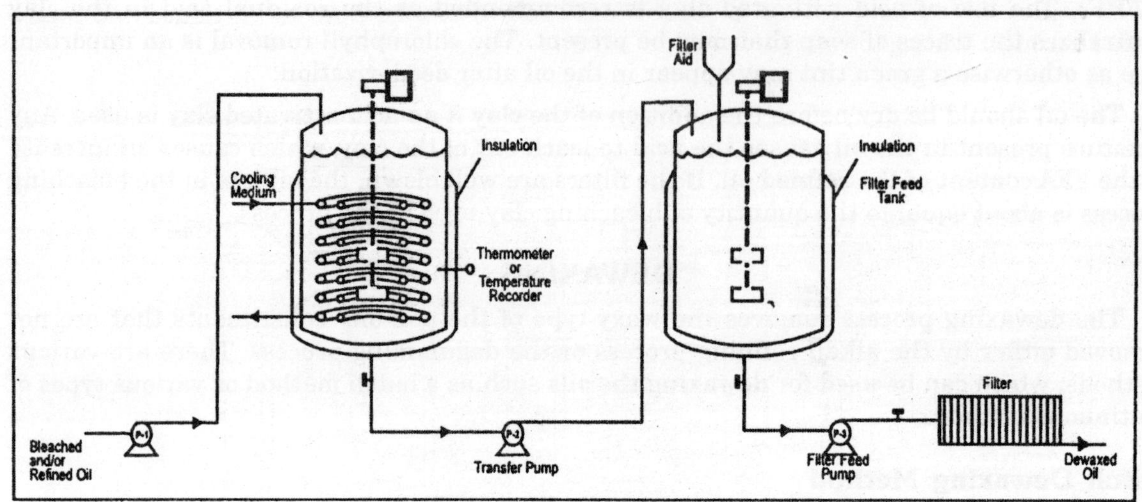

Fig. 9.4. Batch Dewaxing.

The tank used for cooling the oil should also be equipped with a large number of relatively small diameter coils in order to give a large cooling surface. It would be preferable to have a vertical-type tank rather than horizontal, though either is acceptable. However, the agitation should move the oil rather slowly across the cooling coils, yet it should be enough to keep the waxes from settling out of the oil.

Various cooling mediums can be used such as an ammonia expansion system or a cold glycol/water system. Whichever system is used for cooling, the oil should be subjected to the cooling process until the temperature of the oil reaches the desired level which will, after filtration, produce a satisfactorily dewaxed oil.

The oil level in the cooling tank must be high enough to cover the cooling coils (or jacket, if it is jacketed vessel). This is necessary to prevent condensation of moisture on the cooling coils during and after refrigeration.

The oil is cooled from 33°C (91°F) to 0°C (32 °F), which requires about four hours. Following cooling, the refrigerated oil is then pumped to a filter feed tank equipped with a slow speed agitator. There, about 1/4 – 1/2 percent of filter aid base on the weight of oil is added to the oil. When the filter aid is thoroughly mixed with the oil, the mixture is pumped through a filter.

Various filters may be used, but the simplest that can be recommended is a plate and frame filter press. The press should be dressed with flannel, muslin, and a sharkshin-type paper. If the press is so dressed the loss of refined oil during dewaxing is approximately equal to the quantity of filter aid used in the filtration.

The refined and dewaxed oil, using the above procedure, has a chill test of over 36 hours at 0°C, and over 120 hours at 4.5°C.

CONTINUOUS DEWAXING METHOD

A continuous system may use multiple tanks or a continuous chiller. The tanks should be insulated. The oil cools progressively as it is transferred from one tank to the next. The residence time in the system is long enough to permit reducing the temperature of the oil to the desired temperature before it is pumped from the final tank.

The chilled oil is then pumped to a filter feed tank where the filter aid is added and then pumped through a filter to remove the filter aid and waxes.

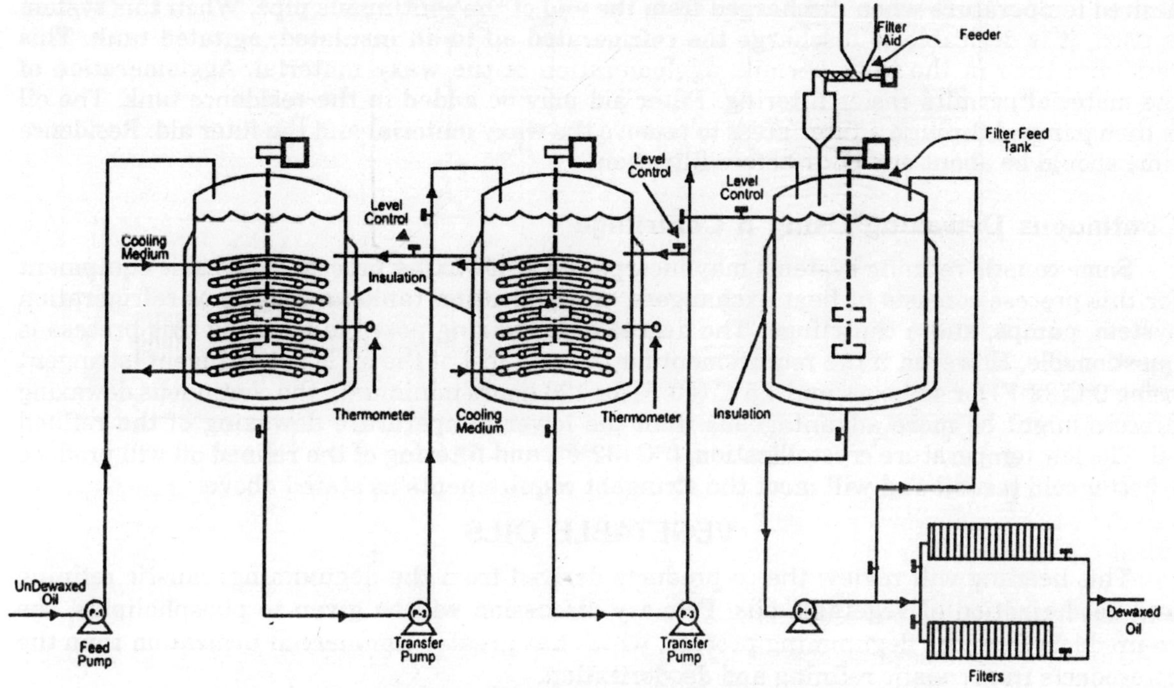

Fig. 9.5. Continuous Dewaxing.

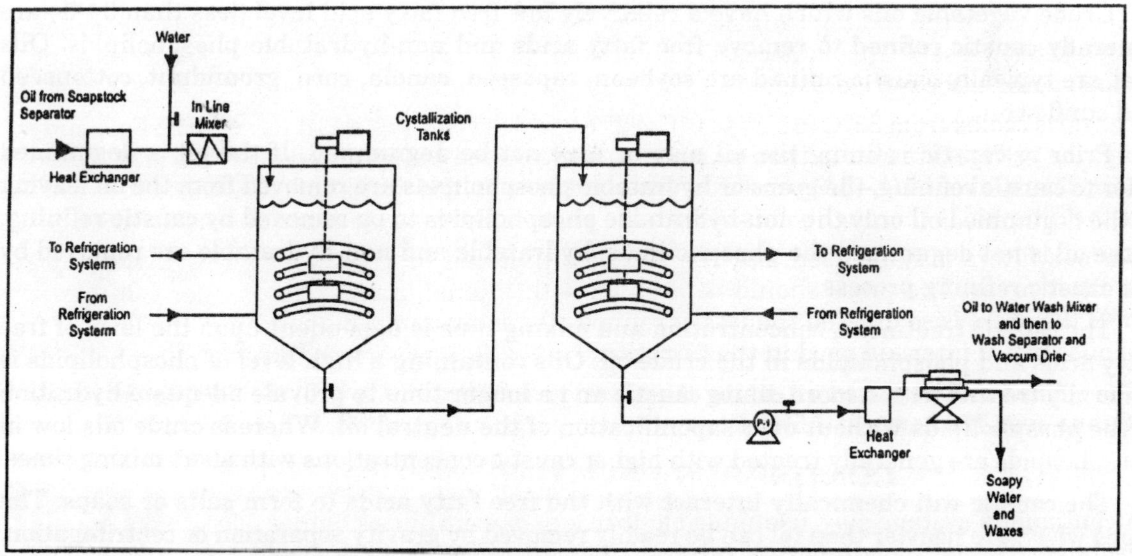

Fig. 9.6. Continuous Dewaxing (Using a centrifuge).

Alternatively, a continuous process may use a continuous chiller. With this process, the oil being chilled passes through a continuous pipe that has a cooling medium on the outside of a jacketed pipe. The flow rate is controlled so that the temperature of the oil is at the desired temperature when discharged from the end of the continuous pipe. When this system is used, it is desirable to discharge the refrigerated oil to an insulated, agitated tank. This residence time in the tank permits agglomeration of the waxy material. Agglomeration of the material permits easier filtering. Filter aid may be added in the residence tank. The oil is then pumped through a filter press to remove the waxy material and the filter aid. Residence time should be about one hour before filtration.

Continuous Dewaxing Using a Centrifuge

Some caustic refining systems may incorporate a dewaxing process. The basic equipment for this process consists of heat exchangers, crystallization tanks, a cooling or refrigeration system, pumps, and a centrifuge. The degree of de-waxing possible by using this precess is questionable. However, if the requirement for the cold test of the oil is not stringent [stringent being 0°C(32°F) for 48 hours and 4.5°C (40°F) for 120 hours minimum], the continuous dewaxing process might be more advantageous than the lower temperature dewaxing of the refined oil. The low temperature crystallization, 0°C (32°F), and filtering of the refined oil will produce a better cold test oil and will meet the stringent requirements as stated above.

VEGETABLE OILS

This heading will review the co-products derived from the degumming, caustic refining and deodorization of vegetable oils. Primary discussion will be given to phospholipids, the co-products from the degumming process, which has greater commercial utilization than the co-products from caustic refining and deodorization.

CO-PRODUCT FROM CAUSTIC REFINING-SOAPSTOCKS

Crude vegetable oils which have a relatively low free fatty acid level (less than 3.0%) are generally caustic refined to remove free fatty acids and non-hydratable phospholipids. Oils that are typically caustic refined are soybean, rapeseed, canola, corn, groundnut, cottonseed and sunflower.

Prior to caustic refining the oil may or may not be degummed. If the oil is degummed prior to caustic refining, the gums or hydratable phospholipids are removed from the oil leaving in the degummed oil only the non-hydratable phospholipids to be removed by caustic refining. If the oil is not degummed the phospholipids, hydratable and non-hydratable are removed by the caustic refining process.

The caustic treatment, concentration and mixing time is dependent upon the level of free fatty acids and phospholipids in the crude oil. Oils containing a high level of phospholipids is generally treated with a more dilute caustic and a longer time to provide adequate hydration of the phospholipids without over saponification of the neutral oil. Whereas crude oils low in phospholipids are generally treated with higher caustic concentrations with short mixing times.

The caustic will chemically interact with the free fatty acids to form salts or soaps. The soaps which are heavier than oil can be readily removed by gravity separation or centrifugation. The hydrated phospholipids and soaps being both more soluble in the water phase are removed collectively in water phase after refining and referred to as the soapstock.

The concentration of soaps (fatty acids) and phospholipids in the soapstock (generally total fatty acid content between 40 to 50%) is related to the strength of caustic (concentration of caustic in water) and additional water washing used for refining. The soapstock is a good source of non-edible fatty acids, which can be used to supplement animal feed or as a source for industrial fatty acids. Because of the high level of water and pH, soapstock can support microbial growth and will ferment.

The "wet" soapstock may be applied directly to animal feed, provided that the additional water from the soapstock does not raise the moisture level of the animal feed above the minimum level required to keep the animal feed stable.

To prevent microbial activity the soapstock can be dried or the pH lowered. Soapstock can be dried using a variety of methods. A common commercial practice of drying soapstock involves neutralizing the soapstock with sulphuric acid to a neutral or slightly acidic pH under mild heat, then removing the moisture via a thin film evaporator to less than 2%.

Soapstock is commonly acidulated with sulphuric acid to an acid pH whereupon the fatty acids are liberated from the metals to form free fatty acids. The acid salts and water are then separated from the free fatty acids via gravity separation or centrifugation to concentrate the total fatty acids to 85-95%.

Co-Products from Deodorization - Free Fatty Acids, Sterols, Tocopherols and Tocotrienols

The final processing step of either physical refining or caustic refining is a high temperature, high vacuum distillation, commonly referred to as deodorization. Of course in physical refining the primary objective of the distillation process is to remove a majority of the free fatty acids and flavor components, and in caustic refining the primary objective is to remove the flavor components and residual fatty acids.

In addition to the free fatty acids, sterols, tocopherols and tocotrienols are also removed by the distillation process along with pesticides, oxidized materials and truncated lipids (a result of oxidation or generated from previous processing steps). The materials along with the water used as sparging steam, are collected and commonly referred to a deodorizer distillate.

The components in deodorizer distillate having greater commercial value (sterols, tocopherols and tocotrienols) must be separated from the components of lessor commercial value by chromatography and additional distillation steps. While the isolated sterols, tocopherols and tocotrienols can command a high value in the market, the processing equipment required to isolate these materials is costly and is commercially practiced by a few oil refiners and oleo chemical manufacturers.

Co-Products from Degumming Phospholipids, Lecithin

Lecithin is a term that collectively refers to all the water-soluble materials or gums that are separated from a crude oil by the degumming process and are retained after the gums are dried.

The gums are removed from vegetable oils prior to caustic or physical refining because the gums will interfere with subsequent processing steps and degrade during the final distillation process imparting dark colors and off-flavors to the final processed oil.

Lecithin contains primarily the following components:

Phosphatidyl ethanol amine (PE), phosphatidyl choline (PC), phosphatidyl inositol (PI), phosphatidyl serine (PS), phosphatic acid (PA), glycolipids (GL) and free fatty acids (FFA), Proportions of these components will vary depending on the source, however the major constituents are generally PE, PC, PI and PA. All PL share a common chemistry. All contain two fatty acids attached at the 1 and 2 position on a glycerol molecule. A phosphate group is attached to the 3 position. If only a phosphate group is attached the molecule is PA. PE has an ethanolamine group attached to the PA; PC has choline attached to PA: PI has an inositol group attached to the PA, and PS a serine group attached to the PA. The functional properties of lecithin are due to the collective chemical properties of all the lecithin constituents.

Lecithins are surface active and have surface activity. The two fatty acid chains are very lipophilic and have a strong affinity to associate with oil. Whereas, the phosphatidyl groups are very hydrophobic, having a strong affinity to associate with water through hydrogen bonding. The surface activity of lecithin provides lecithin with very diverse functionality that is used in a variety of applications and will be discussed later in the paper.

Lecithins also contain small levels of carbohydrates and proteins. Generally the small levels do not affect the lecithin's functional properties. However, lecithin produced from poor quality beans or poor extraction practices will contain higher than normal carbohydrate levels that can alter functional properties.

Because the phospholipids (PL) are insoluble in acetone the purity of lecithin is measured on the basis of the percentage of acetone insolubles or % Al. Most commerçial trading standards require at least 50% Al for a material to be traded as "lecithin." Edible fatty acids and vegetable oils are acceptable diluants to reduce the viscosity or commercial lecithins. Lecithins with greater than 65% Al are extremely viscous. Lecithin with 62% Al has a viscosity similar to honey or molasses.

THE PROCESS

Crude oil is mixed with water (de-mineralized or soft water) at generally an equivalent weight of water to PL's, at approximately 70°C. Because the phospholipids are amphoteric they have an affinity for water and will form gums under mild agitation that can be easily separated by centrifugation. If agitation is too severe the PL's will emulsify the water and oil making separation extremely difficult.

The wet gums are dried using thin film vacuum dryers. To reduce the color lecithin is reacted with hydrogen peroxide. Lecithins may have high peroxide values as consequence of hydrogen peroxide bleaching and is not an indication of excessive oxidation. The % of Al is adjusted using a combination of FFA and oil to obtain a desired viscosity and a lecithin that will not separate during storage. As a final step the lecithin is passed through a fine filter to provide a finished lecithin with less than 1% hexane insolubles and under 2% moisture.

Oil may be removed to make a solid or de-oiled lecithin with a minimum 97% Al by either acetone extraction or ultrafiltration. The de-oiled lecithin is ground or agglomerated to obtain a desired particle size distribution.

Damaged seed or poor crushing and extraction practices will affect the quality of lecithin. Excessive heat will develop non-bleachable colors, reduce oxidative stability and reduce functionality. PL from damaged seed can interact or complex with calcium or magnesium to form non-hydratable phospholipids NHPL will not be removed by the degumming process. Subsequently lecithin yields are reduced and additional effort is required from other oil processing steps (caustic refining or bleaching) to remove the NHPL from the oil.

MODIFICATION

To increase its hydrophobic properties lecithin is commercially modified by the following methods :

Hydroxylation—The fatty acid composition of lecithin is generally slightly more saturated than its source oil. Most commercial lecithins contain a high level of unsaturated fatty acids. The unsaturated double bonds are reacted with hydrogen peroxide using an organic acid catalyst to form hydroxyl groups at the double bond sites. The hydroxyl groups having a strong affinity for water increase the hydrophobic properties of the PL's providing lecithin with greater oil-in water emulsification properties. Approximately 25% of the available double bonds are hydroxylated to provide a functional effect.

Hydrolysis—A lipase enzyme is used to hydrolyze the middle fatty acid at the 2 position commonly referred to as lyso lecithin. Removal of the fatty acid decreases the level of fatty acids on the molecule, therefore reduce the relative ratio between the lipohilic (fatty acid) and hydrophilic (phosphate-based group) and increase the overall hydrophilic properties of the lecithin. A partial hydrolysis will not result in significant change in functionality. Nearly all the fatty acids from the 2 position must be hydrolyzed for lyso lecithin to have a meaningful change in functionality.

The heat stability of lecithin can be enhanced by acetylation of the ethanolamine group in PE. The ethanolamine group can interact with reducing sugars under mild heat to darken the lecithin. By treating the lecithin with acetic anhydride, the amine group is acetylated and will not interact with sugars, and increase the heat resistance of the lecithin. Acetylation also increases the lecithin's hydrophilic properties, increasing its oil-in-water emulsification properties.

Acetylation with hydroxylation produces a lecithin with greater heat resistance and additional hydrophilic properties.

APPLICATIONS OF LECITHIN

Chocolate—Lecithin is used in chocolate to reduce viscosity during tempering to improve the coating and filling properties of the chocolate. Because of lecithins effective lubricating properties the viscosity of chocolate can be significantly reduced with 0.5% lecithin (62% Al), that would otherwise require the addition of 3-5% cocoa butter to achieve an equivalent viscosity.

Margarines and spreads—Lecithin is used in margarines and spreads (margarine products with less than 80% fat-generally 20 to 52% fat) to improve the dispersion of water in fat. Because of lecithin's emulsification properties, lecithin can reduce the surface tension between the water phase and oil phase when they are added together during the production of margarines and spreads. In margarines 0.2 to 0.5% lecithin reduces the size of water droplets dispersed in fat prior to chilling. The reduced water droplets improve mouthfeel characteristics, improve flavor release and decrease microbial activity. In spreads 0.5 to 1.0% lecithin is generally used synergistically with 1 to 2.0% distilled monoglycerides to minimize water droplet size prior to chilling and stabilize the water-in-oil emulsions after chilling and through distribution.

Milk Replacers—The modified lecithins (hydroxylated and hydroxylated and acetylated) are effective oil-in-water emulsifiers to disperse and stabilize 10 to 30% oil-in-water emulsion used as milk replacers.

Baking Shortenings—One percent lecithin in baking shortenings can effectively enhance the emulsifying and aeration properties of other emulsifiers in the baking shortening (such as : distilled monoglycerides, propylene glycol monoesters).

Baking—Standard lecithins and deoiled lecithins are added to baking formulas to improve the lubricity properties of doughs and batters. Doughs with 0.5 to 1.0% lecithin are easier to mix, require less energy, reduce adhesion to equipment and have improved gas retention properties by improve the elasticity of the dough. 0.2 to 0.5% Lecithin in batters reduce batter viscosity improving the batters flow properties and consistent dispensing into baking pans.

Powdered Beverages—Lecithin can be sprayed on powdered beverages that are difficult to disperse in water or milk. Powders with a high fat content, such as a powdered cocoa or chocolate, will not readily disperse cold water or milk because of the high level of surface tension between the fat on the powder and water. Two percent lecithin (30 to 62% AI) sprayed onto the high fat powder will reduce the surface tension between the powder and water and allow easy dispersion of the powder in the water. The lipophilic fatty acids will adsorb on the fat surface, leaving the lecithin's hydrophilic surface to coat the powder and readily associate with the water allowing for easy dispersion.

In beverages that are very hydrophilic, such as, high protein powders, the lecithin is added to reduce the hydrophilic surface. Very hydrophilic powders have a tendency to absorb water very quickly forming capsules or "fish eyes" which are very difficult to disperse. Spraying two percent lecithin on the surface will reduce the hydrophilic surface of the protein powder, reducing water absorption and allowing for improve dispersion.

Anti-stick Sprays—Lecithin is commonly used in anti-stick sprays that are applied to baking pans, trays or belts to reduce the baked product from sticking to the metal. Generally a heat resistant, acetylated lecithin is used for these applications. The viscosity of the lecithin can be reduced for spraying by heating or adding additional vegetable oils.

Lecithins is also used as an anti-sticking agent between packaged cheese slices to prevent cheese slice from sticking to each other.

The anti-stick properties are also useful for mold release applications in the manufacture of ceramics, plastics and concrete.

Dry Mixes—Deoiled lecithins are used in dry mixes (sauces, batters, coatings) to improve dispersion of the powders in water, stabilize the dispersion of the solids in water and reduce sticking in batter dry mixes.

Paints—Lecithin is used to disperse pigments and reduce paint viscosity.

Drilling Muds—Lecithin is added as an effective "environmentally friendly" lubricant used in drilling for oil or water.

The unique chemical properties of lecithin provide lecithin with many multifunctional commercial applications.

BIOTECHNOLOGY APPROACHES TO MODIFICATIONS OF OILSEED AND VEGETABLE OILS

Increasing global human population and specific consumer demands have and will continue to have a direct impact on food supplies. As a major component of many diets definite concern needs to be directed toward oils and fats. This class of compounds not only serves as a major source of dietary energy but also has nutritional implications associated with specific lipids. Particular biological roles exist, for example, for essential fatty acids, sterols (especially cholesterol) and fat-soluble vitamins. Consequently it is imperative that oils and fats be available in both the desired quantity and quality.

There is an essentially universal reliance on plants as the major supplier of human food. In particular, certain herbaceous oilseed crops and cultivated trees are the primary source of edible oils and fats. Projections have been made, however, that greater supplies will be necessary to meet increasing future demands. In addition, health concerns are directing more immediate attention toward the actual composition of dietary oils and fats as related to nutritional requirements.

Several approaches may be adopted to meet the projected supply and demand for edible oils and fats. More efficient agronomic practices can be adopted to ensure greater crop productivity. Use of improved developed plant strains (crop cultivars) can be made to obtain increased oil yields. These improvements in plants can be sought not only from conventional breeding programmes, but also from the adaptation of many of the new techniques of biotechnology and molecular biology. Applications of cell culturing, genetic engineering and modification of enzyme systems have been made with varying degrees of success.

Table 9.1. The source and range of oil and protein yields in major crops (as percentage dry mass)

Crop	*Source*	*Oil*	*Protein*
Cottonseed	seed	18—20	19—24
Groundnut	seed	45—50	21—36
Linseed	seed	40—46	22—26
Rapeseed/canola	seed	40—45	25—29
Safflower	seed	30—35	14—17
Sesame	seed	50—55	18—22
Soybean	seed	19—23	36—44
Sunflower	seed	35—45	14—19
Coconut	endosperm	65—68	
Oilpalm	mesocarp	>60	
Olive	mesocarp	various	

DESIRED GOALS

Several major areas for possible crop improvements may be considered.

Plant

Application of better practices of agronomy should result in increased crop productivity. Cultivation of strains with optimal desired characteristics is, however, ultimately dependent upon the availability of seeds at a reasonable cost. In addition, a definite need exists for the development of further cultivars which can tolerate abnormal soil conditions of high acidity, alkalinity and salinity. Frost tolerance, drought resistance and other environmental conditions are also factors to be considered. Furthermore, scope exists for improvements in plant viability in such areas as germination, maturation and seed-pod shattering propensity.

High crop production is frequently governed by the availability of fertilizers that is accompanied by a significant cost component. A further cost can be incurred from the need to apply pesticides and herbicides which, in turn, can introduce problems of potential environmental concern.

Post-harvest practices can have a major impact on oilseed storage. Protection from conditions of extreme temperatures and moisture, fungal activity and rodent infestation is of prime importance and can be accomplished by procedures that do not require sophisticated approaches such as biotechnology.

Oil Yield

Most of the oilseed crops as presently grown appear to have reached a plateau in oil production. Crops such as rapeseed, sunflower, groundnut and linseed routinely yield seed with an oil content of approximately 40% with a general range of 38-48%. Soybean has a relatively low oil productivity of 19-23%. The very high oil output associated with the oil palm and coconut is associated with the mesocarp of the fruit of these trees.

Subject to a better fundamental understanding of crop physiology, plant genetics and the biochemistry involved with oil production, some definite improvement in the quantity of oil obtainable from oilseeds may be anticipated. Cell culturing techniques and various approaches of biotechnology can be fully expected to be applied in the relevant studies for the development of new crop strains.

Oil Quality

The fatty acid composition of a vegetable oil has a definite significance. Primary concern exists with nutritional requirements and their impact on human health. A second, but none the less important, consideration is with the keeping quality of the oil. Based on the general understanding of the involvement of saturated fatty acids in the aetilogy of cardiac vascular disease interests definitely exist in oils which are deficient or low in this class of fatty acids. Although olive oil and canola oils are characterized by very small contents of saturated fatty acids deliberate lowering of these acids in other commercial edible oils is a major topic of study.

Manipulation of the unsaturated fatty acid component of vegetable oils is variously sought to improve their nutritional value. Oils enriched in oleic acid and certain polyunsaturated fatty acids such as essential fatty acids as well as gamma linolenic acid are considered to be highly desirable.

The polyunsaturated fatty acid content of oils, on the other hand, due to their oxidative instability can present problems of storage and keeping qualities. As a consequence there is a demand by the food industry for the availability of oils with good keeping qualities obtainable from a designed reduction of the polyunsaturated fatty acid content of edible oils.

PLANT ENGINEERING

Trait Detection

Recognition of the occurrence within a plant of a desirable characteristic is essential if the intention is to incorporate that trait into some host crop. The older practice involving selective breeding is handicapped by an inefficient success rate and the length of time generally associated with this approach.

Plant stock evaluation, characterization of the diversity of germplasm and identification of quantitatively inherited traits can be made by one of several techniques of molecular biology. The practice of restriction fragment length polymorphisms (RFLPs) is being replaced by the faster procedures of randomly amplified polymorphic DNAs (RAPDs) or amplified fragment length polymorphisms (AFLPs). Application of expressed sequence tags (ESTs) generated by rapid-throughput automated cDNA sequencing has been made to facilitate the isolation of certain genes associated with the biosynthesis of plant oils.

The basis of these procedures is associated with electrophoretic separation of variously sized fragments of the plant DMA, obtained by the use of restriction endonucleases, and their visualisation by ethidium bromide. The procedures lend themselves to automation. Use of the polymerase chain reaction (PCR) is now being made routinely and at low cost to increase the number of copies of a specific DNA region under investigation. "Finger-printing" or probing of these segments is then made through hybridisation with DNA fragments recognized to be related to some desired trait.

Particular applications of RAPDs and ESTs are anticipated to be most useful for locating and manipulating particular genes as well as in the mapping of polygenic sources of variation such as plant characteristics and oil yields.

There has also been an increasing use of genomic and molecular techniques to assist in the rapid domestication of presently non-commercial plants such as coriander and cuphea.

Cell Culturing

Use of modern cell culturing techniques is continuing to have a major impact on the development of new plant strains. In particular, the time required for the introduction of new commercial cultivars can be shortened significantly from a minimum of approximately fifteen years to about five years. Some further delays, however, may be experienced in obtaining legislative approval for use as a food.

The ready totipotency of individual living cells when isolated in a suitable medium allows for cultured cells to be regenerated into whole plants possessing the desired genomic composition with the ultimate development of a new stable homozygous variety. The general practice is to culture explants (plant cells, tissue or organs) on a chemically defined medium under sterile conditions. Either microspore (immature pollen) or anther culture techniques are commonly used and are so effective that many plants can be produced eventually at a time. Manipulation of the genetics of these isolated haploid (single set of chromosomes) cells is relatively simple and can be attempted by one of several procedures. Diploidisation (creation of a double set of chromosomes) is assisted by colchicine addition or heat treatment and further cell fusion encouraged by electrical stimulation or by the presence of polyethylene glycol. Cell growth and division is then regulated to produce individual cells and callus (clumps of cells) showing little differentiation. Subsequent organised growth is accompanied by embryogenesis and organogenesis of shoots and roots resulting in the ultimate development of plantlets. Following a hardening period the plantlets may be planted in the open field.

The ease of regeneration of plants from cell cultures varies, but a particularly useful system has been established for rapeseed; the system for soybean is more labour-intensive while that for sunflower continues to be developed.

Genetic Modification

Use of haploid cell systems allows for simpler manipulation of the single set of chromosomes than would be the case with diploid cells. As a general practice protoplasts are created by the enzymatic removal of the cell wall to facilitate the introduction of the desired foreign DNA.

Several approaches have been undertaken to induce changes in the plant genome and the incorporation of novel traits. Commercially valuable successes have been limited and varied.

Mutagenesis

An earlier practice with some noted successes in mutation breeding involves the use of a variety of mutagenic agents. The process is associated with point mutations and the attempt to unmask traits normally hidden in the native strain. Compared to targeted procedures, the process trends to have a low success rate due to its random approach.

Exposure of explants, seeds or whole plants can be made to ionizing irradiation or chemical mutagens such as ethyl methanesulphonate (EMS) and sodium azide. The practice is heavily dependent upon the availability of good screening methods. For example, when modification of the fatty acid composition of an oilseed is the aim, good use can be made of gas liquid chromatography to analyse a portion of a developing cotyledon and the remainder can be allowed to develop into a mature plant if found to possess the desired characteristic.

Gene Transfer Technology

Several systems are available for the introduction of selected foreign DNA into plant cells or protoplasts.

1. *Transfection :* Certain bacteria and viruses can be used as vectors in which a desired DNA fragment has been incorporated and then used to infect host, cells leading to alteration of the host genome. Frequent application has been made with the Ti plasmid of *Agrobacterium tumefaciens* as vector to transfect dicots - rapeseed, soybean, sunflower, flax and cotton; this system is ineffective with monocot plants such as corn and oil palm.
2. *Electroporation :* The desired foreign DNA can be introduced into host cells where a transient increase in membrane permeability due to a brief voltage pulse allows for its uptake from the medium.

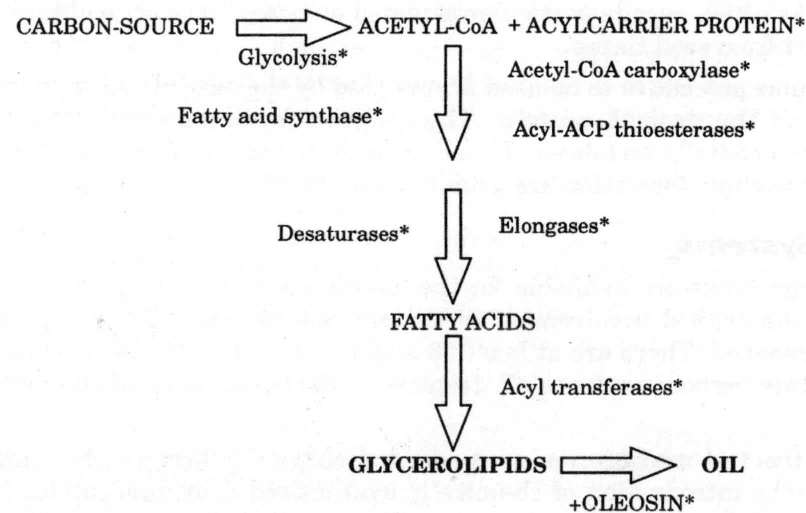

Fig. 9.7. Major target systems* for fatty acid, glycerolipid and oil formation in oilseeds.

3. *Microprojection :* Very small particles of tungsten or gold coated with the desired DNA can be shot into host cells at very high speed using the 'gene gun.'

These procedures are generally limited in their efficiency. Selection of protoplasts and cells where successful uptake of the foreign DNA has occurred frequently involves a known characteristic of the vector system where *e.g.*, the gene for resistance to some antibiotic such as ampicillin or tetracycline resides. Only the transformed systems will survive and grow in medium treated with the appropriate antibiotic.

Successful uptake of the desired genetic material resulting in the creation of a new allele does not guarantee that it will be either expressed or inherited.

cDNA

Specific isolation and .purification of a particular gene is seldom undertaken, although the successful cloning of many of the targeted enzymes of fatty acid biosynthesis and glycerolipid formation has been obtained. Two general practices can be used to prepare a fragment of DNA that contains the desired genetic information. Recombinant DNA libraries representing all of the mRNAs in a given cell can be prepared. Following the isolation and purification of the desired individual single-stranded mRNA double-stranded cDNA can be produced by the action of reverse transcriptase.

Alternatively the desired cDNA may be prepared from synthesis of the mRNA based on the knowledge of the full or partial amino acid sequence of the target protein. A frequently used approach involves the purification of the target protein (enzyme) and raising antibodies against them. Expression gene libraries are then screened, preferably with probes of plant origin, and the desired DNA located. Increased quantities of the desired cDNA can be obtained by cloning in some established system such as *Escherichia coli* or, more effectively, through application of the polymerase chain reaction (PCR). Genetic materials including partial and complete cDNAs which encode particular targeted enzymes have been obtained from several systems derived from seed tissue.

The particular procedure to be used is governed by the ease of isolation, purification and solubilisation of the desired protein. The enzymes systems associated with fatty acid biosynthesis are generally membrane-bound, difficult to isolate and are unstable; the enzymes involved in glycerolipid formation are usually easier to isolate and purify.

Targeted Systems

Several approaches are available for the modification of a single enzyme system. Exact knowledge of the critical involvement of the enzyme in a particular aspect of plant lipid formation is essential. There are at least 50 enzyme systems associated with the conversion of a carbohydrate carbon-source to oil droplets in the seed. Some of the systems that have been targeted.

1. **Site-directed mutagenesis :** A selected enzyme is targeted for modification of its activity by introduction of chemically synthesised deoxynucleotides having discrete changes in sequence from that normally present in the genome. Expression of this altered gene in the host plant is then expected to yield the required protein generally modified at its active site.
2. **Anti-sense RNA technology :** Gene expression for a particular system is effectively reduced by providing a chemically synthesised RNA fragment which is complementary to the mRNA for the targeted enzyme. Subsequent hybridisation of the sense and anti-sense RNA molecules is expected to reduce the quantity of sense mRNA available for translation.

   Application of this technology is particularly useful in obtaining mutants for specific genes when they are members of multigene families *e.g.*, the fatty acid synthase complex and hence more difficult to engineer. Problems, however, may arise when the desired reduction of targeted traits is accompanied with umpredicted accentuation of others. None the less is should be possible, at least in theory, to "fine time" the overall fatty acid composition of plane coils.
3. **Sense gene/co-suppression technology :** Introduction of a hormologous gene into the host plant system can result in suppression of subsequent genetic expression rather than an expected increase in the quantity of the targeted enzyme. Further application of this procedure may prove to be useful ior the modification of the unsaturated fatty acid composition of oils with improved nutritional characteristics.

All of these approaches to modification of the plant genome are heavinly dependent upon accurate knowledge of plant lipid biochemistry, genetics and factors influencing genetic expression. Introduction of foreign DNA into a host plant with the creation of a new allele does not guarantee its expression or its inheritance. Particular requirement exists for the availability of a strong piromoter to initiate the process of transcription.

TRANSGENIC OIL SEED CROPS

Successful modifications of oilseed crops have been made in two general areas and have allowed the introduction of cultivars with desired characteristics.

Table 9.2. Some Available Transgenic Crops and Their Trait Modification

Crop	*Altered Trait*
Canola *(Brassica napus)*	glufosinate ammonium tolerance*
	glyphosate tolerance **
	high-lauric
	high-stearate/low-polyunsaturate
	high-oleic
	high-medium chain
Soybean *(Glycine max)*	glyphosate tolerance
	high-oleic/low-saturate
Corn *(Zeamays)*	Bt-resistant ***
	glufosinate ammonium tolerance
	glyphosate tolerance
Cotton *(Gossypium hirsutum)*	Bt-resistant
	glyphosate tolerance
Flax (Linum usitatissimum	sulphonylurea tolerance
*Liberty Link **Roundup Ready	***Bt = Bacillus thuringiensis

Plant Modifications

Plant protection has been achieved through genetic engineering to provide both tolerance to selected herbicides and resistance to insect infestation.

Strains of corn (maize), flax, rapeseed and soybean are commercially available which are herbicide-tolerant. The presence of weeds can have a major impact on crop yield where growth inhibition of *e.g.*, untreated canola can be as much as 35%. Theoretically plant growth of the crop can be expected to be favoured when weeds sprayed with the herbicide die off. The anticipation is that there will be a lower need to apply the herbicide with an accompanying reduction of costs and potential environmental problems. Preliminary results suggest that herbicide-tolerant cultivars show slightly enhanced crop yield with a lower need for spraying.

Similar increases in crop yields have been obtained with both corn and cotton which have been engineered to contain the gene from Bacillus thuringiensis responsible for the production of a toxin fatal to certain insects.

Selective breeding of cultivars and the application of good agronomic practices have resulted in an apparently optimal range of total oil content for the major oilseed crops.

At present there does not appear to have been any success in developing transgenic crops with enhanced oil productivity. Based on the fact that certain (non-domesticated) plants are capable of accumulating, up to 70% (w/w) oil in the seed an understanding of the mechanism(s) involved might conceivably allow development of higher oil-yielding crops.

Oil Modifications

Some very notable successes have been obtained in the development of transgenic plants in which the seed oil has been modified to yield increased and/or decreased contents of particular fatty acids. Certain of these cultivars are available for commercial use, subject to legislative approval, while others are still in field trials. The impact, if any, of modified fatty acid composition of seed oil on the viability of the plant is not generally known. An engineered high stearic acid content in Brassica rapa seeds resulted in low oil content and low germination.

Further progress in the tailoring of seed oils can be fully anticipated.

COMMENTARY

Successful use of genetic engineering has been made with the incorporation of some desirable traits into commercial crops, particularly canola. The techniques and protocols are available, but more definite applications are heavily dependent upon the availability of more advanced knowledge of plant lipid metabolism-total fatty acid synthesis, unsaturated fatty acid synthesis and glycerolipid (triacylglycerol formation)-gene expression and regulation, and all factors that impact on plant viability.

The first generation of transgenic crops has been primarily associated with input traits associated with protection-herbicide-tolerance and insect-resistance-and with certain modifications of the fatty acid composition of the seed oil. Rapeseed (canola) has proved itself to be particularly amenable to manipulation and further advantages lie in the fact that it is an established crop.

The second generation it is believed will focus on output traits associated with the use of crops modified to produce a wide variety of consumer-desirable products such as nutraceuticals, pharmaceuticals and renewable feedstocks for the oleochemical industry.

Great care will need to be exercised with the use of crop engineering to avoid any possibly negative impacts on the environment and human health. A Luddite philosophy exists in certain segments of society that will have to be overcome by good solid reasoning if further successes with the technology is to be achieved.

Claims that transgenic plants will go a long way to solving the global lack of available food have yet to be substantiated. Emerging nations which generally lack adequate supplies of food may look toward biotechnology and genetic engineering as a solution, but the cost may be prohibitive. It may well be prudent for them to give greater attention to other approaches such as better agronomic practices and post-harvest technologies.

SEPARATION OF OILS AND FATS

Centrifuges are machines from mechanical process engineering for the separation of solids from liquids, the separation of two immiscible liquids or two such liquids and a solid component. The pre-condition is that the media to be separated from each other have a different specific gravity.

Centrifuges are subdivided into different types depending on their application. Disk-type centrifuges are almost exclusively used for oil and fat refining which are in turn classified into solid-wall and self-cleaning centrifuges. At this point let us focus on the self-cleaning centrifuges.

Westfalia Separator has been manufacturing such centrifuges since its foundation in 1893. The firm was taken over by the GEA Group in 1994 and integrated into the Process Technology Division, which is one of the largest suppliers of plant and components for the food industry world-wide.

Westfalia Separator began manufacturing solid-wall disk-type centrifuges for degumming, neutralisation and washing fats and oils back in the thirties. Winterization and fractionation followed later.

In the sixties, when more and more continuous refining lines were being built, Westfalia Separator responded by manufacturing self-cleaning disk-type separators.

WORKHORSES OF THE OIL AND FAT INDUSTRY

In 1970, the first self-cleaning separator type RSA 80 was installed at a customer in Germany. As time progressed, four sizes of this machine were manufactured of which 750 machines had been sold world-wide by 1995.

A cross-section of such a machine. They are straightforwardly designed, extremely robust machines which have been dubbed "workhorses" by a number of our customers.

The machines are characterised by the fact that the product fed practically without pressure from above through a stationary feed pipe into the rotating bowl. In the distributor, a ribbed body, the liquid is accelerated to bowl speed and conveyed into the disk stack. The separated light liquid component, *i.e.*, the oil, is led inwards to the bowl head from where it is discharged under pressure by means of a centripetal pump. The heavy liquid component is forced outwards and fed over a separating disk likewise to a centripetal pump and discharged. The solids are collected in the solids holding space in the periphery of the bowl and can be discharged periodically by partial ejection with an open feed to the machine.

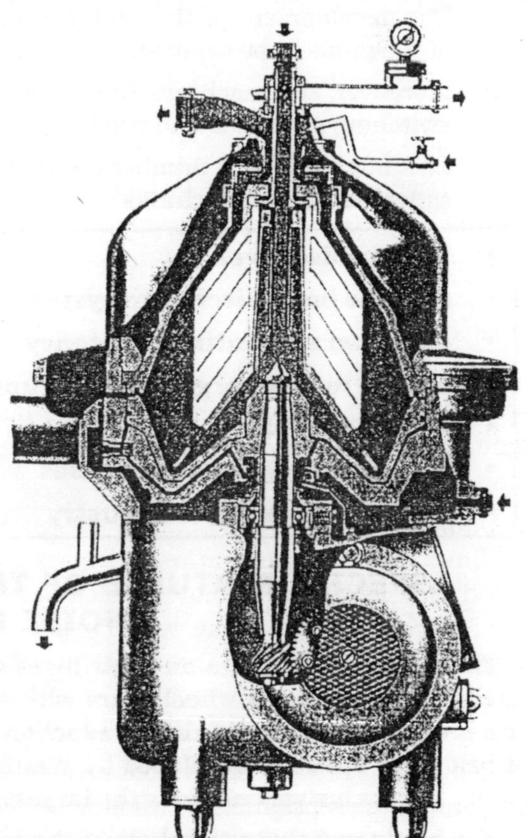

Fig. 9.8. Section through an RSA Separator.

The diameters of the two centripetal pumps are altered to obtain basic adjustment of the separating zone; fine adjustment is carried out by regulating the discharge pressure of the light phase. The basic setting could be altered by exchanging the heavy phase centripetal pump. This was practically only necessary when, for example, winterization was to take palace with a washing separator.

The big advantage of this machine was that no mechanical seals were fitted between rotating and static parts. This made the machines extremely easy to handle and the maintenance costs low.

These RSA separators, which were later replaced by the slightly modified RSB version, have been the state-of-the-art in the oil and fat industry for 25 years. Again the demands of the market required a complete re-design of the machine conception.

REQUIREMENTS OF THE INDUSTRY

- The noise level of centrifuges had to be drastically reduced in response to legal provisions governing industrial safety.
- To ensure operating safety and to avoid unnecessary oil losses during solids ejections, a more reliable bowl ejection system had to be developed.
- Because more and more two-stage neutralization installations with just one washing stage were being built and increasingly efficient acid degumming processes were developed, more importance was attached to an optimum separation efficiency.
- The development of the acid degumming processes resulted in a rise in the viscosity of the gums to be separated, which necessitated special designs.
- Moreover, the machines had to be as flexible as possible to avoid conversions when switching processes and products.
- Last but not least a number of our customers expressed the wish for higher throughput capacities of the machines.

- **lower noise level**
- **reliable bowl discharge system**
- **optimum separating efficiency**
- **ability to discharge viscous gums or soaps**
- **higher flexibility**
- **higher throughput capacities**
- **requirements of the industry**

SPECIAL FEATURES OF THE RSE SEPARATOR SERIES—NOISE REDUCTION

The main noise sources on centrifuges are naturally the rotating bowl and the drive. As regards the drive, worm wheel gears with special tooth design were developed to reduce the noise level emitted. A substantial reduction in noise level can also be achieved by selecting a flat belt drive specially developed by Westfalia Separator. As it can also transmit very high powers, it is exclusively used for the largest refining separator, the RSE 250.

The hoods and sludge catchers of the machines are of double-wall design to reduce bowl noise caused by the air friction in the frame and sludge catchers.

The hollow spaces are filled with a sound-insulating medium, in **addition,** modern separators are today installed fully enclosed.

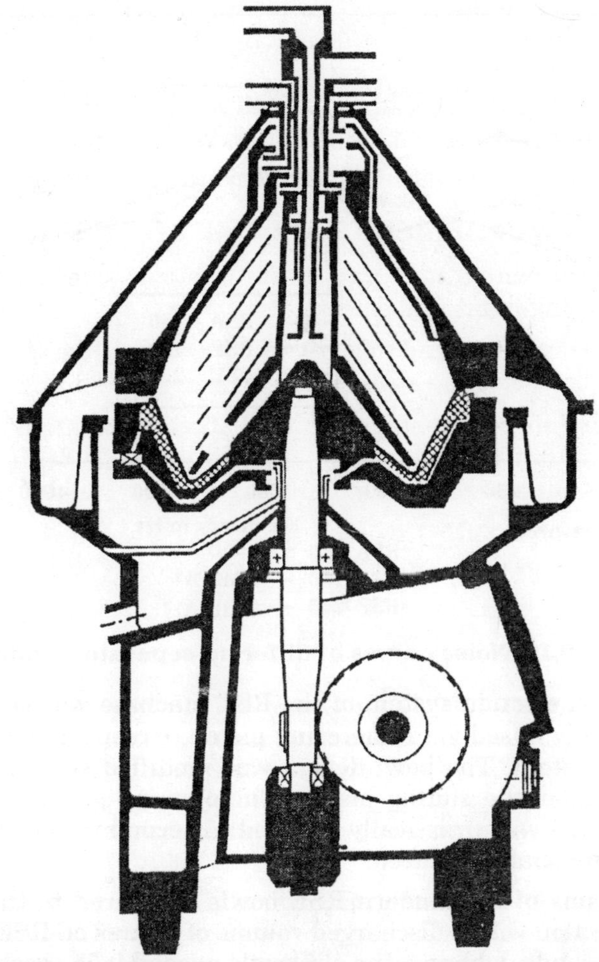

Fig. 9.9. RSE Separator Series—Noise Reduction.

NOISE LEVEL OF DIFFERENT SEPERATORS

Illustrates the noise curves of different separator models. It can be observed that the sound pressure level of the modern RSE machines with worm wheel gear was able to be reduced to less than 80 dB as opposed to approximately 86 dB on the exposed old RSA models. The RSE 250 with flat belt drive shows even better results. Here results below 78 dB can realistically be achieved. All these figures are of course dependent on the environment of the machine.

BOWL EJECTION SYSTEM

Refining seperators are operated almost exclusively with partial ejections. This means that the bowl is opened without interrupting the product feed to the machine until only the accumulated solids in the solids holding space have discharged. A portion of the heavy phase may also discharge, but under no circumstances may light phase, *i.e.* oil, dischage.

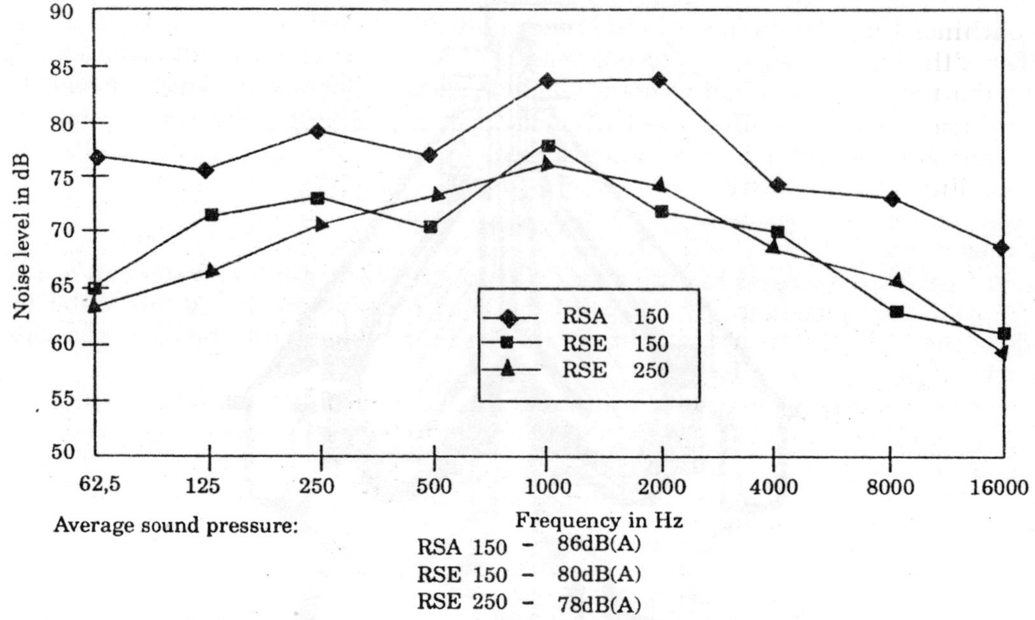

Fig. 9.10. Noise curves of different separator models.

To this end, the bowl ejection system of the RSE machine was completely modified. The opening water is precisely dosed via a metering piston, irrespective of pressure fluctuations in the opening water system. The bowl design was modified so that the closing water can escape more quickly under the sliding piston. The opening gap in the bowl was enlarged, whereby the opening time was drastically reduced. A secondary effect was that the wear to the bowl gasket was appreciably reduced.

The ejection diagrams of the modern RSE bowls compared to the old design. It can be seen that the partial ejection with a discharged volume of 5 litres on RSE machines is practically completed after 0.15 seconds, whereas the old bowls needed 0.55 seconds.

SEPARATION EFFICIENCY

The wish of the industry for lower residual soap contents in neutralisation processes and lower phosphatide contents in the constantly newly developed degumming processes necessitates optimum flow behaviour of the product during the separation process in the rotation bowl.

A specially critical point is the bowl inlet where the product has to be accelerated from zero to the bowl speed of 4000-4850 rpm. The hydrohermetic feed system developed by Westfalia Separator protects the product from exposure to high shearing forces, which would break up the gums or soap particles, consequently making separation a lot more difficult. During washing and winterization, in particular, emulsification is avoided at this point.

The function of this hydrohermetic feed corresponds to that of fully hermetic feed systems with the difference that no mechanical seals are needed.

Especially in the case of very high feed temperatures steam formation can arise in the inlet chamber which can escape up into the light phase on the RSA models from time to time. On the

RSE machines the inlet chamber is hydrohermetically sealed against the centripetal pump chamber of the light phase by means of a sealing disk, *i.e.* likewise without using a mechanical seal. By this means, contamination of the oil phase with unseparated product is avoided.

Soap stock, but especially gums from some acid degumming processes, can be extremely viscous and generate very high flow resistance. To counteract this, special disk stacks were designed. Further, the entire bowl geometry in the area of the heavy phase was adapted so that even extremely viscous phosphatides with a very low oil content can be contnuously discharged reliably.

If downstream processes require a less viscous soapstock, dilution water can be fed direct into the solids holding space by means of the bowl flush device. Since this water is added outside of the disk stack practically at the largest inner diameter of the bowl, the separation process itself is in no way affected.

This water addition can also be utilised to flush out surplus caustic soda which, owing to its high specific gravity, is separated out at the periphery as a third phase and blocks the path of the soap to the separating disk. By this means, the interval between two partial ejections can often be substantially extended, thereby avoiding unnecessary turbulence in the bowl as a result of too frequent ejections.

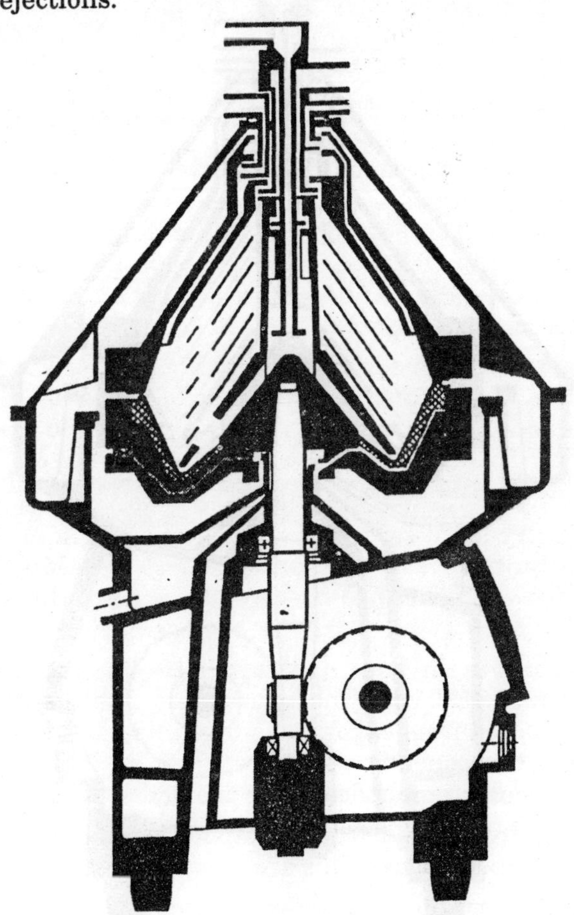

Fig. 9.11.

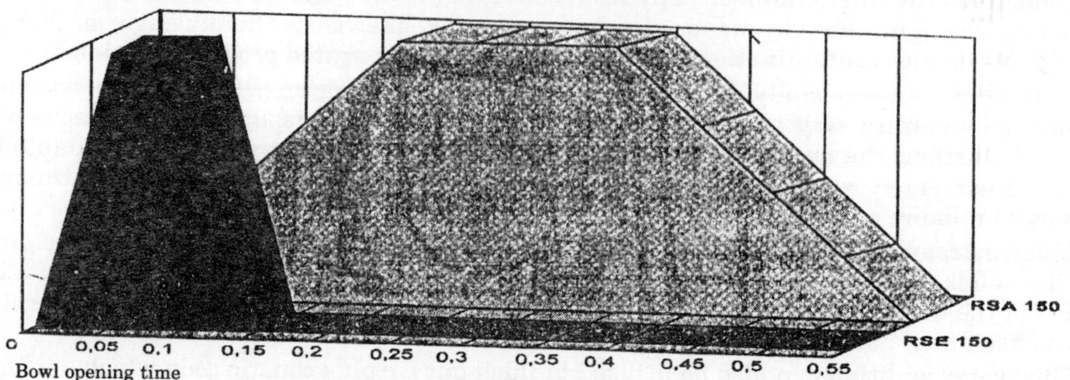

Type of deslydging : Partial / Discharged volume : 5 litre

Fig. 9.12. Ejection diagrams of RSA and RSE Separators.

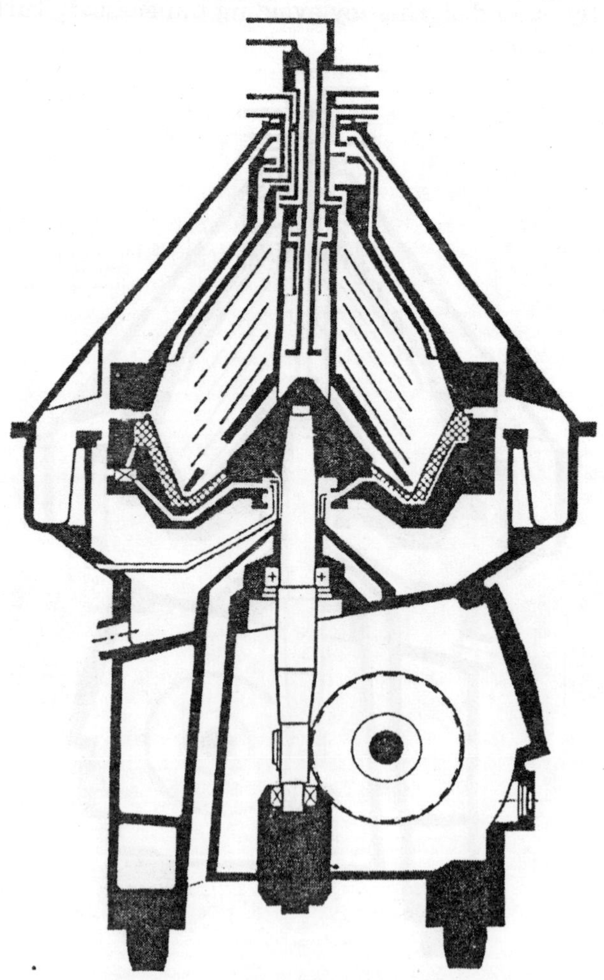

Fig. 9.13. Separation Efficiency.

COMPARISON OF A STANDARD FEED WITH A HYDROHERMETIC FEED

The reduction of the residual soap content with a new RSE 200 compared to the older RSA 200 design. Both machines were run in parallel with the same throughput capacity and the same oil in the neutralisation stage of an installation. The average reduction of the residual soap content was 20%. In other installations even greater improvements were achieved.

FLEXIBILITY

The RSA and RSB separators were equipped with two centripetal pumps for the heavy and light phases. While the lower centripetal pump for the oil phase had a fixed diameter, the diameter of the upper centripetal pump for the soapstock, the gums or the wash water was adapted to the difference between the specific gravity of the light and heavy phase in order to adjust the separating zone between the two phases.

This does not mean that the upper centripetal pump constantly had to be exchanged when changing product or process as the discharge pressure of the light phase could be used to a limited extent to influence the position of the separating zone. However, if one machine is to be used both for hot washing and winterization, there was no option but to replace the centripetal pump. The sometimes extremely high viscosities of the phosphatides form certian acid degumming processes necessitated the use of very large upper centripetal pumps because the flow resistance of these gums forced the separating zone too far inwards.

As such a centripetal pump replacement took anything up to two hours industry's demand for a centripetal pump which could be adjusted externally during operation was quite understandable. The "Fine Tuner" developed by Westfalia Separator is a combination between a centripetal pump and a paring tube and is characterised by a substantially improved efficiency factor compared to conventional centripetal pumps. This is the ratio of the conversion of rotational energy into pressure. In the case of optimally designed centripetal pumps this

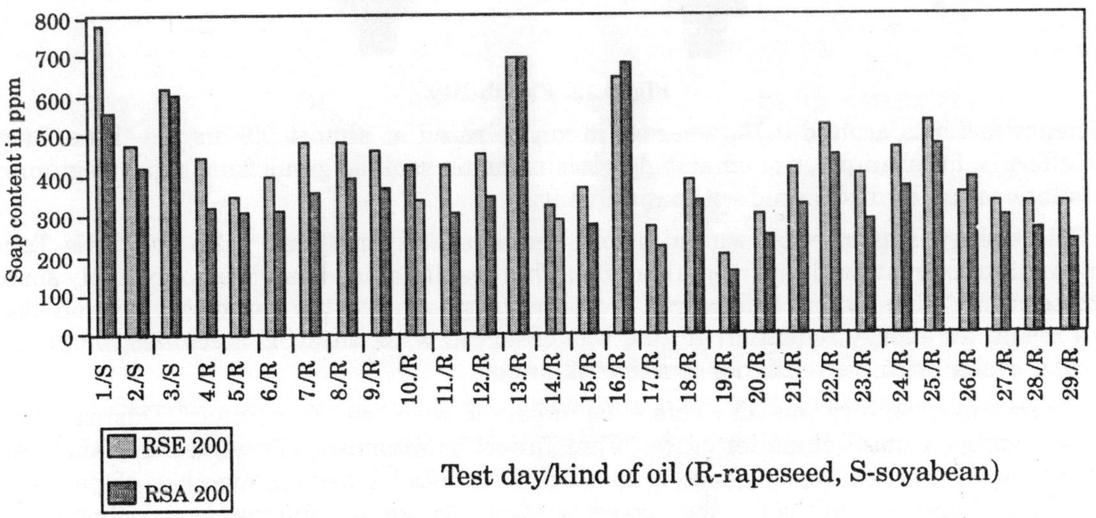

Fig. 9.14. Reducing the residual soap content through an optimum bowl design.

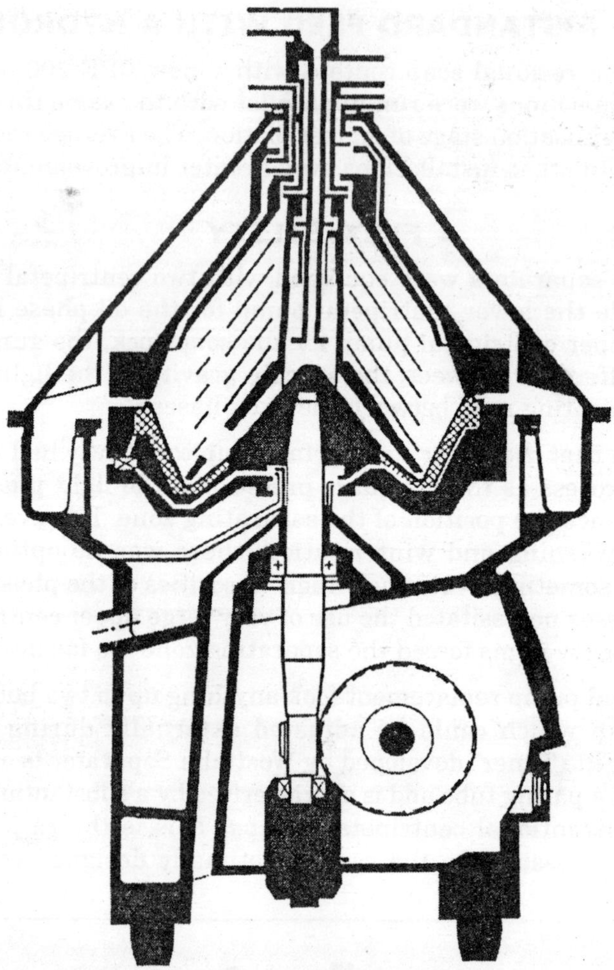

Fig. 9.15. Flexibility.

efficiency factor is around 0.75, whereas it can be rated at almost 0.9 for the "Fine Tuner". The effect is, for example, the amazingly viscous media such as gums from super degumming installations can be discharged without difficulty.

The energy consumption was also to be substantially reduced with the "Fine Tuner" compared to conventional centripetal pumps. This results alone from the appreciably smaller surface of the "Fine Tuner" submerged in the heavy phase and the associated friction losses. As a result, an energy reduction of 10% was observed with the RSE machines compared to an older design with conventional centripetal pump.

Operating reliability was also able to be improved with the "Fine Tuner". During start-up of the machine a small diameter of the "Fine Tuner" is automatically adjusted so that carry-over of the oil into the heavy phase is avoided. Only after a certain time has elapsed is the "Fine Tuner" automatically set to the correct position. The same thing happens during a partial ejection. The "Fine Tuner" is moved inwards immediately before the partial ejection and reset to the correct position after the ejection. These measures virtually eliminate the risk of oil losses.

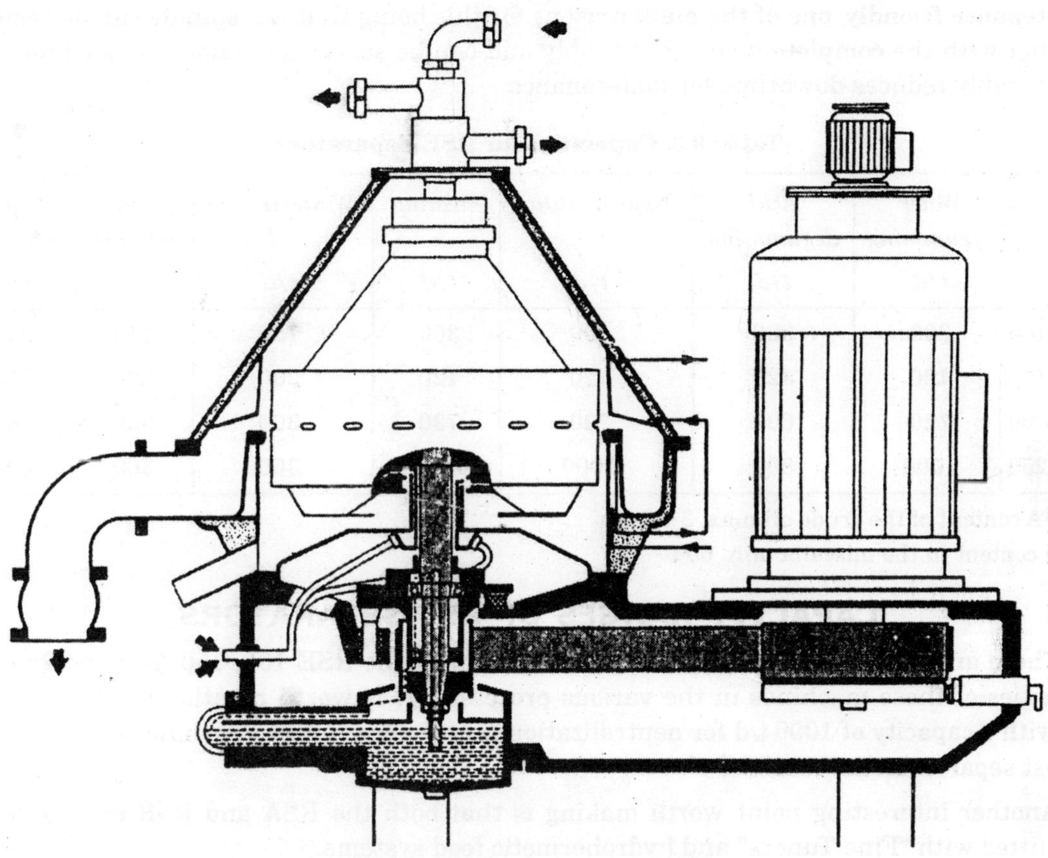

Fig. 9.16. Flat belt drive of the RSE 250.

DRIVE CONCEPTS

The tried and tested and extremely sturdy drive concept of the RSA machines with worm wheel gear, fluid clutch and standard motor with direct switching is likewise used for the RSE separators, However, Westfalia Separator drew on the many years of experience in the chemical industry to build a completely gas-light frame for the explosion-proof machines with a likewise gas-tight neck bearing assembly. In this way, machines that are applied for *e.g.*, miscella refining can be completely blanketed with inert gas under overpressure, by which means optimum safety is achieved.

The frame was re-designed and the toothing of the worm wheel gear optimized to reduce the noise level.

In the case of the heavy bowls "of the largest machines of this series, the RSE 250, and the possible extreme throughput capacities of these machines of 40t/h, the worm wheel gear is pushed to its performance limit. For this reason these machines are delivered exclusively with flat belt drives. Power transmission from the torque-controlled motor to the bowl spindle is via a single flat belt without an intermediary clutch. The design is very simple and maintenance friendly, one of the main reasons for this being that the spindle can be removed together with the complete bearing assembly and can be serviced outside the machine. This considerably reduces downtime for maintenance.

Table 9.3. Capacities for RSE Separators

Type	*Water degumming* *t/d*	*Acid degumming* *t/d*	*Neutralization* *t/d*	*Washing* *t/d*	*Winterization* *t/d*	*Cold refining* *t/d*	*Miscella-refining2* *t/d*
RSE 100	300	300	300	300	150	150	240
RSE 150	420	420	420	420	200	200	300
RSE 200	720	600	720	720	300	300	360
RSE 250	1000	800	1000	1000	360	360	420

(1) FFA content of the crude oil max. 3%

(2) oil content in the miscella min. 60%

CAPACITY RANGES OF RSE SEPARATORS

There are four machine sizes in this series, namely the RSE 100, 150, 200 and 250. The capacities of these machines in the various processes. It is worth mentioning that the RSE 250 with a capacity of 1000 t/d for neutralization, water degumming, washing is presently the largest separtor in the market.

Another interesting point worth making is that both the RSA and RSB models can be retrofitted with "Fine Tuners" and hydrohermetic feed systems.

REFINING SYSTEMS OF VEGETABLE OILS

Refining in itself is a wide subject. I propose to dwell on the current technology involved, the different systems practiced in India and abroad and more particularly, the modern measures for control oil losses.

Oil processing industries abroad have made significant progress in the last thirty years. The continuous system of refining, bleaching and deodorization have replaced the old batch processes. The new process of physical refining is increasingly used for refining oils and fats. More attention is now being devoted in plant designing to control pollution. Developments in laboratory and analytical instrumentation techniques have made it possible for the composition of raw oils and fats more accurately and quickly.

The objective of this presentation is to provide an overview of present refining practices used for refining of vegetable oils in India. Primary processing system for gum-conditioning and refining is described. Crude edible oils and fats contain variable amounts of non-glycerides impurities.

Table 9.4. Non-Glycerides Components

Insolubles	*Solubles*
— Moisture — Oil sludge — Suspended matter	— Phosphatides — Carbohydrates — Chlorophyll's — Proteins — Resinous products — Sterols — Waxes — Coloring and flavoring compounds — Free fatty acids

These impurities must be eliminated by processing before the fats and oils are made suitable for human consumption. These impurities influence the quality of the refined oil.

Table 9.5. Processing of Vegetable Oils

Effect of gums, phosphatides and mucilaginous substances:

- Act as emulsifier increasing loss of oil.
- Decompose at high temperature, increasing color of the refined oil.
- Poison the catalyst, effecting hydrogenation reaction adversely.
- These would effect the self life of deodorized oil adversely.

Effect of Free Fatty Acids:

- Increase foaming and reduce smoke point.
- Diminish keeping properties of oil.
- Erode metals forming metal soaps.
- Interfere with hydrogenation process by forming nickel soap.

A package of processing operation is undertaken to eliminate them.

DEGUMMING OR GUM-CONDITIONING

Water is used for precipitating the phosphatides in oil by hydration. The important parameters for monitoring of degumming are:

(1) Temperature of oil should be 65-70°C.

(2) % soft water should be equal to 75% of phosphatide content.

(3) Retention time should be kept 30 minutes.

Simple water degumming will reduce phosphorus content to about 50 PPM. Phosphoric acid and/or citric acid is used for converting nonhydratable phosphatides into hydratable. Phosphoric acid is added to the tune of about 0.5% to 1% drop by drop to the oil at a temperature of 60°C and mixed for two hours and then soft water is added for hydration of the gums. For best results in both batch and continuous degumming, it is absolutely essential to provide sufficient time for the water and the phosphatides to react. Hydration of phosphatides is not an instantaneous reaction and requires initial intensive mixing, followed by sufficient time to allow hydration and coalescence of the hydrated micelles. The relation between phosphorus and phosphatide content is

% Phosphatides = 30.0 × Phosphorus Content

Refining Operations

The object of refining is to remove the above objectionable impurities with the least possible damage to the neutral oil and minimum loss of oil during processing. The steps are :

- Settling or filtration
- Degumming or gum-conditioning
- Winterization
- Refining
- Bleaching
- Deodorization.

Primarily, the system used for purifying crude edible oils is conventional caustic soda refining. Various processes for vegetable oil refining by batch as well as continuous system are in vogue. The manifold advantages of physical refining over the conventional caustic refining systems.

Table 9.6. Advantages of Physical Refining over Conventional Caustic Refining

Particulars	*% Reduction*
Installed cost	22
Steam	28
Cooking Water	7
Make up Water	85
Waste Water	63
Power	62
Refining loss	60

It should, however, be borne in mind that if the oils are not adequately refined, subsequent operations such as bleaching, hydrogenation, winterization and deodorization would be troublesome and finished product will fail to meet the required quality objectives.

The first step in caustic refining of vegetable oils is the degumming step. The correct dosage of the degumming agent is very essential for effective hydration of gums. Less or more of the degumming solution leads to incomplete hydration and poor separation of oil and aqueous phases. Correct dosage can be checked by a run in a laboratory centrifuge.

The use of caustic solutions for removing unwanted nonglyceride substances and FFAs is effective but can cause some hydrolysis or saponification, which will increase refining losses. This is true for both batch and continuous refining, so the refiner is always required to strike a balance, using enough caustic, based on experience from 30 % to 100% excess and 12 to 24° Be′ depending upon the type of oils, to remove the unwanted substances effectively without causing additional losses. The temperature of refining is also very critical for obtaining better results. The optimum temperature recommended is 80 - 90°C.

The reactions involved in degumming and alkali refining are the following:

1. Phospholipids (hydratable) + water = hydrated gums
2. Phospholipids (nonhydratable) + acid or alkali = hydrated gums
3. Metal/phospholipid complex + acid = hydrated metal gums
4. Free fatty acids + alkali = soaps

The overall goal in caustic refining is simultaneously to maximize quality and minimize loses. Various methods are employed for estimating practical refining losses.

Table 9.7. Refining Loss Methods

Fatty acid method,	Loss % = loss factor × ffa
Acetone insoluble method,	Loss % = % phosphatides + %ffa + H_2O
Wesson loss method,	$\text{Loss }\% = \frac{Wc - Wn}{Wc} \times 100$
Chromatographic method,	$\text{Loss }\% = \frac{Wc - Wn}{Wc} \times 100$
Cup refining method,	$\text{Loss }\% = \frac{1}{10}\left[(Wc - Wr) + (Ws - Wa) + \text{evap loss}\right]$

Where, Wc = weight of crude oil

Wn = weight of neutral oil

Wr = weight of the residue (refined oil)

Ws = weight of soap stock

Wa = weight of alkali solution used.

The Chromatographic and Wesson methods are replacing the traditional refining loss estimation method based on F.F.A. Some equipment manufacturers like Sharpies and Alfa Laval base their refining loss guarantee figures on Wesson Loss.

Table 9.8. Guaranteed Refining Loss

Sharples process,	Loss % = 0.5 + 1.25 × W + W2 /80
Alfa Laval process,	Loss % = 0.3 + 1.35 × W

Where, W = Wesson loss (theoretical refining loss).

Loss Monitoring Systems

Because of the steeply increasing prices of edible oils, it has become more essential to curtail losses during refining. Several developments to minimize losses have been made abroad employing pH measurement, sodium balance and loss monitoring system etc.

It has been found that the neutralization of oils at a particular pH helps in better separation of soap stock and lower refining losses. The optimum pH values of caustic - oil reaction mixtures for various oils.

Table 9.9. pH Monitoring System (Optimum pH of caustic - oil reaction mixtures)

Crude oil	*Optimum pH*
Palm	10.2—10.4
Soybean (degummed)	10.3—10.4
Soybean (nondegummed)	10.5—10.8
Groundnut	10.4—10.6
Cottonseed	10.8—11.0

A continuous pH monitoring and caustic soda feed adjustment system can be installed in line with continuous refining plants.

Sodium balance is another technique for measurement of refining losses. It helps in measuring the losses by analyzing the content of sodium in various streams.

Table 9.10. Determination of Refining Losses (Sodium Balance Method)

$$\% \text{ Refining Loss} = \frac{(\text{Nar-Nas})\text{-}100(\text{Nao})}{\text{Nas-Nao}}$$

Where, r= % caustic reagent used per weight of crude oil.

Nar =% sodium in caustic reagent.

Nas =% sodium in soap stock.

Nao =% sodium in refined oil.

The method does not require actual weighment or flow measurement. Analysis required is only volumetric and can be completed within 2 hours.

Yet another development in controlling losses in refining is the Refining Loss Monitor. This instrument measures the inflow and outflow at any particular time, calculates and directly indicates % loss at that time. On altering the various factors temperature, % excess and strength of caustic, one by one, their effect on % oil loss is immediately known. Thus refining conditions can then be set for minimum loss.

Further, factors important to an efficient refining operation include the source of oils supply, receiving and storage techniques, impurities in the oil, type of crude oil processing system, processing conditions and control systems used for yield and quality.

The major factors which could influence quality of crude oil before processing is a topic worthy of discussion by itself but briefly these factors can be listed as follows:

1. Conditions at harvest, maturity of seeds, etc.
2. Damage during harvesting
3. Percentage of moisture
4. Storage conditions
5. Foreign matter in seed

Now there is a need to have a relook into our refining practices and on our oil processing industry. We should go for physical refining. This will enable saving in energy consumption and pollution problems. The newer practices in controlling refining losses during alkali refining should be adopted, so that oil losses are minimized thus making the process more economical. The traditional methods of monitoring oil efficiency by calculating FFA of soap stock during refining is time consuming and less accurate. Automation and instrumentation would minimize measuring time and improve accuracy by eliminating personal errors.

PROCESSING OF VEGETABLE OILS

Vegetable oils are refined in order to remove the impurities that are present in the crude oil. Some of these impurities are naturally present in the seeds [or oil -bearing fruits], a great majority of these impurities are also formed during harvesting and storage of the seeds [fruits], or during extraction of crude oil and, subsequently, during its refining.

Oil refining processes for vegetable oils are designed to remove these impurities from the oil or reduce them to a level where their deleterious effects on oil stability are minimum. An oil processor needs to be familiar with the characteristics of these impurities, how they are formed in the oil and how to minimize their formation during oil processing.

MODES OF OIL BREAKDOWN

Vegetable oil undergoes degradation almost immediately after the seed is crushed. The oil becomes more prone to oxidation as it is processed further from crude to the finished stage. Deodorized oil has got the highest tendency towards oxidation.

The oil starts to show the sign of primary oxidation as measured by its peroxide value. Under certain circumstances the oil may develop a darker colour or higher free fatty acids and eventually an unpleasant flavor obviously, all of these reactions require certain catalysts that are very specific for these reactions. Presence of oxygen, moisture and high temperatures simply accelerate the reaction process.

Following are the three principal modes in which a vegetable oil undergoes degradation.

1. Oxidation

(*a*) Autoxidation

(*b*) Photooxidation

2. Hydrolysis

3. Thermal decomposition

(*a*) oxidative polymerization

(*b*) Thermal polymerization

OXIDATION

Oils get oxidized in two different ways, such as:

(1) Autoxidation, and

(2) Photooxidation

There are two types of photooxidation, namely:

(1) Photochemical reaction and

(2) Photosensitized reaction.

Autoxidation and photochemical reactions are somewhat similar in nature. Each reaction starts with the help of an initiator. In case of autoxidation, the initiator is a metal. In the other reaction, the initiator is Ultra Violet (UV) light. The reaction is propagated in both cases by group of catalysts, called the radicals, that are formed from the unsaturated fatty acids in the oil in presence of the initiators.

Autoxidation

As mentioned above, autoxidation in the oil is initiated by metal ion that produces a free radical from a molecule of unsaturated fatty acid. This occurs when the oil is subjected to high temperature and in presence of a metal ion, nickel, aluminum, etc, and is called the "initiation" phase of the reaction. Unsaturated fatty acid breaks down to form a peroxy radical, Which is also a free radical. The peroxy radical then reacts with an unsaturated fatty acid, forming hydroperoxide and another free radical, which continues to catalyze the reaction. This is known as the "propagation" phase. The reaction proceeds very rapidly in presence of oxygen or air.

Unsaturated fatty Acid (RH) + Initiator	=	R* (Free Radical) + H*-Initiation
R*+O_2	=	ROO*(Peroxy Radical) -Propagation
ROO*+RH	=	ROOH (Hydroperoxide) +R*-Propagation
R* + R*	=	RR (Dimer**)-Termination
R*+ROO*	=	ROOR-Termination
ROO* + ROO*	=	ROOOOR-Termination
** These are oxidative dimers and polymers		

Fig. 9.17. Reactions during autoxidation of oil.

During reaction the peroxide value in the oil increases, reaches a peak and then it begins to decrease because hydroperoxide molecules decompose readily due to their fragile nature (10). The hydroperoxide molecules break down to produce a whole host of secondary and tertiary oxidation products, such as aldehydes, ketones, acids, alcohol, epoxides, polymers, etc.

Ultimately, under low oxygen supply situation and also as the available unsaturated fatty acids begin to decrease in the substrate, reaction takes place between the free radicals, which eventually terminates the autoxidation process. This is referred to as the "termintion" phase of the reaction. The oil becomes inedible, long before this state is reached. The reaction in the autoxidation process.

The above process can occur in the oilseeds and fruit palm even before the oil is extracted. This can also occur during extraction as well as during the subsequent steps of oil processing.

Sources of Free Radicals

Free radicals are formed in the oilseeds, palm fruits and the oils derived from them at various stages of the production as listed below:

1. Poor quality oilseeds
2. Improper handling and storage of the oilseeds
3. Improper conditions during crushing
4. Poor handling and storage of the crude oil
5. Over-heating of the oil during processing

6. Undue exposure of the oil to oxygen during processing
7. Atmospheric bleaching
8. Deodorization under poor vacuum
9. Storage and handling of the finished oil without nitrogen.

Poor quality oilseeds, which are field damaged or damaged due to drought or excessive rainfall, are known to have poor stability. Poor quality oilseeds require excess caustic treatment and extra bleaching. This also produces higher than normal level of free radicals in the oil. When the soybeans are stored at temperatures higher than 45°C and at a moisture content of 14% or higher, lipase and lipoxygenase in the beans are activated, resulting in free radical formation. In addition, another enzyme, known as phospholipase-D produces a higher level of non-hydratable phospholipids in the crude oil This requires extra caustic in refining. As mentioned earlier, extra caustic treatment can increase the formation of free radicals in the crude oil. Like the oilseeds, fruit palm also shows an increase in free fatty acid content, when they are damaged during harvest or they are not processed within 24 hours after harvest.

Therefore, one needs to be aware of the source of free radicals in crude oil as it is received. The degree of oxidation in the crude oil must be monitored. Peroxide value indicates the state of primary oxidation of the oil. However, hydroperoxides are very unstable. These compounds break down to produce secondary oxidation products such as aldehydes and ketones. Some of these secondary oxidation compounds can be measured by analyzing the oil for para anisidine (or commonly termed as anisidine value, AV).

To extract oil from soybean, the beans are cracked, heated (tempered) in horizontal or vertical cookers, flaked and then the oil is extracted with hexane. This process allows ample opportunity for the free radicals to be formed because the cracked seeds are heated in presence of air. The cracked pieces of soybean go through a temperature range where the enzymes are most active. This creates an opportune environment for free redicals in the oil while it is still in the cracked seeds.

Normally the cracked seeds get heated faster in a horizontal cooker than in a vertical cooker. This offers some advantage of a horizontal cooker over a vertical one. Another method used in seed exraction process is known as expander process. Expanders are extruders, where the soybean flakes are subject to extrusion in presence of live steam. The temperature of the feed reaches a high temperature very rapidly. This aids in the process of enzyme deactivation. The author has experienced reduced level of non-hydratable phospholipids in the crude soybean oil obtained from the expander process. In addition, the crude oil from the expander process, after degumming to less then 5 parts per million of phosphorus, produced soybean oil that had superior flavor stability than the oil obtained from the standard process. It is an indirect evidence of the fact that reduced enzyme activity during crushing can produce improved oil stability due to reduced free radical activity in the crude oil.

As for fruit palm, the fruit bunches are autoclaved with steam at 45 to 50 pounds per square inch pressure. Fruit bunches are placed in the cages (roughly 1-1.5 ton capacity, each). The cages are rolled into an autoclave and live steam is injected into the autoclave. After 15 to 20 minutes of heating, the steam from the autoclave is vented. The purpose is to remove the air from the inside of the autoclave. The autoclave is repressurized with steam and the fruits are heated for another 60 to 75 minutes. The mode of operation of the autoclave makes the conditions conducive to the formation of free radicals and free fatty acids. The fruits go through a heating process where they are in contact with the air but are not heated rapidly enough to deactivate the enzymes and to prevent enzymatic reaction.

Crude oil produced from oilseeds or fruit palm must be protected against exposure to oxygen. It is common knowledge that peroxide value in the oil is reduced to zero when an oil with high peroxide value is bleached and deodorized. This does not necessarily imply that the quality of the finished oil is improved as the peroxide value drops to zero. This is because when the peroxide is reduced to zero, the anisidine value in the oil goes up. These are secondary oxidation products and contain free radicals that can propagate autoxidation in the oil. In addition, peroxide formation can occur while the fatty acid is still attached to the glycerol molecules. In the deodorization process 90 per cent of these molecules still remain in the oil. These partial molecules are free radicals. They can continue to break down more unsaturated fatty acids while the finished oil is in storage. Researchers have demonstrated that when crude soybean oil is heavily preoxidized, the freshly deodorized oil may show good flavor but it develops oxidized flavor and can become rancid very rapidly. Over-heating the oil during processing or undue exposure to oxygen increases the formation of peroxides, which leads to poor oil stability.

It is recommended that the degree of oxidation in the incoming crude oil must be monitored at the time of receipt and during its storage, prior to refining. Peroxide value (PV) alone does not provide the true picture of the state of oxidation in the crude oil. Besides PV, one must also measure the anisidine value (AV) in the crude oil. These analyses must also be conducted in the crude oil while it is in storage and waiting to be refined. The goal should be to keep both of these values as low as possible. The crude oil must be refined immediately, if these values begin to rise.

Many processors of seed oils do not cool the crude oil before it is stored. Many soybean crushers send the crude soybean oil into the storage tank at 70°C, after the oil leaves the de-solventizer. Crude palm oil, in most cases is stored at 75°C to 90°C by the fruit palm crushers. This practice is highly damaging to the oil.

It is recommended that for both seed oils and palm oil processes, an oil cooler must be installed to cool the oil to less than 40°C, before the crude oil is sent to storage

In many third-world countries oil is bleached in atmospheric kettles. The oil is heated to 110°C to 120°C, bleaching clay *is* added and the oil is agitated for 30 to 45 minutes and then filtered. This is an ideal situation for the oil to react with oxygen. Generally the peroxide value in the freshly bleached oil would be very low or zero but the anisidine value of the bleached oil from the atmospheric bleaching process is always high. This is because the oil produces hydroperoxides during bleaching. Hydroperoxides break down in the process soon after they are formed. Therefore, determination of only peroxide value in the bleached oil is not sufficient to establish its state of oxidation in the oil.

It is recommended that anisidine value should also be determined in the bleached oil along with peroxide value to establish the exact state of oxidation in the oil. Bleached oil must be analyzed for the trace impurities such as phosphorus, calcium, magnesium, iron, soap, chlorophylls and lovibond Red color.

In order to minimize the formation of excess free radicals in the bleaching process, one must avoid using the atmospheric process and adopt the vacuum bleaching method. In addition, use of hydrogel silica, such as Trisyl, can reduce soap, trace metals, phospholipids and free radicals that are present in the oil and improved its stability.

The final step in oil processing is deodorization. In this process the bleached oil is steam distilled under vacuum .The object is to remove the volatile impurities from the oil. The red color of the oil is reduced due to heat bleaching of the carotenes

In the deodorization process it is necessary that the absolute pressure in the deodorizer system is maintained as specified by the manufacturer. The author has experienced in the United States and around the world that the production personnel as well as the management in an oil processing operation tend to operate the deodorizers as described below:

1. A throughput far beyond the designed limit.
2. Maintain an absolute pressure in the deodorizer much higher than the designed operating limit.
3. Deodorize oils at a temperature that is too high in order to compensate for poor vacuum in the deodorizer and/or higher deodorizer throughput.

All of the above practices produce either incompletely deodorized oil or damage the oil in deodorization process.

A higher absolute pressure in the system, generally, indicates presence of air. At a deodorizer temperature of 242° to 260°C, the oil can undergo very rapid oxidation and polymerization if the system is not maintained at the designed operating pressure. Researchers have shown that as dimer content in the deodorized oil goes up, flavor acceptability in fresh deodorized oil goes down. At the same time, the oil containing higher levels of dimers, demonstrates poor flavor stability. Poor vacuum in the deodorizer can produce high amounts of oxidation polymers in the oil. Some of these polymers contain as much as nine times the amount of oxygen present in a triglyceride molecule. These compounds break down in the absence of light, oxygen and even at very low temperatures, releasing more free radicals that cause flavor degradation in the oil in storage or in the food processed with this oil. This is known as hidden oxidation in the oil. Standard analytical tools that are normally used to measure oil oxidation, such as peroxide value, cannot explain this phenomenon.

The oil in the deodorizer must be cooled down to 150°C or less, before it is taken out of the deodorizer for external cooling. Addition of citric acid is critical to chelate the metals that can cause autoxidation of the oil in storage. Citric acid must be added at a temperature less than 145°C to prevent its decomposition.

The oil, leaving the deodorizer must be cooled down to a temperature not more than 5°C above its melt point before it is stored. The liquid oils must be cooled down to less than 40°C. It is highly recommended that all deodorized oils must be saturated with nitrogen as they leave the deodorizer. The oils must be stored under nitrogen atmosphere with a maiximum oxygen content of 0.5 per cent in the head-space.

Oils stored under these conditions have shown no rise in the peroxide value for 6-8 weeks, without using antioxidants.

Photooxidation

It has been mentioned earlier that there are two types of photooxidation reactions:

1. Photochemical reaction

2. Photosensitized reaction

Photochemical Reaction

This reaction is initiated by the ultraviolet (UV) light. Unsaturated fatty acids form free radicals in presence of the UV light. The rest of the reaction proceeds in the same manner as the autoxidation process. Photochemical reaction proceeds at the same rate as autoxidation.

Photosensitized Reaction

Photosensitized reaction occurs when the oil is exposed to the visible light in presence of a photosensitizer. This reaction proceeds 1500 times faster than autoxidation or photochemical reaction.

Chlorophylls, naturally present in soybean and canola oils, are photosensitizers. This is why excess chlorophylls, present in these oils must be removed in order to reduce photosensitivity of these oils in following applications:

1. When the oils are packaged and stored in clear bottles.
2. Processed foods are prepared with these oils and packaged in clear bags or containers.

Chlorophylls (A and B) are reduced in the bleaching process. Acid activated clay is used to remove chlorophylls from vegetable oils. In the United States and Canada, the chlorophyll content is maintained at less than 30 parts per billion in both soybean and canola oils.

It has been found that chlorophylls undergo decomposition under acidic pH in presence of oxygen. The breakdown products of chlorophylls are known as pheophytins, pheophorbides and pyropheophorbides. These components are ten times stronger photosensitizers than their parent compounds. Therefore, measurement of only chlorophylls in the bleached oil may not provide total information regarding its photosensitivity. It has also been demonstrated that chlorophyll B and their decomposition products are stronger photosensitizers compared to chlorophyll A. Therefore it is important to measure the breakdown products of chlorophylls in the bleached oil to establish its photosensitivity.

Atmospheric bleaching process provides an ideal opportunity for the chlorophylls to break down and make the oil more prone to photooxidation.

Silica treatment of oil in both conventional refining process and modified chemical refining process (28) has shown improved oil stability. This is because silica is a very good adsorbent for soap, trace impurities and including free radicals. This can improve the autoxidative stability of the oil. In addition, use of silica reduced the dosage of bleaching clay required to achieve the desired green color in soybean and canola oils. This may provide the opportunity to reduce chlorophyll breakdown during bleaching and improve photo-oxidative stability of the oil.

1. Atmospheric bleaching process is damaging to the oil. Vacuum bleaching process is a preferred option for oil bleaching.
2. Silica pre-treatment, prior to the acid activated clay treatment of the oil is recommended for bleaching all types of vegetable oils.
3. In addition to the chlorophyll measurement, it is prudent to check for the chlorophyll break down compounds in soybean and canola oils. This can provide a better idea about photo-oxidative stability of the oil.

HYDROLYSIS

When a molecule of water reacts with a molecule of triglyceride, the resultant products are diglyceride and fatty acid (commonly termed as free fatty acid). As more water molecules react with the oil, more free fatty acids are formed together with diglycerides and monoglycerides. Eventually the reaction products are glycerol and fatty acids. This reaction can occur only when the oil and water are in perfect solution. Oil and water do not form a solution under normal processing conditions, because both temperature and pressure are insufficient to create a solution. However, the presence of a small amount of surfactant in the oil increases inter-solubility between the two liquids and promotes hydrolysis of the oil during processing. This occurs because the surfactant reduces the interfacial tension between oil and water.

In commercial fat-splitting reactors, a small amount of caustic is used to produce soap through the reaction between the oil and caustic. The soap thus produced act as surfactants for the reaction. The crude oil contains natural surfactants, such as phospholipids. The other source of surfactants are:

1. Soap left in the oil from the caustic refining process or from neutralized (unbleached) oil.
2. Some products of oil oxidation have surface activity that can promote hydrolysis in the oil.
3. Mono and diglycerides produced in the oil due to over refining of the crude, or excess caustic treat needed for refining poor quality crude oil.

In chemical refining process of seed oils, the final phosphorus content in the oil is recommended to be less than 1 part per million, it has been experienced by the author that when the phosphorus content in the vegetable oil is 3 parts per million or higher, the free fatty acid in the finished oil begins to rise while the oil is in storage. This is commonly experienced in physically refined palm oil or palmolein, where due to the distance between the source and the end-users, the oil has to be purchased in large quantities and stored for months at a time. In many instances, the oil needs to be re-deodorized to reduce the increased free fatty acids produced during storage. Physically refined palm oil or palmolein contains 3 to 5 parts per million of phosphorus in the finished oil. This is believed to be one of the major drivers for the rise of free fatty acid in these oils during storage.

High phosphorus, left in the oil, can cause some hidden losses in oil yield. High phosphorus in the oil poisons the nickel catalyst in hydrogenation. This increases the consumption of catalyst. Additional amount of diatomaceous earth is needed to remove the extra nickel catalyst

Hydrogenated oil, produced with poisoned catalyst tends to contain high level of colloidal nickel. This is removed from the oil in the post-bleaching process where the hydrogenated oil is treated with acid activated bleaching clay together with citric or phosphoric acid under vacuum, post-bleaching process always increases the free fatty acid content in hydrogenated oil, which increases oil loss in the deodorization. Process, A higher free fatty acid in deodorizer feed reduces the rate of deodorization. This means reduced productivity due to slower deodorizer throughput. Although the excess free fatty acids are removed in the deodorizer, there is always some additional loss of neutral oil in the distillate when the deodorizer feed contains high concentration of free fatty acids. If the phosphorus content is high in the deodorizer feed, the oil can undergo hydrolysis during deodorization. This can be verified by measuring the diglyceride content of the feed to the deodorizer and the deodorized oil. The additional diglycerides formed in the deodorized oil do not get distilled and can be analyzed.

There is a tendency to deodorize physically refined palmolein at a temperature of 260°C to 265°C. This is generally done to reduce the deodorization time and increase deodorizer throughput. However, at this higher temperature the oil can undergo hydrolysis, creating a condition where reduction of the free fatty acid to less than 0.05 per cent becomes difficult. One must also recognize that surfactants, such as phospholipids reduce interfacial tension between the oil and air. Therefore, under poor vacuum and with high level of phosphorus, there is an opportunity for the oxygen to react with the oil in the deodorizer and form free radicals that lead to the formation of oxidative dimers and polmers.

Soap can be left in the refined oil due to inadequate bleaching. This can produce similar reaction as phospholipids in the oil . Some cottonseed oil processors believe that prime Bleach summer yellow (PBSY) cottonseed oil does not require bleaching because of the low color of the oil. This is not correct because PBSY cottonseed oil contains some residual soap that must be removed from the oil to prevent hydrolysis and oxidation of the oil in subsequent processing and handling. Neutrallized oils also suffer form the same risk as the PBSY cottonseed oil since it contains soap.

Crude palm oil may contain soap. This normally comes from improperly cleaned shipping vessels. Even a small amount of soap content must be eliminated from the oil before it is deodorized. Some unscrupulous palm oil processors have also delivered crude palm oil that has been partially neutralized in order to meet the 5 per cent limit on the free fatty acid, set forth by PORAM. This practice causes long-term instability in the oil.

Table 9.11. Recommended Oil Quality Parameters for Seed Oils

Analysis	*Crude Oil*	*Refined Oil*	*Water Washed Oil*	*Bleached Oil*
% FFA	—	0.01—0.03	0.02—0.05	0.04—0.1
Soap, ppm	—	< 500 Not > 1000	< 50 Not > 100	0
PV, maximum	4*	4	4	0
PAV	4	4	4	6
Phosphorus, ppm	—	3	3	< 1
Calcium, pm	—	—	—	< 0.2
Magnesium, ppm	—	—	—	<0.2
Iron, ppm	—	—	—	< 0.2

*At PV higher than 8, seed oil may produce unacceptable flavor stability.

It is recommended that the soap content in all palm oil and all types of partially neutralized oil receipts is measured and the oil is bleached to remove the soap before the oil is further processed. The oil can be treated either with silica or bleaching clay to accomplish this. The recommended oil analytical parameters for crude, refined, water washed and bleached oils for best performance.

THERMAL DECOMPOSITION

Vegetable oils undergo decomposition when they are subjected to high heat. When the oil is heated to 150°C in air, even the saturated fatty acids break down to produce compounds, such as homologous series of carboxylic acid, 2-alkanones, 2-alkanals, lactones, *n*-alkanes and *n*-alkenes. When triglycerides containing even numbered fatty acids, ranging from 6 to 16 carbon atoms are heated in air to 180°C to 250°C for 1 hour, they all produce homologous hydrocarbons, aldehydes, keton lactones.

Thermolytic non-oxidative products from saturated fatty acids can be detected even when the oil is heated to 180°C under vacuum. Volatile compounds detected under this condition include normal alkanes, fatty acid, oxypropyl ester, propene, propanediol esterdiglycerol and acrolein.

In case of unsaturated fatty acids, formations of dimeric and cyclic compunds appear to be the predominant thermolytic breakdown product. Degrees of complexity of these compounds increase with the degree of unsaturation of the fatty acids present. Heat treatment also leads to the formation of several aromatic compounds. Thermal dimers and cyclic monomers have been isolated from several types of heated vegetable oils.

It can be seen from the above dicussion that exposure of oil to high heat can lead to both oxidative and thermolytic degradation of the oil.

RECOMMENDATIONS

The recommended analytical paramcters for deodorized seed oil is listed. The general information provided in this table is applicable to all types of seed oils and palm oil (palmolein).

Table 9.12. Recommended Anlysis for Various Vegetable Oils.

Analysis	*Target*	*Maximum / minimum*	*Remarkes*
%FFA	< 0.03	0.05, max.	Higher FFA may indicate lower self-life stability
PV	0	0.5 max... 1.0 max...	As loaded in trucks As received by Customer
AV	2	4	Never to exceed 6
%Dimers	< 0.2	0.4 max	Never to exceed 0.5%
%Polymers	< 1	2 max.	Higher polymers may impart bitter flavor in prepared foods
Chloropyhll	—	30 max.	Increases photosensitivity in the oil
Phosphorus, ppm	< 0.5	1.0 max.	Baking shortening and margaring can tolerate higher phosphorus level
Calcium, ppm	—	0.2 max.	Higher than 0.5 ppm implies poor quality oil
Iron, ppm	—	0.2 max.	Higher than 0.5 ppm implies poor quality oil
Magnesium, ppm	—	0.2 max.	Higher than 0.5 ppm implies Poor Quality Oil
Lovibond Red color	< 0.1, for Soybeean Oil, Canola Oil, Sunflower Oil	1.5 max. for Soybeean OIl, Canola Oil, Sunflower Oil	> 1.5 Lovibod Red color Implies Poor Quality Crude or Poor Oil Processing
% Triglycerides	98 minimum	-	A Lower Value Implies Poor Quality Crude In Case of Seed Oils
% Monoglyceride*	—	0.3	A Lower Value Implies Poor Quality Crude or Over Refining In Case of Seed Oils
% Diglyceride**	—	0.6	A Lower Value Implies Poor Quality Crude or Over Refining In Case of Seed Oils

*Typical values in commerical palm and palmolein : 0.5 – 1% monoglycerides

**Typical values in commerical palm oil and palmolein : 3 – 11% diglyceride

Lovibond Red color is different for different oils. Mono and diglyceride contents in palm oil and palmolein are always higher than in the seed oils.

PALM OIL INDUSTRY

Composition

Palm oil is obtained from the flesh surrounding the seed, simply by cgoking, meshing and pressing. In this process the seeds are separated and after cracking and removing the shell, the kernel can be processed to yield palm kernel oil. Palm kernel oil forms about 10% of the total oil yield and has different composition and uses from palm oil.

As regards to fatty acid composition, palm oil occupies an intermediate position among natural fats. Its saturated fatty acid content is 50% and the rest is monounsaturated and polyunsaturated. By virtue of the fatty acid composition, palm oil is semi solid containing mainly palmitic, oleic and linoleic acids, which are the most common components of all food fats. Palm kernel oil, on the other hand is rich in lauric acid and is interchangeable with coconut oil in most of its applications but is more flexible because it is more unsaturated compared to coconut oil.

Crude palm oil, besides its fatty acid composition also contains 1% minor components. A diagrammatic presentation of crude palm oil minor components.

CAROTENOIDS AND TOCOLS

Carotenoids form a class of C40 polyunsaturated hydrocarbons (carotenes) and their oxygenated derivatives (xanthophylls). These compounds impart a rich orange-red colour to the oil.

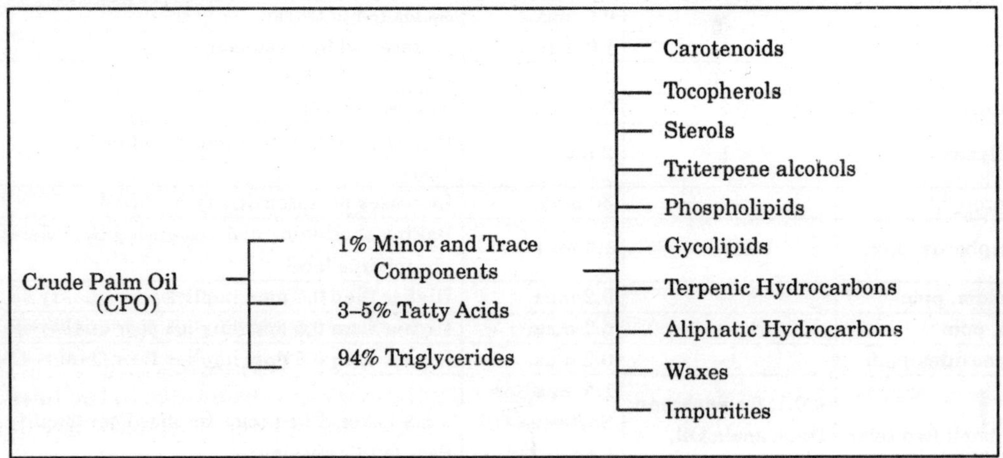

Fig. 9.18. Constituents of Crude Palm Oil

Crude palm oil is one of the richest natural plant sources of carotenes with the concentration of 500-700 ppm. It has 15 times more retinol equivalents than carrots and 300 times more than tomatoes. Analysis shows that alpha and beta carotenes constitute approximately 90% of the total carotenoid content and the rest are delta carotenes, gamma carotene, phytofluene, phytoene, zea-carotene, lycopene, neurosporene, alpha and beta zeacarotenes.

The production of palm oil from Malaysia is projected, to reach ever 8 million tonnes by the year 2000. Hence, the potential availability of carotenes in the year 2000 is about 4800 tonnes.

The tocopherols and tocotrienols are found in cude palm oil at 600-1000 ppm concentrations. The main constituents are 44% gamma tocotrienol, 22% alpha tocophercel, 12% delta tocotrienol, the rest being alpha and beta tocotrienols, gamma and delta tocopherols.

NUTRITIONAL BENEFITS OF CAROTENOIDS AND TOCOLS IN PALM OIL

In human nutrition, carotenoids play important role as a source of provitamin A. Beta-carotene, especially, is the most nutritionally active carotene as provitamin A. It is converted in the gastrointestinal tract into vitamin A. Vitamin A of course, plays an important role in regulating vision growth and reproduction. It is essential for the normal cellular differentiation of most epitilia, including those of the skin, bronchi and trachea, stomach, intestine, uterus, kidney and other organs.

The FAO/WHO recommendations on fats and oils in human nutrition stated that "in countries where vitamin A deficiency is a health problem, the use of rcd palm oil, wherever readily available should be encouraged."

Human studies have also indicated strongly that carotenoids possess antioxidant properties and may protect against certain types of cancer. It was postulated that beta carotene has significant protection against cancer. This has sparked much interest and numerous reports have now appeared to shed new light on this question. An inverse association between serum beta carotene and the risk of squamous cell carcinoma of the lung has been found. It was also reported that beta carotene inhibited oral tumors when applied topically or given as a dietary supplement.

The antitumor effects of palm carotenes had been studied on the developments of chemically-induced skin cancer and viral-induced lymph cancer in HRS/J female hairless mice. Palm carotene was found to possess antitumorigenic effects and the inhibition of subcutaneous lymphoma was concentration dependent: 0:3% palm carotenes (PC) > 0.1% PC > 0.05% PC. All the palm carotenes group inhibited epithelial pappilloma formation as well.

Vitamin E functions primarily as an antioxidant. especially to prevent oxidation and peroxidation of polyunsaturated fatty acid units of membrance phospholipid (within and on the plasma membrane of cells). This works against injury to cell membranes as in red blood cell fragility in man. It is also implicated in preventing photooxidative damage to the lens and thus in preventing cataract formation. The antioxidant properties of vitamin E have been implicated to prevent/delay the onset of various degenerative diseases such as aging, cancer, arthritis, damage to cells caused by air pollution and also in preventing coronary heart diseases.

The nutritional benefits of palm vitamin E (palm vitee) has been studied on human as well as animal models. The hypocholesterolemic effect of palmvitee has been studied by various researchers. It was found in one study that the concentrations of serum total cholesterol (–15%), LDL cholesterol (–8%), apoprotein B(–10%) and thromboxane (–25%) decreased significantly in 15 hypercholesterolemic subjects fed 200 mg palmvitee capsules/day . Another study on 9 healthy volunteers taking one capsule of palmvitee perday also showed a reduction of serum cholesterol (reduction rage –5% to –35.9%) and LDL cholesterol (reduction range 0.9% to 37%).

The effect of palm oil vitamin E on isolated Langendorff perfused rat hearts has also been studied and results indicated that palmvitee containing a mixtue of tocotrienols and tocopherols may be more efficient than alpha tocopherol alone in the protection of the heart against oxidative stress induced by ischemia-reperfusion.

Large population studies have analysed the relationship between antioxidant nutrients such as carotenoids and vitamin E and degenerative diseases. A continuing study among 22,000 male physicians in the US yielded an interim finding that those taking a supplement of beta carotene reduced the occurrence of major cardiovascular events by almost 50%. It was reported that an inverse correlation between plasma vitamin E levels and mortality from ischemic heart disease in a cross cultural epidemiology.

CAROTENOIDS

Sources of Carotenoids

Various analytical methods have been used for the determination of the carotenoid profile of crude palm oil. Column chromatography has been used in earlier studies using different adsorbents and reverse phase HPLC has been shown to have several advantages for the separation of carotenoids in the oil.

Table 9.13. Carotenoid contents of Palm Oil Extracts

Palm Oil Extract	*Carotenoid Content (ppm)*
Crude palm oil	630—700
Crude palm olein	680—760
Crude palm stearin	380—540
Second pressed oil	1800—2400
Residual oil from fiber	4000—6000
Total carotenoids estimated at 446 nm	

The carotenoid contents of various palm oil extracts. A carotenoid rich oil can be obtained from palm pressed fiber, a palm oil by product which is presently burnt as a fuel in palm oil mills. The pressed fiber was found to contain about 5-6% of residual oil with a carotenoid concentration of 4000-6000 ppm, and the carotenoid content of the residual oil in the pressed fiber of the hybrid oil palm fruit is even higher, at 6000-7000 ppm.

Carotenoids from the commercial crude palm oil are concentrated during the extraction and fractionation process. A system of palm oil extraction based on a double pressing technique has been implemented by several mills in Malaysia. The expected advantages of double pressing over the conventional single-stage pressing are lower oil loss in fiber, higher kernel extraction rate, less wear on the screw worm and cages, and reduction of contamination of the kernel oil with crude palm oil. More interestingly, the oil produced from the second pressing has a higher concentration of carotenoids. This could be due to the fact that the first pressing in a double pressing process is carried out at a lower pressure (to avoid cracking of the nut) and oil relatively higher in carotenoids from the second pressing is extracted. After removal of the nuts the fiber is then subjected to a higher pressure and more carotenoids are extracted out of the mesocarp, together with the residual oil from the first pressing.

Carotenoids are also being concentrated in the industrial process called fractionation. Palm oil is a semi-solid fat at ordinary room temperature. Fractionation is carried out to extend the uses of palm oil and the products obtained are the liquid oil (olein, 70-80%) and the solid fat (stearin, 20-30%). The carotenoid content in the palm olein (lower melting) fraction is enriched by 10-20%.

Recovery and Extraction of Carotenoids

The carotenes on the market at present are chemically synthetic beta carotene, carotene extracted from carrots and carotene extracted from algae Dunaliella. The carotenoid composition of the various carotenes. The demand for natural carotenes is such that customers are willing to pay a premium for these, which they value for their perceived safety. The major commercial uses of beta carotene in the global market in which they are widely used as a food colorant.

Table 9.14. Carotenoid Compositions of Various Carotenes.

	alpha-C	*beta-C*	*Others*
Carrot carotene	30—50%	40—60%	10—15%
Synthetic beta-carotene		100%	
Dunaliella cartonene	6—10%	84—90%	
Palm oil carotene	30—35%	60—65%	5—10%

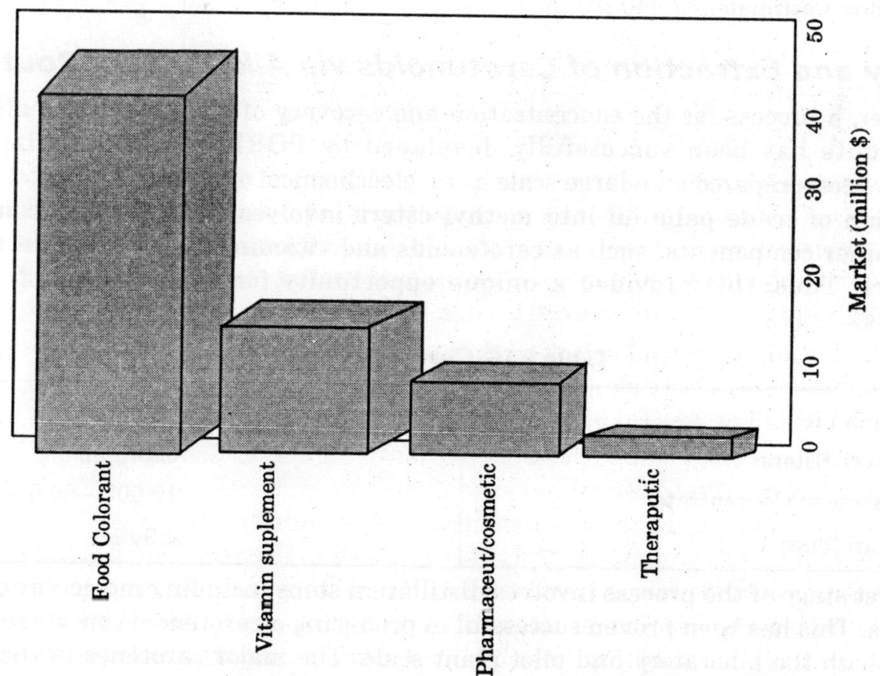

Fig. 9.19. Major commercial uses of beta carotene.

Because carotenoids are likely to grow in importance and value, the recovery of carotenoids from palm oil and its by-products is important. In addition, some studies are also being carried out to recover and concentrate palm-based carotenoids for pharmaceutical purposes.

Numerous extraction methods have been developed to recover the carotenoids from crude palm oil. Most of the reported methods are difficult, costly or inefficient.

Table 9.15. Carotenoid Concentrates from Various Methods.

Method	*Carotenoid content (ppm)*	*Recovery*
Vacuum distillation*	>20,784	<46
Molecular distillation*	>80,000	>80
Adsorption*		
C18	8000—9000	>90
Carbon	5000—7000	<50
Molecular distillation of crude palm oil	1290—1990	
Adsorption from crude palm oil (activated carbon)	3700—5600	<80

* through methyl ester route
Total carotenoids estimated at 440 nm

Recovery and Extraction of Carotenoids via Alkyl Esters Route

However, a process for the concentration and recovery of carotenoids via alkyl esters of palm oil route has been successfully developed by PORIM. Volatile palm oil methyl esters have been prepared on a large scale as an oleochemical or diesel substitute. This process of conversion of crude palm oil into methyl esters involves mild reaction conditions, the valuable minor components, such as carotenoids and vitamin E would remain intact in the alkyl esters. Thus, this provided a unique opportunity for the recovery of these minor components.

Table 9.16. Carotene Content

Palm oil	600 ppm
Alkyl Esters	600 ppm
Carotene Concentrate	10,000—80,000 ppm
Carotenes	< 90%

The first stage of the process involves distillation steps including molecular distillation of alkyl esters. This has been proven successful in producing a carotenoid concentrate of >80,000 ppm from both the laboratory and pilot plant scale. The major carotenes in the concentrate are alpha and beta carotene. The second stage involves column chromatography with an absorbent to give a carotene fraction. The concentration of this carotene fraction ranged from

90-100%. The carotene fraction can then be diluted to produce the desired commercial carotene concentrate in various concentrations (1-30%) for various applications.

Table 9.17. Carotene composition (%)

Carotene	*Concentrate*	*Crude Palm Oil*
Phytoene	1.5	1.3
Phytofluene	0.3	0.1
Cis-β-Carotene	0.9	0.7
β-Carotene	49.9	56.0
α-Carotene	33.3	35.1
Cis-α-Carotene	5.5	2.5
γ-Carotene	1.7	0.7
ζ-Carotene	1.3	0.3
δ-Carotene	0.6	0.8
Neurosporene	0.1	0.3
β-Zeacarotene	1.3	0.7
α-Zeacarotene	0.4	0.2
Lycopene	3.4	1.3
Total (ppm)	80560	673

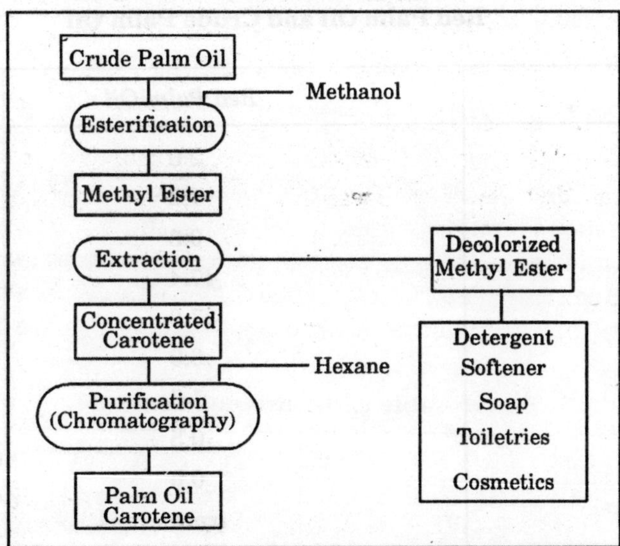

Fig. 9.20. Lion Procedure for Producing Palm Oil Carotene

Lion Corporation of Japan has made efforts to recover carotene from palm oil since the 1970s. They have developed a process which is also via the methyl ester route. Crude palm oil is converted into methyl esters by transesterification with methanol at low temperature

which results in little loss of carotenes, if any. In the second step methyl esters are mixed in an organic solvent, and then the mixture is separated into a carotene rich layer and a decolourizsed methyl ester layer. The methyl ester is used for oleochemical purposes while the carotene may be further concentrated from the carotene rich layer by ordinary means. Finally, the carotene is purified by liquid chromatography with hexane. The palm oil carotens contains more than 95% of carotenoids (60-70% of beta carotenes, 30-35% alpha carotenes and small amounts of gamma and delta carotenes).

Carotech has also developed an integrated propess for recovery of carotenoids and tocotrienols from crude palm oil, again via the alkyl esters route. Fatty acids in a vegetable oil are subjected to alcoholic esterification to form an ester-rich layer including fatty acid alkyl esters, carotenoids and tocotrienols. The ester rich layer is exposed to solvolytic micellization to form a carotenoid rich layer. The ester rich layer is separated from the carotenoid-rich layer. The carotenoids in the carotenoid-rich layer are concentrated and can be adsorptively separated from the carotenoid-rich layer. Fatty acid alkyl esters are separated from the ester-rich layer to form a tocotrienol-rich layer.

Red Palm Oil

A process has also been developed by PORIM to produce deacidified and deodorized red palm oil from the degummed palm oil with more than 80% of the original carotenes still intact as well as the vitamin E. This red palm oil is of similar excellent quality to the refined, bleached and deodorized (RBD) palm oil that is normally traded but contains no carotenes.

Table 9.18. Carotene Composition (%) of Deacidified and Deodorized Red Palm Oil and Crude Palm Oil

Carotene	*Red Palm Oil*	*Crude Palm Oil*
Phytoene	2.0	1.3
Phytofluene	1.2	0.1
Cis-β-Carotene	0.8	0.7
β-Carotene	47.4	56.0
α-Carotene	37.0	35.1
Cis-α-Carotene	6.9	2.5
Carotene	1.3	0.7
ζ-Carotene	0.5	0.3
γ-Carotene	0.6	0.8
Neurosporene	trace	0.3
β-Zeacarotene	0.5	0.7
α-Zeacarotene	0.3	0.2
Lycopene	1.5	1.3
Total (ppm)	545	673

The process involves two stages *i.e.*, pretreatment of the crude palm oil followed by deodorization and deacidification by molecular distillation. The pretreatment is carried out in a conventional manner using phosphoric acid followed by bleaching earth. This allows impurities and oxidative products in the crude oil to be removed. The deodorization and deacidification is carried out using molecular distillation unit at temperatures <165°C and a pressure of 20-35 X 10-3 Torr.

Analysis by HPLC shows that the profile of carotenes are not destroyed during the process and the quality of red palm oil has also been found to be good. Sensory evaluation carried out on red palm oil showed that it is of very good quality and comparable to RBD palm oil.

Table 9.19. Quality Parameters of Red Palm Oil

Carotenes	>80% intact
Tocopherols and Tocotrienols	>80% intact
Free fatty acids	<0.1%
Peroxide value	<0.2
Phosphorus content	<2 ppm
Moisture and impurities	<0.1%

The commercial production of red palm oil involves the capital investment in a molecular distillation unit, which includes a vacuum system. The other costs involved are the power supply, water supply for cooling and storage tanks. The cost of production, based on a 5-year investment plan is given below.

Plant capacity (T/yr)	10,000	30,000
Pretreatment (RM/T)	17.00	17.00
Molecular distillation (RM/T)	99.00	96.30
Total cost of Red Palm Oil (RM/kg) (includes cost of crude palm olein)	1.2529	1.2374

Unitata has also developed a process to produce carotene enriched red palm oil. This is a chemical process which involves chemical neutralisation of the free fatty acids, followed by pretreatment, then deodorization using conventional method to remove the odour. The quality of the product obtained from this process is also similar to the RBD palm oil, containing not less than 250 ppm of carotenes and 800 ppm of vitamin E. The oil produced has been marketed which bears the tradename of NUTROLEIN.

TOCOLS

Source of Tocols

Table 8 shows the various sources of tocopherols from palm oil and its products. Tocopherols accumulated in the palm fatty acid distillate (PFAD) after physical refining can be as high as

0.8% or greater so that the PFAD has been looked upon as a good source of natural tocopherols and tocotrienols. In fractionation or fractional crystallizations tocopherols tend to remain in the liquid or olein fractions thus rendering the solid fractions deficient in antioxidants.

Table 9.20. Tocopherols in Palm Oil (ppm)

Crude Palm Oil	**600_1000**
Crude Palm Olein	**800—1000**
Crude Palm Stearin	**250—350**
PFAD	**150—8500**
RBD Palm Oil	**356—630**
RBD Palm Olein	**468—673**
Palm Kernel Oil	**(80—100)***

The vitamin E content of palm oil and other edible oils. Palm oil has an abundance of tocotrienols as compared to other oils.

Palm Vitamin E Isolation

The palm fatty acid distillate (PFAD) contains 0.4%-0.8% tocopherols and tocotrienols with the tocotrienols present as a major component. The bulk of the PFAD is free fatty acids.

The concept of isolating the tocopherols and tocotrienols (palm vitamin E) is based on the conversion of free fatty acids, partial glycerides and triglycerides into methyl esters after removal of which by distillation a concentrate of palmvitee is obtained. Purification of the concentrate will yield tocopherols and tocotrienols.

The purification of the crude mixture is effected by crystallization during which sterols are separated, followed by ion exchange chromatography, in which the tocopherols and tocotrienols are separated from squalene. Palmvitee is then washed dried and purified by molecular distillation. Finally it is deodorized before being dissolved in super olein prior to encapsulation.

The process flow of isolation of palmvitee. The concentration of vitamin E at different stages of the process.

This pilot plant is capable of producing vitamin E with the following results:

purity	95—99%
overall yield	65%
production	5—6 kg/batch

Table 9.21. Total Tocopherol and Tocotrienol of Oils and Fats

Oil and fats	*Tocopherol*					*Tocotrienol*					
	α – T (ppm)	*β – T (ppm)*	*γ – T (ppm)*	*δ – T (ppm)*	*%T (ppm)*	*α – T3 (ppm)*	*β – T3 (ppm)*	*γ – T3 (ppm)*	*δ – T3*	*%T3 T + T3*	*Total (ppm)*
Palm Oil	256	—	316	70	55%	143	32	286	69	45%	1172
Palm Oil	279	—	61	—	31%	274	—	38	69	69%	1081
Palm Oil	152	—	—	—	17%	205	—	439	94	83%	890
Soyabean Oil	101	—	593	264	100%	—	—	—	—	0%	958
Safflower Oil	387	—	174	240	100%	—	—	—	—	0%	801
Corn Oil	112	50	602	18	100%	—	—	—	—	0%	782
Cottonseed Oil	389	—	387	—	100%	—	—	—	—	0%	776
Sunflower Oil	487	—	51	8	100%	—	—	—	—	0%	546
Peanut Oil	130	—	216	21	100%	—	—	—	—	0%	367
Cocoa Butter	11	—	170	17	99%	2	—	—	—	1%	200
Olive Oil	51	—	—	—	100%	—	—	—	—	0%	51
Coconut Oil	5	—	—	6	31%	5	1	19	—	69%	36
Palm Kernel Oil	13	—	—	—	38%	21	—	—		62%	34
Lard	12	—	7	—	73%	7	—	—	—	27%	26

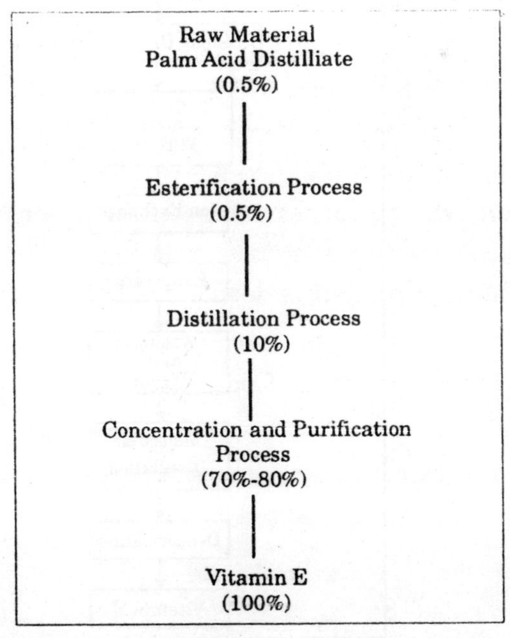

Fig. 9.21. Concentration of Vitamin E at Different Stages.

ECONOMIC VIABILITY EVALUATION

It has been shown that it is technically feasible to produce tocopherols and tocotrienols from PFAD and preliminary economic assessment on the process has been attempted.

Vitamin E extraction is a concentration process from the 0.5%. Vitamin E content of the feed material, PFAD to 99% purity. There are 8 processing steps, and based on two capacities of 25000 tonnes and 50000 tonnes PFAD per year, the plant investment amounts to RM 49.86 and RM 75.50 million respectively (1989 estimates).

The material, utility and labour costs are estimated on the pilot plant experience. The effect of variation of selling prices on project feasibility is analysed.

The economic analysis indicated that for 50000 tonnes of plant capacity for PFAD or 162500 kg of vitamin E, the project is feasible when the selling price is maintained at about RM200 per kilogram for an investment of 75.5 million ringgit. For 25000 tonnes of plant capacity for PFAD or 81250 kg of vitamin E the project is marginally feasible if selling price of vitamin E can be maintained at RM200 and more, some of the costs may be further reduced such as depreciation of plant and material (PFAD).

Nevertheless a detailed analysis, including market study is required before these processes could be fully commercialized.

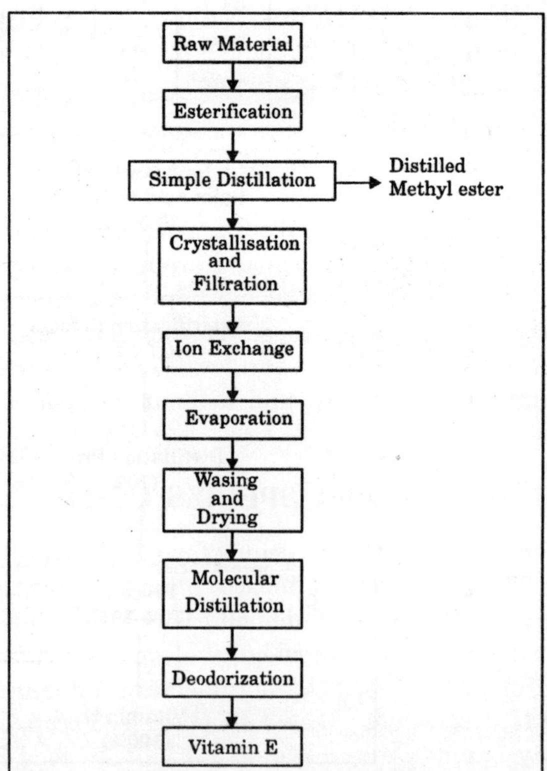

Fig. 3.22. Vitamin E pilot plant production process.

CHAPTER 10 Pharmaceutical Proteins in Plants

Molecular farming or *pharming*™is the production of pharmaceutically important and commercially valuable proteins in plants. Since the first production of a pharmaceutical protein in transgenic tobacco a broad range of proteins with potential medical applications has been expressed in plants. These range from small peptides such as enkephalins to complex, multisubunit molecules such as secreted antibodies of the IgA type. In this chapter we will address the decisions that must be made during the production and marketing of such proteins. This begins with the choice of production host, which can range from cellular green algae to a variety of crop species. For pharmaceutical applications, it is necessary to work under current good manufacturing practice (cGMP) conditions and the use of plant suspension cultures must therefore be considered as an alternative to whole plants.

Another crucial consideration is the design of the expression cassette, which mus optimized to achieve high-level accumulation of functional recombinant protein. Intrinsic factors such as how the protein is processed in the plant need to be taken into account, as well as the ability of the host species to carry out specific forms of posttranslational modification. When sufficiently high levels of expression and protein accumulation are achieved, efficient downstream processing protocols must be developed to insure product quality and the economic feasibility of production. As the demand for safe, recombinant pharmaceutical proteins continues to expand, the market potential of plant-produced recombinant proteins is considerable. Molecular farming can produce iccombinant proteins at a lower cost than traditional expression systems based on microbial or animal cell culture, and without the risk of contamination with human pathogens.

PLANT SPECIES USED

Many species can be exploited for molecular farming with tobacco being the most widely used host plant to date. The advantages of tobacco plants include the high biomass yield (which can reach 50 – 100 tonnes ha^{-1} in a high-density population depending on the cultivar), the ease of stable transformation either by co-cultivation with *Agrobacterium tumefaciens*, or transiently by infiltration with transgenic agrobacteria or transfection With viral vectors. Another benefit is that tobacco is not used as a food or feed crop, ensuring that a transformed line expressing a highly potent drug will not contaminate food resources by outcrossing or during the processing steps. A drawback of this species is its content of nicotine and other alkaloids. Although there are cultivars available with reduced alkaloid content it is necessary to remove all traces of these toxic compounds during downstream processing, especially if the recombinant protein is intended for clinical applications.

An additional disadvantage is the limited shelf life of transgenic leaf material after harvest, a problem that can be solved in other plant species by targeting the desired protein to natural storage organs like seeds or tubers. For tobacco leaves, downstream processing must commence immediately after harvest to ensure the stability of the recombinant protein, although Fiedler *et* al. reported no losses of scFv (single chain fragment variable) antibody specificity or antigen binding capacity in dried leaves three weeks after harvest. In the quest to find other plants that are suitable as bioreactors, various monocotyledonous and dicotyledonous species have been tested. These include corn, rice and wheat, alfalfa, potato, oilseed rape, pea, tomato and soybean.

The major advantage of cereal crops is that recombinant proteins can be directed to accumulate in seeds, which are evolutionar specialized for storage and thus protect proteins from proteolytic degradation. Recombinant proteins are reported to remain stable in seeds for up to five months at room temperature and for at least three years at refrigerator temperature without significant loss of activity. In addition, the seed proteome is less complex than the leaf proteome, which makes purification quicker and more economical. lt has been shown recently that the yields of recombinant protein in transgenic plants vary according to intrinsic factors such as the developmental stage and extrinsic factors such as the climate. Such factors also affect the precise nature of post-translational modification, particularly glycosylation.

Elbers and colleagues reported that the glycosylation profile, expressed as the ratio of complex to high mannose type glycans, increased with the age of tobacco leaves. Therefore, it is necessary to consider measures that ensure protein homogeneity, e.g. restricting expression to defined plant organs or specific compartments. Alternatively, it is possible to use an inducible expression system to restrict protein expression to a defined time point or period. Examples of such production methods include the postharvest system developed by Croptech and the amylase-based system for controlled expression in suspension cells.

ALTERNATIVE EXPRESSION SYSTEM

Plant cell cultures have not been used as frequently as intact plants for molecular farming but there are several examples that demonstrate their suitability as hosts for the production of recombinant proteins. Free cell suspension is regarded as the most suitable cultivation system for large-scale biotechnology applications, but tentative experiments have also been performed with hairy roots and shooty teratomas. Examples of foreign proteins successfully expressed in suspension cells include monoclonal antibodies and their derivatives, cytokines, hormones and enzymes. The tobacco suspension cell line BY-2 appears to be a particularly suitable host cell line because of its high growth ratejtfl]. Like plant tissues, BY-2 suspension cells can be transformed easily by cocultivation with *A. tumefaciens.* In addition to this tobacco cell line, soybean cell culture has been used for the production of hepatitis B virus surface antigens and rice suspension cells have been employed for the expression of human α-1-antitrypsin and human lysozyme.

In contrast to the tobacco transformation process, gene transfer to rice is often performed by particle bombardment. Table 10.2 provides an overview of pharmaceutical proteins that have

Table 10.1. Pharmaceutical Proteins Produced in Transgenic plant

Protein	*Regulatory elements*	*Additional elements of the construct*		*Species/Tissue*	*Accumulation level*
α creatine kinase, MAK33 mAb (Fab)	Ca MV 35S promoter/ different termination sequence (ocs and rbcS)	LP of 2S2 storage protein of Arabidopsis *thaliana*	Fd and k chain with 3'ocs	*A. thaliana* (leaves)	3.6% of TSP
			Fd chain with 3'ocs, κ chain with 3'rbcS	*A thalina* (leaves)	6.5% of TSP
			Fd chain with 3'ocs, κ chain-KDEL with 3'rbcS	*A. Thaliana* (leaves)	5.9 % of TSP
α human creatine kinase (fab, lgG)	CaMV 35S promoter/ ocs terminator	LP of the 2S2 storage Protein of A. *thaliana*	Fab	*Nicotiana tabacum* (callus)	0.128% of TSP
			IgG	*N. tabacum* (callus)	0.477% of TSP
			Fab	*A. thaliana* (callus)	0.188% of TSP
			IgC	*A. thaliana* (callus)	0.217% of TSP
α substance P (single-domain Ab)	CaMV 35S promoter/ nos terminator	*pel*B LP		*Nicotiana benthamiana*	1 % of TSP
IG of 38C13 mouse B Cell lympoma (scFv)	Viral vector TTO1A, a hybrid fusion of TMV and tomato mosaic virus, N-terminal: rice α-amylase LP			*N. benthamiana* (leaves)	12.3–30.2 μg g^{-1} FW
α Streprococcus mutans abhesin (IgA, IgG)	Ca MV 35S promoter/ nos terminator	Native mouse immunoglobulin LP The γ chain constant region was replaced by an α chain constant region		*N. tabacum*	IgG: 7.7. α g mL^{-1} Plant extract IgG1/A: 1.5 μg mL–
1					plant extact lgG2/A:2.1 μg mL^{-1} plant extract
α *Streptococcus mutans* adhesin (slgA-G)	CaMV 35S promoter/ Nos terminator	Native mouse immunoglobulin LP, four transgenic lines expressing either the k, the hybrid lgA-G heavy chain, the murine J chain or the secretory component were crossed to obtain a sIgA experessing line		*N. tabacum*	200-500 μ g^{-1} FW
α human (IgG))	CaMV 35S promoter/ T7 terminator	5′murine LP of HC and LC LC anc HC separrately transformed		*Mddicago sativa*	1% TSP (full-size IGG)

Table 10.1 Continued....

Protein	*Regulatory elements*	*Additional elements of the construct*	*Species/Tissue*	*Accumulation level*
α HCV-2 glycoprotein B (IgG)	CaMV 35S promoter/ nos termineator	LP of tobacco extensin	*Glycine max*	Not mentioned
Human hemoglobin (α + β)	CaMV enhanced 35S promoter/ CaMV 35S	Transit peptide of small subunit of Rubis CO of *Pisum sativum*	*N. tabacum*	0.05% of seed protein
EGF (synathetic gene)	terminator CaMV 35S promoter/ nos terminator	Codon-optimized for high ,evel peptide expression in Escherichia coli	*N. tabcum* (leaves)	0.001% of TSP
Human glutamic acid decarboxylase (hGAD)	CaMV enhanced 35S promoter/ TEV5′ UTR/nos terminator		Low alkaloid tobacco *Salanum* *tuberosum*	0.4% of TSP of leave or tuber
Isoform of hGAD (hGAD65)	CaMV 35S promoter/ nos terminator		*N. tabacum* (leaves)	0.01–0.04% of TSP
			Daucus carota (tap root)	0.01% of TSP
			N. Tabacum (leaves)	0.04% of TSP
		$GAD67_{1-87}$ (rat)/$hGAD65_{88-585}$ Fusion		0.19%ofTSP
	Expression mediated by potato virus X infection		*N. benthamiana* (leaves)	2.2% of TSP
Human collagen Pra α (1) chain	CaMV 35S promoter/ CaMV 35S terminator	Pathogenesis-related protein (PR) LP or native IP	tobacco	>100 μg g^{-1} tissue
Hepatitis B surface antigen	CaMV enhanced 355S promoter/ TEV 5′UTR/nos terminator		*N. tabacum* (leaves)	66 μg g^{-1} TSP
	Patatin promoter		*S. tuberosum* (tuber)	1.1 μg g^{-1} FW
	CaMV 35S promoter/TEV 5′ UTR/nos terminator			0.33 μg g^{-1} FW
		Soybean vegetative storage protein LP (vspαS)		0.8 μg g^{-1} FW
		vspαS plus putative vacuolar targeting signal (vspαL)		2.4 μg g^{-1} FW
		Transit peptide of small subunit of RubisCO		n.d.
		KDEL		1.25 μg g^{-1} FW
	CaMV 35S promoter/ΩTMV 5′UTR/nos terminator			0.33 μg g^{-1} FW

Table 10.1 Continued....

Protein	*Regulatory elements*	*Additional elements of the construct*	*Species/Tissue*	*Accumulation level*
	CaMV35S promoter/ Ω TMV 5′UTR/nos B terminator			6.5 μg g^{-1} FW
	CaMV/35S promoter/TEV 5′UTR/vsp B gene terminator			16.0 μg g^{-1} FW
Binding subunit of heat-labile enterotoxin B	CaMV35S promoter/TEV 5′UTR/vsp B terminator	With and without KDEL	*N. tabacum* (leaves)	5 μg g^{-1} TSP (without KDEL) 14 μg g^{-1} TSP (with KDEL)
			S. tuberosum	30 μg g^{-1} TSP (without KDEL) 110 μg g^{-1} TSP (with KDEL)
Cholera toxin B subunit	CaMV35S promoter nos terminator	Kozak seqence, KDEL	*Lycopersicon esculentum*	Leaves: 0.02% of TSP Fruits: 0.04% of TSP
IL-10	CaMV 35S promoter/ nos terminator	Native LP, with and without KDEL	*N. tabacum*	800 ng g^{-1} TSP (without KDEL) 55 μg g^{-1} TSP (with KDEL)
IL-2	Patatin promoter	Native LP	*S. tuberosum*	115 units g^{-1} FW
Enkephalins	Promoter and terminator of 2S1 albumin gene of *A. thaliana*	Integrated in 2S1 albumin of *A. thaliana* and flanked by tryptic cleavage sites	*A thaliana* (seed) *Brassica napus* (seed)	2.9% of TSP (200 nmol g^{-1}) 50 nmol g^{-1}
T84.66 α CEA (scFv)	Legumin A seed specific promoter/Ω sequence TMV 5′UTR	Light chain LP of mAb24, KDEL	*P. sativum* (seed)	9 μg g^{-1} FW
T84.66 α CEA (diabody)	CaMV enhanced 35S promoter/ CHS 5′UTR/Pw 3′UTR of TMV/ CaMV 35S terminator	Light chain LP of mAb24, and without KDEL	*N. tabacum* (leaves)	1 – 5 μg g^{-1} (apoplast/transient) 4 – 12 μg g^{-1} (ER/ transient) 0.44 – 0.93 μg g^{-1} (apoplast/stable) 3 – 9.3 μg g^{-1} (ER/stable)

Table 10.1 Continued....

Protein	*Regulatory elements*	*Additional elements of the construct*	*Species/Tissue*	*Accumulation level*
Aprotinin	Maize ubiquitin promoter/ pinII terminator	Maize ubiquitin 5′intron, barley α-amylase LP	Extracellular matrix of embryo (maize seeds)	In T5: 0.35% of extractable seed proteins
Human serum albumin	CaMV enhanced 35S promoter/ AIMV RNA synthetic leader sequence/nos terminator	Native or PR-S leader peptide	*S. tuberosum* (leaves)	0.02% of TSP
Hirudin	Oleosin promoter/ nos terminators	Fused to native oleosin	*B. napus*	1% of seed protein
β-Interferon	CaMV 35S promoter/ CaMV 35S terminator	Native LP	*N. tabacum*	170 ng g^{-1} FW
α human Rhesus D (IgG)	CaMV 35S promoter (LC, HC and tandem construct)/E9 PolyA terminator of *P. sativum*	Murine immunoglobulin LP	*A. thaliana*	Cross of LC + HC: 0.6% of TSP, tandem : 0.3% of TSP
	*mas*1′2′ dual promoter (bidirectional)/nos and E9 terminator			Bidirectional: 0.12% of TSP
Dog gastric lipase	CaMV enhanced 35S promoter/ CaMV 35S terminator	LP of rabbit gastric lipase precursor	*N. tabacum* (leaves;	7% of acid extractable proteins
		LP of sweet potato sporamin precursor	*N. tabacum* (leaves) (vacuole)	5% of acid extractable proteins
α Hepatitis B surface antigen (scFv)	CaMV 35S promoter/TMV Ω 5′UTR/nos terminator		*N. tabacum* (cytosol)	n.d.
		Sporamin LP	*N. tabacum* (apoplast)	0.31 % of TSP
		Sporamin pre-pro-peptide LP	*N. tabacum* (vacuole)	0.032% of TSP

Table 10.1 Continued....

Protein	*Regulatory elements*	*Additional elements of the construct*	*Species/Tissue*	*Accumulation level*
		Sporamin LP, KDEL	*N. tabacum* (ER)	0.22% of TSP
α Hepatitis B surface antigen (scFV)	CaMV 35S promoter/TMV Ω 5'UTR/nos terminator	KDEL	*N. tabacum* (root exudates)	630 – 760 ng g^{-1} dry weight root/day
		N-terminal ER LP from calreticulin		
Human interferon (hIF-α2b and hIF-α8)	CaMV 35S promoter/ nos terminator)	Native LP of HIF-α2b and hIF-α8 resp.	*S. tuberosum*	560 IU g^{-1} tissue

Abbreviations used in table 6.1: AIMV: alfalfa mosaic virus; CaMV: cauliflower mosaic virus; CEA: carcinoembryogenic antigen; CHS: chalcone synthase; Fab: antigen binding fragment; Fd: variable domain and first constant domain of the heavy chain; FW: freshweight; HC: heavy chain; hIF: human interferon; HSV: herpes simplex virus; IL: interleukin; LC: light chain; LP: leader peptide; mAb: monoclonal antibody; mas: mannopine synthase; n.d.: not detectable; nos: nopaline synthase; ocs: octopine synthase; *pel*B: pectate lyase; pin II: potato proteinase inhibitor II; PR-S: pathogenesis related protein S; Pw: pseudoknot wildtype of TMV; rbcS: ribulose-1, 5-bisphosphate carboxylase small subunit; RubisCO: Ribulose-1, 5-bisphosphate Carboxylase Oxigenase; scFv: single chain Fragment variable; TEV: tobacco etch virus; TMV: tobacco mosaic virus; TSP: total soluble protein; UTR: untranslated region; vspB: vegetative storage protein B.

been produced in plant cell cultures so far and the expression cassette elements used in each case. Possible contamination by chemical or biological substances is one of the most important concerns when producing pharmaceutical proteins. Plant cell cultures ensure the production of the desired protein in a controlled, sterile and sealed environment and can be adapted to cGMP conditions. Therefore, the risk of contamination is minimized and the production conditions can be modified more easily in a contained reactor than in the field.

Another advantage is the ability to freeze plant suspension cells in liquid nitrogen. So that master and working cell banks can be established, aprerequisitefor cGMP procedures. Although plant cell culture is not as cost effective as plant cultivation in the open field, it will become an economical process if higher protein yields can be achieved. The cultivation medium of plants is chemically defined, consisting of a carbon source, minerals, vitamins and phytohormones. Furthermore, it is protein-free and relatively inexpensive. In contrast, animal cells often require complex supplements such as fetal calf serum and/or expensive growth factors, although serum-free cultivation is possible in case of Chinese hamster ovary (CHO) cells. When trying to increase the yields of recombinant protein inplant suspension cultures, one should consider optimizing the nutrient supply and including product stabilizing agents.

Sharp and Doran showed that antibody accumulation in hairy root cultures was improved by increasing the dissolved oxygen tension to 150% air saturation and that loss of the antibody could be minimized by inhibiting protein transport in the secretory pathway with the antibiotic brefeldin A. The beneficial effects of stabilizing solutes were demonstrated among others by Magnuson *et al.* who achieved a 35-fold increase in scFv yield by adding polyvinylpyrrolidone (PVP) to the culture medium. Other proteins were stabilized to a lesser degree by supplementing the medium with the protease inhibitor bacitracin, bovine serum albumin (BSA), gelatin or dimethylsulfoxide (DMSO).

Fischer *et al.* reported a transient threefold increase in recombinant antibody levels after supplementing cell cultures with amino acids prior to harvest. All these studies show, however, that the benefits obtained by certain expression cassette elements or defined culture conditions cannot be generalized for all types of pharmaceutical proteins. These factors have to be investigated and optimized individually for every product. In order to facilitate purification of a recombinant protein from plant cell culture, the recombinant protein can be directed to the extracellular space (apoplast and culture medium) by attaching an appropriate signal sequence or using a native signal sequence if one is present. Only about 100-1000 endogenous proteins are thought to be secreted from cultured plant cells, so the purification of a recombinant protein from the culture medium is much easier than purification from whole cell extracts.

The plant cell wall is a natural barrier to protein secretion. Proteins >20 kDa tend to be retained although higher-molecular-weight proteins such as monoclonal antibodies are secreted. Secretion can be further enhanced by the addition of DMSO. Although cell cultures produce a much lower amount of biomass than plants cultivated in the open field, tobacco suspension cells have been cultivated at volumes of up to 20 m^3 and references therein).

PROCESSING STEPS IN PLANTS

In order to make molecular farming commercially profitable, recombinant proteins must be produced at a sufficiently high yield and in an active form. It has become clear that, for

Table 10.2. Pharmaceutical Proteins Produced in Plant Suspension Cultures

Protein	*Regulatory elements*	*Additional elements of the construct*	*Species*	*Accumulation level*
Human α_1 antitrypsin	Promoter of rice a amylase gene *RAmy3D*/terminator of *RAmy3D*	LP of rice a amylase gene *RAmy3D*	*Orza sativa*	4.6–5.7 mg g^{-1} dry cell weight
		Additional first intron of *RAmy3D*		No significant increase
		LP of rice α amylase gen *RAmy3D*		18.2–24.2 mg g^{-1} dry cell weight
Human erythropoietin	CA MV 35S Promoter/CaMV35S terminator	Native LP	*N. tabacum* cv By-2	26ng^{-1} TSP
α hepatitis B surface antigen-(scFv)	CaMV 35S Promoter/ nos terminator	Sporamin LP	*N. Tabacum* cv Havana SR	1 mg L^{-1} in culture supernatant and 5 mg kg^{-1} wet weight from cell extracts
hIL-12	CaMV enhaced 35S promoter	Native LP	*N tabcum* cv Havana SR	175µ L^{-1}
hIL-2, hIL-4	CaMV 35S promoter/ T7 terminator	Native LP	*N. Tabacum* NT-1	0.1 mg L^{-1} (hIL-2) and 0.18 mg L^{-1} (hIL4)
Human Serum albumin	CaMV enhyanced 35S promoter/ AIMV RNA synthetic leader sequence/nos terminator	Native or PR-s LP	*N. tabacum*	250 µg g^{-1} TSP
Bryodin 1	CaMV 35S promoter/ nos terminator	Extensin LP	*N. tabacum* NT-1	30mgL^{-1}
hG-CSF	CaMV enhanced 35S promoter/ TMV Ω 5'UTR/nos terminator		*N. tabacum*	105 µg L^{-1} extracellular
GM-CSF	CaMV35Spromoter/TEV5' UTR/T7 terminator		*N. tabacum* cv BY-2	150 µg L^{-1} intracellular and 250 ug L^{-1} extracellular
	CaMV enhanced 35 S promoter/ TMV Ω 5'UTR/nos terminator	Native GM-CSF LP	*N. tabacum* cv Havana SR	180 ug L^{-1} extracellular, after adding of gelatin increase to 738 µgL^{-1}

Abbreviations used in table 2: AIMV: alfalfa mosaic virus; BY-2: *Nkotiana tabacum* cv Bright Yellow 2; CaMV: cauliflower mosaic virus; FW: freshweight; GM-CSF: granulocyte macrophage colony stimulating factor; hG-CSF: human granulocyte colony stimulating factor; hIL: human interleukin; LP: leader peptide; nos: nopaline synthase; PR-S: pathogenesis related protein S; RAmy 3D: rice amylase gene promoter 3D; scFv: single chain Fragment variable; TEV: tobacco etch virus; TMV: tobacco mosaic virus; TSP: total soluble protein; UTR: untranslated region.

high-level protein accumulation, the stability of transgene expression can be as important as the expression level itself. The quantity of protein is determined by the rate of protein synthesis, assembly as well as proteolytic degradation. Processes that influence the production of a stable and functional end product in plants and plant cell suspension cultures include transgene integration, transcriptional and translational activity, posttranslational modification. The first hurdle is stable transgene integration and expression in the genome of the transgenic plants or transformed cells, since transgenes are often silenced either immediately after integration or in subsequent generations.

A novel and very interesting method for dealing with this problem in transient gene expression is the use of tomato bushy stunt virus P19 protein. This is a known suppressor of posttranscriptional gene silencing (PTGS) and transient expression of a range of proteins was increased 50-fold in the presence of p19. Another approach is the inclusion of scaffold attachment regions (SARs) or matrix attachment regions (MARs) within the expression construct. Such elements, which attach to the nuclear matrix and may help divide interphase chromatin into functional domains, have been shown to enhance transformation efficiency to achieve stable gene expression and to increase transgene expression levels up to 650-fold although the level of enhancement appears to be dependent on the promoter. Increased transcription levels are assumed to result in increased protein synthesis. One approach to reach this goal is to raise the transgene copy number by the use of amplification-promoting sequences derived from a spacer sequence of tobacco ribosomal DNA. Posttranscriptional processes such as capping, splicing and polyadenylation are important for high protein yields, and it is also important to maximize mRNA stability. The synthesis of a protein may be hampered by suboptimal translation initiation or termination sites. Lukaszewicz *et al.* studied the context sequence of the initiation codon and Sawant *et al.* were able to increase the expression level of a reporter gene by inserting a sequence commonly found in highly expressed plant proteins.

Other problems can include the presence of cryptic splice or polyadenylation sites, and unfavourable codon bias (i.e. clusters of rare codons). If problems of this type are encountered, the construction of a synthetic gene can be considered. The artificial synthesis of a whole gene is expensive, but codon optimization at the 5′ end of the open reading frame can be sufficient to enhance gene expression. Batard *et al.* demonstrated an increase in wheat P450 enzyme activity in transgenic tobacco after optimization of the first 111 bp of the wheat open reading frame. This corresponds to 7% of the entire open reading frame. An useful alternative to nuclear transformation is the direct transformation of plastids, since plastid genomes are not affected by epigenetic silencing and they have a codon bias different to that of nuclear genes.

By creating transplastomic plants, very high levels of recombinant proteins have been obtained. In the case of pharmaceutical proteins, 7% total soluble protein (TSP) has been achieved for human somatotropin and 10% TSP has been achieved for human serum albumin (HSA). This exceeds the levels obtained through nuclear transformation, although the production of homoplastic plants is unfortunately time consuming and has only been achieved for a few species. Posttranslational modifications can have dramatic effects on the accumulation of a recombinant protein. If proteins are not properly cleaved, pro-proteins will accumulate instead of mature proteins.

For example, it has been shown that one of the causes for improper processing of soybean glycinins in potato tubers is the absence of the necessary proteases in the host plant. Some proteins need specific modifications to become active, but these are not carried out in plants. These include the addition of certain sugar molecules or the hydroxylation of proline for the production of native-like collagen in tobacco plants. Differences in glycan structures between plants and mammals (e.g. the presence of xylose and fucose residues in plants) have raised concerns regarding the potential immunogenicity of recombinant proteins, especially those used as human therapeutics. In order to obtain an antibody with a glycan profile similar to the mammalian type, Bakker *et al.* expressed human β-1, 4 galactosyltransferase in tobacco plants and crossed them with plants expressing a murine antibody. The result was a plantibody with partially galactosylated N-glycans. The folding of polypeptide chains and the assembly of multiple subunits arc critical requirements when complex and multimeric proteins such as full size antibodies are expressed in plants.

Correct folding of the protein may be favoured by retention in the endoplasmic reticulum (ER), which can be achieved by adding a C-terminal K/HDEL ER-retention signal. This prolongs the exposure of the protein to endogenous chaperones like BiP (binding protein). Bouquin *et al.* expressed the heavy and light chains of an antibody individually in *Arabidopsis thaliana* plants. Strongly expressing parental lines were selected and crossed to obtain a hybrid line producing the full-size antibody. The multimeric antibody accumulated to higher levels than either of the individual chains in the parental lines, indicating that assembly of the mature protein had a stabilizing effect in comparison to the individual subunits. Expressing recombinant proteins as N-terminal fusions with ubiquitin is another strategy that can help to achieve proper folding.

Additionally, this is an elegant way to obtain proteins that do not start with a methionine residue, since the ubiquitin moiety is cleaved off by endogenous ubiquitin-specific proteases. If the desired protein is prone to proteolytic degradation, the use of appropriate targeting signals to direct the protein to different cellular compartments (e.g. the ER, chloroplast, vacuole or apoplast) or the use of tissue specific promoters (e. g. the hordein promoter for protein expression in barley grains) could be considered. Targeting is especially important if the recombinant protein is toxic to the production host. For example, targeting the non-pharmaceutical protein avidin to the cytosol in transgenic tobacco plants was toxic, but plants were regenerated successfully when this molecule was targeted to the vacuole. The expression of a secreted version of avidin in maize *(Zea mays)* resulted in partial or complete male sterility

IMPROVED PROTEIN YIELDS

We will focus on recombinant antibodies (rAbs) and their derivatives (e.g scFvs, dia-bodies) in this section because these classic plant-produced pharmaceuticals have been expressed in many diverse production systems using a variety of regulatory elements. However, even in the case of this extensively studied class of proteins, direct comparisons between systems are difficult to make, since different genes, regulatory elements and production hosts have been used. Some investigators have compared the expression of a particular rAb in different plant species, or have targeted the antibody to different tissues or subcellular compartments in one particular host. Generally, such comparisons have revealed significant variation in expression and accumulation levels. Stoger *et al.* expressed the scFv antibody T 84.66, which recognizes the carcinoembryogenic antigen (CEA), and systematically compared several plant systems, regulatory elements and subcellular targeting signals.They concluded that the major factor determining overall accumulation levels was the stability of the antibody, which could be controlled in part by subcellular targetig.

In agreement with others, the highest levels of active antibody were obtained by targeting the protein to the secretory pathway using its native signal peptide, and then retaining the antibody in the ER lumen using a KDEL tag. Presumably because of the favourable biochemical environment, the lack of proteases and the presence of chaperones, rAbs accumulate to higher levels in the ER than in any other compartment. Furthermore, this strategy can limit the extent to which core glycans are replaced by plant-specific carbohydrates like α(l,3)-fucose and β(1, 2)-xylose.

Without ER retention, antibodies are secreted to the apoplast where accumulation *i* levels are typically 10 – 100-fold lower than in the ER. The cytosol appears to be an inappropriate compartment for antibodies, resulting in very low accumulation levels. An exception is described by De Jaeger *et al.* who targeted a phage display-derived scFv to the cytosol and achieved a level of 1% TSP. Stoger *et al.* also observed that between species, the amounts of scFv per unit fresh weight were in the same range, and did not correlate with the total protein content in the plant. For example, even though pea is a much more proteinaceous crop than rice, the amounts of antibody measured as a percentage of TSP were considerably leach species, individual promoters resulted in distinct, tissue-dependent accumulation patterns. The cauliflower mosaic virus (CaMV) 35S promoter, for example, led to high-level accumulation in callus and leaves whereas the maize ubiquitin-1 promoter was the best choice for producing recombinant proteins in cereal seeds even though it is not in itself seed-specific.

The lack of such comparative studies for proteins other than rAbs makes_it_ difficult to generalize an optimal expression strategy for all proteins. List recombinant proteins expressed in plants and provide details of the production system, promoters and other regulatory elements used in each case.

DOWNSTREAM PROCESSING

When choosing a plant expression system, downstream processing should be taken into account since this will contribute a substantial proportion of the overall costs. A cost distribution for the production of β-glucuronidase in maize has been made by Evangelista *et al.* This is not a pharmaceutical protein, but the principles of cost analysis should hold true anyway. It was stated that protein extraction and purification would account for 40% and 48%, respectively, of the total annual operating costs, assuming a product purity of 83%. When higher purity is needed, as for pharmaceutical proteins, downstream processing could account for an even larger proportion of the total costs. Various strategies have been proposed to simplify downstream processing. We have already mentioned the use of tissue-specific promoters or targeting signals that allow protein secretion or accumulation in a particular organelle.

A useful example is protein targeting to oil bodies. which is used in combination with the oleosin-partitioning technique for protein isolation from oilseed crops. For new products that have yet to undergo clinical testing, the addition of an affinity tag can be considered to simplify protein isolation. Such tags have unique properties, which allow fusion proteins to be isolated from cellular extracts or media by affinity to particular ligands.

Alternatively, the tag may confer a novel physical property on the protein. An example is the synthetic elastin-like polypeptide (ELP) tag. When a temperature transition occurs, the tags fold into spirals that aggregate. These aggregates can easily be isolated by centrifugation. However, if such tags are used, they must not adversely affect the function of the protein. Importantly, the presence of affinity tags on pharmaceutical proteins may result in non-compliance with regulatory issues. A minimalist extraction buffer is often recommended for protein purification since most additives provide only a marginal improvement in yields, but will increase costs significantly.

The removal of phenolics, for example by tangential-flow ultrafiltration/diafiltration is an important step that should be carried out as early as possible in the purification procedure, since these molecules can become covalently linked to amino acid side chains and can oxidize certain residual groups. When using plant cell cultures for the production of biopharmaceuticals, targeting the protein to the apoplast (and from there to the culture medium) is beneficial for later purification. Plant cell culture media are protein-free, which makes the purification of the product relatively easy. Unfortunately, secreted recombinant proteins are often unstable in the cell culture medium. One strategy to overcome this problem was presented by James *et al* in form of an affinity chromatography bioreactor. By initiating protein purification while the cells are still growing, the exposure of the protein to destabilizing agents in the culture medium is minimized.

PLANT-DERIVED PHARMACEUTICALS

The production of pharmaceutical proteins in plants has clear advantages over traditional systems in terms of cost-efficiency and product safety, since there is no risk of contamination with human pathogens. Furthermore plants are much less likely than mammalian cells to be affected by the expression of certain human proteins, such as growth factors and cell cycle inhibitors. Therefore, plants provide a strategic complement to existing microbial and animal production systems. Antibodies account for more than 20% of all biotechnology-derived molecules, and the market is expected to reach $5 billion by 2005. More than 250 companies are working on over 700 therapeutic antibodies, of which 220 are in clinicaltrials. These numbers demonstrate the ongoing commercial expansion of pharmaceutical protein production and indicate a growing demand for production capacity.

It has been estimated that plant-derived antibodies can be produced in a 250 m^2 greenhouse at about one tenth of the cost of hybridoma-derived antibodies. Planet Biotechnology Inc. presents an even rosier picture, estimating that the cost of plant-derived antibodies will be 5 % of the cost of antibodies produced in animal cells. This calculation, however, was based on yields of 500 $\mu g\ g^{-1}$ leaf fresh weight, a rather ambitious level for antibody production in plants. Once again, the greatest proportion of the cost reflects downstream processing and purification. When purification is needed, it is estimated that a recombinant protein must be expressed at greater than 0.01% of fresh weight or 1% of TSP for molecular farming to be commercially feasible. To date, several antibodies with therapeutic potential have been produced in plants. Among them are the Guy's 13 antibody that recognizes *Streptococcus mutans* (the causative agent of dental caries), humanized antibodies that recognize viral antigens (herpes simplex and respiratory syncitial virus), an antibody against CEA and a tumor-specific vaccine for the treatment of lymphoma.

Although none of these antibodies is produced commercially at the current time, the lymphoma scFv has completed phase I clinical trials and the Guy's 13 antibody has reached phase II clinical trials in the USA *(www.planetbiotechnology.com/products.html)*. Regarding other pharmaceutical proteins, enterotoxin B and aprotinin have both reached phase I trials, while gastric lipase is undergoing clinical phase IIa trials in France and in Germany *(www.meristem-therapeutics.com/GB/intro.htm)*. Although no plant-derived pharmaceutical protein is yet available on the market, two technical proteins (avidin and β-glucuronidase) are successfully produced in corn for commercial purposes.

CONTAINMENT STRATEGIES

The issue that concerns people the most, with respect to the safety of plant biotechnology, is the possibility that transgenes could spread in the environment by out-crossing. Methods to contain foreign genes within transgenic plants are reviewed by Daniell and include maternal inheritance, male or seed sterility and induced transgene excision before flowering. Chloroplast genetic engineering takes advantage of the fact that the chloroplasts of most crops are maternally inherited. Male sterility can be achieved by crossing plants that produce the pharmaceutical protein of interest with plants that are male sterile or through the use of genetic methods to prevent the production of functional pollen.

Seed sterility can be achieved using Monsanto's Terminator Technology™or the recoverable block of function systern. The Cre-loxP svstein has been used for transgene excision. Another strategy is the cultivation of transgenic plants in greenhouses. Although this increases production costs considerably, it could still be the cheapest solution overall. Khoudi *et al.* calculated the cost of 1 g of purified antibody produced in greenhouses, and found that it was only 10% of the cost of hybridoma technology. A further advantage of molecular farming in greenhouses is that the product itself would be contained, a desirable outcome in the case of biopharmaceuticals that are potentially harmful, capable of persisting in the environment or of accumulating in non-target organisms. Such safety issues came to the fore after the ProdiGene fiasco in late 2002, prompting the US Department of Agriculture in March 2003 to tighten its permit conditions relating to the field-testing of plants engineered to produce pharmaceutical and industrial compounds.

Another means to improve safety standards is the use of inducible promoters, which could be used to delay protein expression until the crop has reached the factory. This would also eliminate protein degradation problems caused by the delay between harvesting and processing. Selectable markers are an essential part of the transformation construct, since they allow the selective propagation of transformed cells.

Antibiotic resistance is the most widely used selectable phenotype, but there are concerns that resistance genes could be transferred into bacteria and might spread in the environment. For example Mercer *et al.* reported that oral bacteria can be transformed with naked DNA in human saliva. The persistence of DNA in the field was reviewed recently by Dale *et al.* who stated that there is no evidence for the incorporation of functional plant (trans)genes into the meat products of animals fed with genetically-modified feed. Nevertheless, the use of such markers has been legally challenged in Europe, and the commercial use of antibiotic resistance markers will be prohibited from 2005 (EU Directive 2001/18/EC released 21/03/2001). Other types of marker gene may be used, such as metabolic markers that allow the use of an alternative carbon source or growth-regulation markers that express enzymes involved in the biosynthesis of plant hormones. Another approach is the use of removable markers. For a review on these selection procedures see. Also, a recent publication proves the feasibility of marker-free transformation, which must be regarded as the best possible technique to prevent an unintended spread of antibiotic resistance genes.

CONCLUDING REMARKS

Numerous issues must be taken into account in order to make molecular farming a commercial success. These issues include the yield and quality of the target protein, downstream processing methods, storage and purification, transgene containment and the cost of production. Currently, there are few examples of plant-derived pharmaceutical proteins that have been produced in a manner that addresses such concerns. We are facing a growing demand for therapeutics and diagnostics, but almost the entire capacity for cell-based production is already in use, leaving little room for the estimated 50-60 new protein pharmaceuticals that will be approved over the next 6-7 years. A shift to plant-bioreactors could therefore become a necessity. Although molecular farming still faces a number of challenges, one can clearly see the benefits and possibilities of the technology. Further research is needed to meet these challenges and facilitate the cost-effective and safe production of pharmaceuticals in plants.

Index

❑❑❑